Advances in monitoring of native and invasive insect pests of crops

It is widely recognised that agriculture is a significant contributor to global warming and climate change. Agriculture needs to reduce its environmental impact and adapt to current climate change whilst still feeding a growing population, i.e. become more 'climate-smart'. Burleigh Dodds Science Publishing is playing its part in achieving this by bringing together key research on making the production of the world's most important crops and livestock products more sustainable.

Based on extensive research, our publications specifically target the challenge of climate-smart agriculture. In this way we are using 'smart publishing' to help achieve climate-smart agriculture.

Burleigh Dodds Science Publishing is an independent and innovative publisher delivering high quality customer-focused agricultural science content in both print and online formats for the academic and research communities. Our aim is to build a foundation of knowledge on which researchers can build to meet the challenge of climate-smart agriculture.

For more information about Burleigh Dodds Science Publishing simply call us on +44 (0) 1223 839365, email info@bdspublishing.com or alternatively please visit our website at www.bdspublishing.com.

Related titles:

Integrated management of diseases and insect pests of tree fruit
Print (ISBN 978-1-78676-256-6); Online (ISBN 978-1-78676-258-0; 978-1-78676-259-7)

Integrated management of insect pests: Current and future developments
Print (ISBN 978-1-78676-260-3); Online (ISBN 978-1-78676-262-7; 978-1-78676-263-4)

Biopesticides for sustainable agriculture
Print (ISBN 978-1-78676-356-3); Online (ISBN 978-1-78676-358-7; 978-1-78676-359-4)

Improving integrated pest management in horticulture
Print (ISBN 978-1-78676-753-0); Online (ISBN 978-1-78676-755-4; 978-1-78676-756-1)

Chapters are available individually from our online bookshop:
https://shop.bdspublishing.com

climate•SMART•publishing
IN AGRICULTURAL SCIENCE

BURLEIGH DODDS SERIES IN AGRICULTURAL SCIENCE
NUMBER 128

Advances in monitoring of native and invasive insect pests of crops

Edited by Dr Michelle Fountain, NIAB-EMR, UK and Dr Tom Pope, Harper Adams University, UK

burleigh dodds
SCIENCE PUBLISHING

Published by Burleigh Dodds Science Publishing Limited
82 High Street, Sawston, Cambridge CB22 3HJ, UK
www.bdspublishing.com

Burleigh Dodds Science Publishing, 1518 Walnut Street, Suite 900, Philadelphia, PA 19102-3406, USA

First published 2023 by Burleigh Dodds Science Publishing Limited

Library of Congress Control Number: 2022949262

British Library Cataloguing in Publication Data
A catalogue record for this book is available from the British Library

ISBN 978-1-80146-107-8 (print)
ISBN 978-1-80146-110-8 (PDF)
ISBN 978-1-80146-109-2 (ePub)
ISSN 2059-6936 (print)
ISSN 2059-6944 (online)

DOI 10.19103/AS.2022.0113

Typeset by Deanta Global Publishing Services, Dublin, Ireland

Contents

Part 2 Identification, modelling and risk assessment

Series list

Title	Series number
Achieving sustainable cultivation of maize - Vol 1 From improved varieties to local applications Edited by: Dr Dave Watson, CGIAR Maize Research Program Manager, CIMMYT, Mexico	001
Achieving sustainable cultivation of maize - Vol 2 Cultivation techniques, pest and disease control Edited by: Dr Dave Watson, CGIAR Maize Research Program Manager, CIMMYT, Mexico	002
Achieving sustainable cultivation of rice - Vol 1 Breeding for higher yield and quality Edited by: Prof. Takuji Sasaki, Tokyo University of Agriculture, Japan	003
Achieving sustainable cultivation of rice - Vol 2 Cultivation, pest and disease management Edited by: Prof. Takuji Sasaki, Tokyo University of Agriculture, Japan	004
Achieving sustainable cultivation of wheat - Vol 1 Breeding, quality traits, pests and diseases Edited by: Prof. Peter Langridge, The University of Adelaide, Australia	005
Achieving sustainable cultivation of wheat - Vol 2 Cultivation techniques Edited by: Prof. Peter Langridge, The University of Adelaide, Australia	006
Achieving sustainable cultivation of tomatoes Edited by: Dr Autar Mattoo, USDA-ARS, USA and Prof. Avtar Handa, Purdue University, USA	007
Achieving sustainable production of milk - Vol 1 Milk composition, genetics and breeding Edited by: Dr Nico van Belzen, International Dairy Federation (IDF), Belgium	008
Achieving sustainable production of milk - Vol 2 Safety, quality and sustainability Edited by: Dr Nico van Belzen, International Dairy Federation (IDF), Belgium	009
Achieving sustainable production of milk - Vol 3 Dairy herd management and welfare Edited by: Prof. John Webster, University of Bristol, UK	010
Ensuring safety and quality in the production of beef - Vol 1 Safety Edited by: Prof. Gary Acuff, Texas A&M University, USA and Prof. James Dickson, Iowa State University, USA	011
Ensuring safety and quality in the production of beef - Vol 2 Quality Edited by: Prof. Michael Dikeman, Kansas State University, USA	012
Achieving sustainable production of poultry meat - Vol 1 Safety, quality and sustainability Edited by: Prof. Steven C. Ricke, University of Arkansas, USA	013
Achieving sustainable production of poultry meat - Vol 2 Breeding and nutrition Edited by: Prof. Todd Applegate, University of Georgia, USA	014

Introduction

Insect pests remain a major threat to crop production primarily because of their ability to inflict severe damage on crop yields, as well as their role as key vectors of disease. Early identification of pests is critical to the success of integrated pest management (IPM) programmes and essential for the development of phytosanitary/quarantine regimes to prevent the introduction of invasive insect pests to new environments.

This volume reviews the wealth of research on techniques to monitor and thus prevent threats from both native and invasive insect pests. The book is split into three parts: Part 1 chapters draw attention to detection methods for crop insect pests, focusing specifically on techniques to trap and monitor them as well as developments in sampling and survey design, detection techniques and ways to monitor airborne movement of crop insect pests. Chapters in Part 2 focus on identification, modelling and risk assessment, highlighting advances in image-based identification and analysis, population growth models and pest risk assessment techniques. Part 3 chapters review invasive species and how their potential economic impact can be assessed, including the development of effective phytosanitary measures to prevent invasive insect pests and successful IPM programmes to control or eradicate them.

Part 1 Detection

The first chapter of the book draws attention to advances in techniques for trapping crop insect pests. Chapter 1 begins by describing the basic components of an insect trap. It then moves on to discuss methods to make traps more attractive for an insect, emphasising the use of methods such as visual and olfactory cues. The chapter also highlights common trap types for catching insect pests, focusing specifically on water traps, sticky traps, suction traps, interception traps and funnel and cone traps. Case studies of their usage and an evaluation of their merits and limitations are also included. Automated traps for insect pests are also reviewed, such as time-sorting traps, automated counting and identification and automated visual identification.

The next chapter reviews current insect pest monitoring systems for native and invasive pest species of corn, soybean, and cotton in the U.S. Chapter 2 presents case studies of European corn borer, corn earworm, western bean cutworm and western (*Diabrotica virgifera* LeConte) corn rootworm as pests of corn; soybean aphid, stink bugs, and Japanese beetle as pests of soybean; and tobacco thrips, plant bugs and bollworm as pests of cotton. It discusses how new technology has driven recent improvements in monitoring of these pests from the lab, field and areawide levels. It also discusses recent trends and tools for insect resistance management.

The subject of Chapter 3 is interpreting data from insect pest trap networks. The chapter begins by highlighting models to analyse trap networks, focusing primarily on the development of the TrapGrid network. TrapGrid can be used to quantify the probability of capturing insects instantaneously or over time using a function that relates distance from a given trap to probability of capture and two models of insect dispersal. Brief descriptions of other modeling approaches to these questions, some of which have seen application outside of research, are discussed followed by ideas for the application of TrapGrid, including a way to determine trap attraction (the parameter λ in the model) and how to compare alternative trap layouts on a landscape scale. A working example is also provided, comparing two alternative trap layouts in a 1 km^2 area via quantification of capture probability instantaneously and over 30 days.

Chapter 4 focuses on developments in crop insect pest detection techniques. The chapter first discusses the use of camera systems for pest detection at micro-scale ranges, then moves on to review how drone/camera systems can be used at meso-scale detection ranges. This is followed by an overview of the use of Landsat systems for pest detection and macro-scale detection ranges. A section on sound- and vibration-sensors is also included. The chapter provides a case study focused on augmenting traditional pest detection and biological control with nanoscale- and micro-scale sensor technologies, before concluding with an overview of future goals and developments for insect pest detection technologies in crops.

The final chapter of Part 1 explores monitoring airborne movement of crop insect pests and beneficials. Chapter 5 highlights the importance of trapping methods when it comes to monitoring specimens. This is then followed by a review of the visual observation of insects and a discussion of various forms of radar that can be used, such as scanning entomological radars, vertical-looking and weather surveillance radars. A section on optoelectronics is also provided, focusing specifically on short-range and long-range monitoring as well as imagers and the strengths, limitations and prospects of these devices. The chapter also provides a review of radiotelemetry and harmonic radar as well as a case study on moths over a cotton crop.

Part 2 Identification, modelling and risk assessment

Part 2 opens with a chapter that reviews progress in developing automated image-based systems for identifying crop insect pests. Chapter 6 identifies the challenges in distinguishing insect pests in field conditions and ways they can be addressed. The chapter outlines key steps in image-based identification (image capture, processing, segmentation, feature extraction and classification) and the growing use of artificial intelligence to increase accuracy and reliability. It also provides examples of commercially-available systems and considers future developments. The chapter is aimed at practitioners and scientists new to the topic and as a useful reference on the pros and cons of different monitoring strategies for those already in the field.

Chapter 7 looks at advances in insect pest monitoring using pest population growth and geospatial data for pest risk management. The chapter reviews approaches and provides case studies of current advances that are in development or have been applied in operational IPM systems. It first analyses incorporating knowledge of expected pest population growth to adjust economic thresholds for use in applying remedial IPM tactics to individual fields. The chapter then turns to principles and advancements in using spatial data, including spatially-referenced pest, crop and remote sensing data. Spatial data linked to insect pest monitoring may improve strategic use of remedial IPM tactics at the field level and use of preventive IPM tactics at larger scales.

The subject of Chapter 8 is advances in pest risk assessment techniques focusing on invertebrate pests of European outdoor crops. The chapter breaks down the processes and tools available to undertake a pest risk assessment, first focusing on the ways to assess the likelihood and hazard of pest infestations. It then highlights the potential influence of the perception of risk on decision making and how risk can be reduced. The chapter also provides a worked example for a priority pest in wheat.

Part 3 Invasive species

Chapter 9 addresses the potential economic impact of invasive plant pests. The chapter evaluates techniques such as partial budgeting, partial equilibrium models, input-output models and computable general equilibrium models in terms of their economic concepts, scope and required data and skill. This is then followed by a general reflection on the question of how to choose the most appropriate technique whilst considering the trade-off between completeness in economic scope and required resources. The chapter concludes with two case studies to illustrate the potential of the methods that are most often applied.

The subject of Chapter 10 is developing effective phytosanitary measures to prevent the introduction of invasive insect pests. The chapter provides a definition of phytosanitary measures, why they are needed, how standards for phytosanitary measures developed through international cooperation and how they relate to international agreements. This chapter then explains how policy decision making and the management of invasive insect pests using phytosanitary measures is supported by the structured framework of pest risk analysis which follows international standards. Outlines of some of the more commonly used phytosanitary measures used to manage pest risk are provided. Case studies are used to illustrate the impact of phytosanitary measures. The chapter closes by identifying some of the key challenges the industry faces to inhibit the introduction and spread of invasive insect pests with the aim of stimulating future research.

The final chapter of the book is focused on mitigating invasive insect species, focusing specifically on eradication, long-term management and also highlights the importance of sampling and monitoring. Chapter 11 introduces major concepts associated with the management of invasive insects. It then focuses more deeply on eradication and long-term management, especially IPM, describing common factors associated with successful programmes and potential challenges of each strategy. Brief case studies are provided to illustrate applications of eradication and IPM to various invasive insects in different regions. Lastly, the chapter summarises recent research and technology that have progressed effective management of invasive insects and highlight areas where further research is needed.

Part 1

Detection

Chapter 1

Advances in techniques for trapping crop insect pests

Archie K. Murchie, Agri-Food & Biosciences Institute, UK

1 Introduction

Monitoring of insect pests is an essential component of integrated pest management. To determine whether pest densities justify an intervention, such as an insecticide application or other control measures, their numbers need to be assessed and compared to economic thresholds (Pedigo and Buntin, 1994; Ramsden et al., 2017; Stern, 1973). Such assessments can be done by physical sampling of the crop and visual inspection or passively using traps. The method used will depend on the biology of the pest species, the ease of assessment, the availability of labour and the relationship between the stage monitored and the damage done. For example, traps may monitor the adults, whilst the damaging stage is the larva. The closer the temporal and spatial association between the monitored stage and the damaging stage, the more accurate the assessment but the less advanced the warning provided.

For traps, cost and ease of use are important practical elements for in-crop sampling. For all trapping systems, the processing times and costs are major considerations. Some flight interception traps such as Malaise traps catch a

http://dx.doi.org/10.19103/AS.2022.0113.01

wide variety of flying insects, with several thousand specimens collected over a trapping period (Skvarla et al., 2021). For biodiversity assessment, non-selective traps like this are ideal; however, for monitoring crop pests, selective traps are usually required that produce as targeted a sample as possible with little bycatch. This makes for speedy and accurate processing, as well as minimising impacts on non-target species. In-crop trapping systems therefore aim to be attractive to the target pest species but not to other insects within the crop. The exceptions to this may be where two species interact, such as, two pest species [e.g. cabbage seed weevil, *Ceutorhynchus obstrictus* (Marsham), and brassica pod midge, *Dasineura brassicae* (Winnertz)] (Smart et al., 1992), or where there is an opportunity to monitor a pest and its natural enemy at the same time (Murchie et al., 1997).

The first challenge in successful pest monitoring is to produce a cheap, selective trap that is highly attractive to the pest species. The second challenge is that the trap must be effective enough to provide an accurate population assessment. The third challenge is to reduce the trap sample processing time. Correspondingly, there is increasing interest in automatic methods of sorting and identification of trap catches. The fourth challenge is to reduce the environmental impact of trapping. Indiscriminate traps such as sticky traps are notorious for their large bycatch, even including small birds, bats and lizards, whilst water traps catch beneficial insects such as bees, hoverflies, lacewings and parasitoids. There is also the difficulty of disposal. Sticky traps are awkward to handle and invariably made of plastic, which once contaminated by the sticky glue cannot be easily recycled (Solorzano et al., 2015). The handling and disposal of any chemical preservative involved also needs to be considered.

2 Basic trapping elements

There are four basic components to an insect trap. The first is a method to make the trap attractive to the pest. This is commonly achieved using an attractive colour, light (for nocturnal insects) or more selectively volatile chemicals, in particular pheromone lures. Combinations of colour and chemical attractants are common, and trap design can use both visual and olfactory cues to enhance collection (Blight and Smart, 1999). An exception to this requirement would be where the trap is intended to sample specifically background insect populations, for example with suction traps, pitfall traps or clear flight interception traps. The second component is a mechanism to retain the catch. Insects may be retained on sticky glue, water with a dash of detergent to break the surface tension, or caught into a preservative such as alcohol. Some traps may be run for the purposes of live capture, particularly for biodiversity assessment (e.g. moth trapping) or to collect specimens for further work. In such cases, the design of the trap will allow entry but minimise the likelihood

Published by Burleigh Dodds Science Publishing Limited, 2023.

of exit. Often traps will be left for several days or longer before being emptied, so some method of preserving the catch must be used, if identification of specimens is required. Common preservatives are alcohol, if evaporation can be minimised, and ethylene or propylene glycol (anti-freeze), where the trap is more open to the elements. Alternatively, some traps operate dry and use insecticides to rapidly kill the catch. Lastly, the trap must be mounted in such a way as to maximise catch. Often for flying crop pests, this is at the crop canopy height and unobscured by plant foliage, which changes as the crop grows; therefore, height mounts for the trap are adjustable.

3 Making the trap attractive

3.1 Visual cues - colour and light

Colour is the simplest method of increasing trap attractiveness. Insects have diverse and highly developed colour vision, with some species having up to six different spectral receptors and a visual range of <300 nm [ultraviolet (UV)] to >700 nm (Briscoe and Chittka, 2001; Fennell et al., 2019). Specific wavelengths are attractive to certain insect species. Yellow in particular has been used for attracting phytophagous insects and has been recommended for monitoring glasshouse pests since the 1920s (Lloyd, 1922). This may be because the peak colour reflectance of plants is in the yellow band at 500–580 nm (Prokopy and Owens, 1983), due to brightness/intensity effects (Döring and Chittka, 2007) or attraction of phytophagous insects to stressed/diseased plants (Hodge et al., 2011). However, other colours have also proven attractive depending on the specific pest. For example, Kirk (1984) demonstrated that within the thrips species caught in coloured pan traps in an English sports-field, there were strong species-specific colour preferences for yellow, white, or blue, but there were also some grass-feeding thrips that were caught equally in all colours. The author suggested that yellow was in general the most attractive for non-grass foliage feeders, dark colours (black or red) for biting flies and wood borers, and white or blue for predators and parasites not associated with foliage. Responses to colour are therefore specific to the ecology of the pest species and can be subtle. For example, the contrast between the trap colour and background can affect the response of insects to traps, as can the sex and physiological state of the insect (Blackshaw, 1983; Košťál and Finch, 1996; Murchie et al., 2018). Due to the sensitivity of insects to colour, when conducting choice experiments to aid in designing traps, spectral reflectance measurements in visible and UV using spectrophotometers are preferable to human-vision assessment of colour hues.

Light trapping is used extensively for monitoring flying nocturnal insects. Many species have strong attractions towards artificial light sources particularly those emitting UV wavelengths. The reasons for this are not entirely known but seems to do with innate phototactic responses to natural nocturnal light sources such as moonlight and starlight for orientation (Nowinszky, 2003; Spalding, 2019), with the artificial light source acting as a super-stimulant. A wide variety of light-trap designs for catching moths are available, and there are networks of moth recorders in many countries running these for biodiversity assessment (Fox et al., 2011; Woiwod and Harrington, 1994). The other major use of light traps is to monitor biting insects, predominantly mosquitoes but also biting midges and sand flies. Light-trapping techniques for biting insects of medical importance are well established, and standard protocols are often available (European Commission, 2007; Silver, 2007). As with attraction to colour, attraction to light is dependent on the light source wavelength and intensity and is often species and sex specific. Consequently, whilst a general issue with light traps is that they can be non-specific and result in large bycatches, this can be mitigated against by use of very specific wavelengths, such as those available through wavelength-specific light emitting diodes (LEDs), which are highly attractive to the target pest (Kim et al., 2019). However, for many major lepidopteran pests, species-specific pheromone traps have been developed, replacing light traps for monitoring. For crop pests, light trapping has mainly been used for scientific studies or surveillance at the national level, rather than in-crop monitoring. This is due to the cost of the traps and the need for a power source, although solar-powered light traps are becoming more common as technology develops (Mohammed et al., 2018). Light-trap catch data of cotton bollworm, *Helicoverpa armigera* (Hübner), and the native budworm, *Helicoverpa punctigera* (Wallengren), were used to develop weather-based regression models to forecast pest populations in Australia (Maelzer et al., 1996; Maelzer and Zalucki, 1999; Zalucki and Furlong, 2005). Recently, wavelength-specific light traps have been developed for monitoring invasive stink bugs (Pentatomidae), utilising LEDs and pheromone lures, with the specific aims of reducing bycatch and lowering trap power consumption (Endo et al., 2022; Rondoni et al., 2022). Perhaps the most extensive current use of light traps for monitoring pests is to analyse and predict the migrations of the rice pests, the brown planthopper [*Nilaparvata lugens* (Stål)] and white-backed planthopper [*Sogatella furcifera* (Horváth)], into Japan, China and the Korean Peninsula. These migratory pests are monitored using networks of light traps run by plant protection authorities (Hu et al., 2011; Hu et al., 2014; Hu et al., 2017; Lu et al., 2017; Ma et al., 2017; Matsumura, 2001).

3.2 Olfactory cues - semiochemicals

Insects have well-developed olfactory senses and mostly detect volatile chemicals through their antennae. Correspondingly, most insect pests respond

Published by Burleigh Dodds Science Publishing Limited, 2023.

strongly to olfactory cues derived from their host crop or from conspecifics. The term 'semiochemical' refers to a range of behaviour-altering chemicals. For the purposes of pest trapping, these are usually classed as kairomones and pheromones. Kairomones are chemicals that benefit the receiver but not the sender, for example, plant-volatile chemicals that attract pests, whilst pheromones are chemicals released by organisms that illicit a specific response in the receiving organism of the same species (Dicke and Sabelis, 1988), for example, sex pheromones. Due to their specificity and attractiveness, pheromone-based trapping provides an ideal mechanism for monitoring insect pests on crops, and hundreds of chemicals have been identified for this purpose (Witzgall et al., 2010). However, in cases where pheromones are not suitable, not identified, or synthesised, plant-volatile chemicals are also used as trap lures, although these often lack the specificity of pheromones. The online database, Pherobase, details pheromones and other semiochemicals associated with more than 7500 species (El-Sayed, 2022).

Pheromone trapping commonly utilises either sex pheromones or aggregation pheromones. The former are usually released by the female to attract the male, whilst aggregation pheromones are released by conspecifics often to mass attack a plant in order to overcome its defences as well as initiating courtship and mating. Isolation of sex pheromones usually starts with behavioural observations, such as the mating behaviour of a pest, or the attraction of multiple males to virgin females. Next extracts of the natural pheromone are extracted from the pest. For example, brassica pod midge (*D. brassicae*) males were attracted to live virgin females, crushed females, hexane washes of females and hexane washes of ovipositors but not to washes of female bodies without ovipositors (Williams and Martin, 1986). Natural pheromones can be collected either by solvent washes of females or by air entrainment, with the latter providing a purer sample for subsequent analyses (Hall et al., 2012). Gas chromatography coupled with electroantennography studies are usually conducted to determine the active components of the pheromone. In this process, the raw pheromone source is passed over the target insect's antenna whilst at the same time passing through a gas chromatograph. Electrical signals generated by the antenna in response to the chemicals can be compared with the gas chromatograph trace and the components that elicit most responses identified. There is a great diversity in pheromone chemistry. In the Lepidoptera, which is the most studied group, pheromone components are typically alcohols, acetates or aldehydes derived from fatty acids; however, insect pheromones also include, amongst others, polyenic hydrocarbons, methyl-branched hydrocarbons, epoxides, ketones, terpenes, terpenoids, iridoids and alkaloids (Jacquin-Joly and Groot, 2018; Rizvi et al., 2021). Furthermore, identification and synthesis of pheromones can be complex due to the sensitivity of insects to precise multi-blend formulations and isomeric compounds including chiral molecules. On the other hand, pheromone components can be active across several species, and there is often similarity in pheromone components within taxa, which can aid in identification and synthesis (Xu et al., 2020).

Once synthesised and tested, the synthetic pheromone is embedded in a suitable substrate that allows for a measured and consistent release of the volatile chemicals over weeks or months. This is usually a rubber septum, polyethylene vial or cellulose sponge enclosed in a sealed polyethylene pouch. Fortunately, insects are highly sensitive to pheromones, and therefore small to minute quantities of chemical are required. The optimum release rate of the pheromone lure and its positioning on the trap is derived from field trials. Good examples of the development process for pheromone trapping of apple leaf midge (*Dasineura mali* Kieffer) and strawberry blossom weevil (*Anthonomus rubi* Herbst) from pheromone synthesis through to field use are given in the previous studies (Cross and Hall 2009; Cross et al., 2006a; Cross et al., 2009; Cross et al., 2006b).

4 Common trap types for collecting pest insects

There are a multitude of different insect trap designs reflecting the great diversity of insect pests, the crops on which they are found, the behaviour of the pest in question and the influence of local factors such as the availability of materials, power sources and labour. Trap types and the names given to them are not absolute, and some traps may use a combination of techniques to trap pests, so, for example, what might be termed a funnel trap in one cropping system may be called a cross-vane trap in another. However, there are some basic commonalities, and these are reflected in the following general categories of crop pest trap design. Some traps which are predominantly used for biodiversity studies are not covered in detail here. This includes common designs such as pitfall traps, where a container is dug into the ground to soil level and crawling insects fall into it. Pitfall traps are commonly used in agro-ecology studies to assess the prevalence and activity of ground-dwelling predators (Brown and Matthews, 2016; Hohbein and Conway, 2018; Woodcock, 2005) or occasionally pests (Hanula, 1990) but are not typically used for routine pest assessment.

4.1 Water traps

Water traps are simple generalist traps, also known as Moericke or pan traps. They consist of coloured bowls filled with water and a little detergent, which serves to break the surface tension as the insect lands on the water and then sinks. Water traps are normally mounted at crop canopy height. The most common design is for the trap to be mounted on a central pole that passes through the centre of the bowl, through an appropriate conduit, which in commercial traps is a moulded part of the bowl. The trap can then be easily raised and lowered by moving it up and down the pole. The other feature is mesh-covered drain holes to prevent overflow if it rains. The attractiveness of the trap is due to the colour of the bowl. For phytophagous insects, yellow is the most common attractive colour (Kirk, 1984), whilst white and blue are also used widely. The effectiveness of the colour is dependent on the taxa (Disney et al., 1982) and other visual issues such as contrast with the background. In some cases, the attractiveness of the trap will alter as the crop matures, e.g. on oilseed rape a yellow trap becomes more conspicuous and attractive to pests as the yellow flowers senesce and the crop becomes green, whilst *Drosophila suzukii* Matsumura are caught less in baited traps when the surrounding crops ripen and thus become more attractive to the flies. Sodium benzoate, 50/50 ethylene glycol/water or saturated salt solution will aid in preserving the catch, although after a few days in water, specimens will become macerated and start to decay. Trap catches can be sieved and preserved in alcohol before sorting. Due to the lack of specificity, water-trap catches are usually sorted under the microscope.

4.1.1 Case study of a water-trapping regime – aphids on potatoes

Yellow water traps are used to monitor pest aphid species on potato crops in the UK (AHDB, 2021; AHDB Potatoes, 2021) (Fig. 1). Aphids can vector potato viruses, and this monitoring system is concerned with transmission of potato

Figure 1 Yellow water trap mounted on a central pole within a potato crop and used to monitor pest aphid species. Source: AFBI image.

Published by Burleigh Dodds Science Publishing Limited, 2023.

virus Y by peach-potato aphid [*Myzus persicae* (Sulzer)] and other aphid species. A sampling kit consisting of yellow water traps, mounts and sample pots is sent to growers or crop inspectors. Growers maintain the traps on their crops, and weekly trap samples are sieved and posted back into the laboratory for identification. The results consist of aphid species counts and a vector pressure index that is calculated as the number of aphid species present multiplied by their relative transmission efficiency (Northing et al., 2004). Growers receive their individual results which are also incorporated into an online national risk map. Growers and advisers can then use this information for timing of haulm burn down and insecticide application if necessary. For seed potato producers, EEC directive 66/403/EEC states that the virus incidence must be less than 4% in the direct progeny of the crop. The service is funded through a levy board, the Agriculture and Horticulture Development Board (AHDB), on a semi-commercial basis in England, Wales and Scotland, and through Government agricultural advisory bodies in Northern Ireland (AFBI 2022).

4.1.2 How water traps meet the key challenges of pest monitoring (e.g. selectivity, accuracy, processing and cost)

Water traps are widely used for crop pest and biodiversity assessment. They are poor to moderately selective as they rely on colour to attract insects, so trap catch is likely to be diverse. Semiochemical lures can be mounted over water traps to improve catch of target species (Blight and Smart, 1999; Hardie et al., 1996), but without a barrier to incursion of other insects (Conrad et al., 2016) or the use of clear or plant-coloured traps with reduced visual attractiveness, the problem of an indiscriminate catch remains. The accuracy of water traps relies on visual attraction and is good where there is contrast between background foliage and the trap colour (Döring et al., 2004). This is particularly so for migrating insects, such as aphids, which are attracted to conspicuous yellow traps as they fly onto the green crop (Döring, 2014). Absolute accuracy for any pest monitoring trap is difficult to determine, and on-plant counts remain the most accurate assessment. Water-trap catches mostly correlate with on-plant counts. However, this varies with cropping system and pest species (Kisimoto, 1968; Lykouressis et al., 2017; Metspalu et al., 2015; Trumble et al., 1982). Nonetheless, trap catches, at the very least, provide an early indicator of pest species within the crop. Processing is straightforward, and water-trap catches can be sieved and stored in alcohol. However, sorting trap catches to species requires significant microscope time, and only cursory assessment can be done by eye or using a hand lens in the field. Although, possibly some distinctive species such as cabbage stem flea beetle [*Psylliodes chrysocephala* (L.)] may be identified in the field especially if they are sampled in the autumn/winter, when bycatch is low. The great advantages of water traps are that they are cheap and simple, using ready-available containers and water, plus household chemicals. They do catch a wide range of non-target species, but other than that environmental impact is low as they are reusable.

4.2 Sticky traps

Sticky traps are one of the simplest and cheapest methods of monitoring insects. At the basic level, they consist of a thin plastic sheet coated in non-drying, water-resistant glue. The plastic sheet is usually of an attractive colour, with yellow, blue or white being the most common commercial colours. Sticky traps are usually mounted vertically, although horizontal and 45°-mounted traps can be more effective depending on the target pest (Blight and Smart, 1999; Holland et al., 2021; Smart et al., 1997). Some traps utilise a cross-vane design, with vertical panels set perpendicular to each other, e.g. the Rebell® trap (Katsoyannos et al., 2000; Remund and Boller, 1978). Insects alighting on the trap are caught in the sticky glue. The glue used on commercial sticky traps is petroleum derived such as polyisobutylene, but other sticky substances such as petroleum jelly, tree resins, boiled linseed oil, corn syrup or honey have also been used (Murphy, 1985). The type of glue can change the reflectance of the trap colour and alter its attractiveness to pests (van Tol et al., 2021). A coloured sticky trap is non-specific and will catch a wide range of flying insects. If the target species for monitoring is easy to identify or if the trap is operated in a protected environment with a single pest species present, counts can be derived directly from the sticky trap by eye, using a hand lens or dissecting microscope. This can be facilitated by covering the sticky trap with a transparent plastic film (e.g. cling film). In this manner, sticky traps can be easily stored and, provided the trap was dry, kept indefinitely. Otherwise handling sticky traps can be messy due to the glue.

The disadvantages of sticky traps are that specimens are often distorted and can be damaged by bird feeding and features may be obscured due to the glue. Various methods have been tried to dislodge and clean specimens for more accurate identification through microscopy or molecular techniques. Murphy (1985) lists several solvents which can be used to dissolve polyisobutylene, such as toluene, heptane, hexane, xylene, ethyl acetate and others. More recently, for health and safety reasons, solvents such as the commercially available De-Solv-it® (Davidson et al., 2015) and orange oil have been recommended. However, the use of solvents can interfere with polymerase chain reaction procedures for identification of pest species or detection of pathogens within insect vectors (Bextine et al., 2008; Marshall et al., 2010).

Sticky traps are often baited with pheromone lures in a delta trap or prism trap design. The body of the trap typically consists of corrugated plastic or waxed cardboard formed into a triangular prism with a sticky trap insert placed glue-side up on the base of the trap. A pheromone lure is mounted inside the prism, either hanging from a thin wire or directly onto the sticky trap insert. This is a cheap and simple design. The outer delta housing protects the pheromone lure and catch from rain and sunlight and helps minimise bycatch of non-target organisms. Choice of colour and other visual cues can enhance catch (Domingue et al., 2016) or dissuade non-target organisms (Clare et al., 2000).

4.2.1 Case study of sticky traps - pea moths

Pea moth [*Cydia nigricana* (F.)] is an economic pest of peas (*Pisum sativum* L.). The larvae feed inside the pea pod, and, as well as direct feeding damage, contamination can lead to the downgrading or rejection of crops for human consumption. This is particularly so with vining peas (harvested green for freezing or canning) where the moth larvae cannot be removed physically during processing, and the threshold for rejection by processing factories can be very low, e.g. 0.5% damage of the consignment in organic crops in Germany (Schultz and Saucke, 2005; Thöming et al., 2011). Female pea moth lay eggs on the leaves, and the first-instar larvae move to and penetrate the developing pod to feed on the peas. The optimum time for any insecticide treatment is when the larvae are exposed and prior to them entering the pod. Depending on whether the crop type is combining or vining, there is a need to determine the requirement for - and timing of - any control measures (Wall, 1988). This problem was addressed by the development of one of the first commercial examples of a pheromone-based pest monitoring system using economic thresholds (Witzgall et al., 2010).

Pheromone-baited delta traps are used to monitor pea moths within crops (Fig. 2). Originally, two compounds were found to be highly attractive to male moths in the field, (*E*,*E*)-8,10-dodecadien-l-yl acetate and (*E*)-10-dodecen-l-yl acetate, with the former subsequently identified as the female sex pheromone (Greenway, 1984). However, this compound isomerised in sunlight to form a repellent (Witzgall et al., 1993). Therefore, the less attractive pheromone analogue [(*E*)-10-dodecen-l-yl acetate] was the practical choice as a more persistent lure, although the pheromone itself combined with an antioxidant to prevent isomerisation was recommended for the more precise detection of moths required for vining crops (Wall, 1988). This is a good illustration of the unforeseen and subtle responses of insects to semiochemicals and demonstrates the fine-tuning and practical compromises required in developing pheromone lures for field trapping. The pheromone lure is impregnated into a rubber septum and hung inside a delta trap with a sticky, removable basal plate. The current recommendation is that two pheromone traps are set per field and examined every 2 days. For combining crops for human consumption, the

Figure 2 Delta trap with sticky trap insert. The trap is baited with a pheromone analogue to attract male pea moths (*Cydia nigricana*). Source: Image courtesy of the PGRO - Processors & Growers Research Organisation, Peterborough, UK.

threshold is 10 or more moths in either trap during two consecutive occasions, with spray date determined by a computer model of moth egg development. For vining crops, the presence of any male moths suggests that control will be necessary (PGRO, 2022; Ramsden et al., 2017; Wall, 1988).

4.2.2 How sticky traps meet the key challenges of pest monitoring (e.g. selectivity, accuracy, processing and cost)

The advantages of sticky traps are that they are a cheap and flexible method of monitoring pest insect numbers. However, the selectivity, accuracy and processing time of sticky traps depends on whether they are relying on colour or semiochemical lures as an attractant and consequently the visual and olfactory sensory ranges of the target insects. A non-conspicuous delta trap baited with a sex pheromone can be highly specific and allow direct counts by eye for pest monitoring; whereas a yellow sticky trap mounted in an open field situation will catch many insect species. Even within a protected environment dealing with a single pest species, there are many factors influencing the accuracy of sticky traps. Pinto-Zevallos and Vänninen (2013) provide an in-depth review of the use, placement and processing of yellow sticky trap in glasshouses to monitor whiteflies. They comment that there are relatively few studies quantifying yellow sticky trap efficacy but of those that do, yellow sticky traps in glasshouses are comparable in terms of efficiency to other sampling methods, including direct on-plant counts.

4.3 Suction traps

Suction traps are non-attractive and sample flying insects from a defined stratum of air. At their simplest, they consist of an impeller fan that sucks air plus insects through a conical mesh sieve to concentrate the insect sample into a collecting jar. Depending on the design, the fan may be mounted at the air intake or at the bottom of a vertical tube that allows higher strata to be sampled. The advantages of suction traps are that they measure absolute abundance per unit volume of air (Harrington et al., 2007b; Taylor, 1982) and are not reliant on trap attractiveness, which may vary depending on the pest's physiological state or background colours or odours. The main disadvantages of such traps are that they require a mains or generator electric power source (Johnson, 1950), are large, are comparatively expensive pieces of equipment and are therefore less suitable for multiple on-crop use. The first design for a suction trap for use on crops was that of Johnson (1950), with improved designs in subsequent versions (Taylor, 1951) and therefore known as the Johnson–Taylor suction trap. This design of trap is still commercially available through Burkard Manufacturing Co Ltd (Rickmansworth, UK) (www.burkard.co.uk). Due to the expense of the trap and the requirement for a power source, most work on crop pests using the Johnson–Taylor suction traps has been for scientific studies or for national surveillance of pests and not monitoring by commercial growers. A smaller battery-powered suction trap was developed and used by Wainhouse (1980) to sample first-instar larvae of the felted beech scale, *Cryptococcus fagisuga* Lindinger, within a forest environment, and Belding et al. (1991) used a solar-powered suction trap for sampling aphids in arable crops in Illinois.

The most widespread use of suction traps has been the sampling of aphid pests of crops. A network of 12.2-m stationary suction traps throughout the UK (Harrington and Woiwod, 2007; Woiwod and Harrington, 1994) was initiated in 1964 to monitor aphid populations ostensibly but has subsequently been used for much more besides this (see Case study in Section 4.3.1). Developed at Rothamsted Research in Harpenden, UK, the Rothamsted 12.2 m suction trap consists of a 9.2 m plastic pipe (internal diameter 244 mm) sitting on a 3 m tall, sealed box containing an electric fan (Fig. 3). The height of 12.2 m (40 feet) matches the mean logarithmic height for aphid flight and is sufficiently high to minimise the effects of immediate surroundings (Harrington and Woiwod, 2007; Taylor, 1974). Air sucked down the pipe passes into an expansion chamber to reduce its speed, and the catch is then sieved through a stainless steel mesh cone into sample jars containing preservative (Macaulay et al., 1988). Similar networks of aphid-monitoring suction traps have been developed across Europe (Harrington et al., 2007a; Woiwod and Harrington, 1994) and also in the USA since 1983, although they are largely based on the design of Allison and Pike (1988). More recently, suction trap networks have been used as forecasting tools for soybean aphid,

Figure 3 Rothamsted 12.2 m suction trap used to sample aphids. The trap is permanently sited and mains powered with an impeller fan housed in the stainless steel body and a long, vertical sample pipe to collect aphids migrating to and from crops. The door of the trap is open showing the collecting mesh and four sample containers, which rotate every 24 h. Source: AKM image.

Aphis glycines Matsumura, an invasive pest first found in North America in 2000 (Rhainds et al., 2010; Schmidt et al., 2012). There are also suction trap networks in New Zealand (Teulon et al., 2004), South Africa (Krüger and Laubscher, 2013) and China (Qin et al., 2013). Individual traps are operated in Northern Ireland by the Agri-Food and Biosciences Institute and in the Republic of Ireland by Teagasc.

4.3.1 Case study suction traps – Rothamsted trap network

A suction trap network for sampling pest aphid species in Great Britain currently consists of 16 12.2 m Rothamsted traps, 12 in England and 4 in Scotland, managed by the Rothamsted Insect Survey (RIS), England, and Science and Advice for Scottish Agriculture (SASA), Scotland (Bell et al., 2015; Harrington and Woiwod

2007). Trap catches are sent to the RIS or SASA for expert identification. The traps provide daily records for a range of aphid species from April to November and weekly aphid counts in the winter (Rothamsted Research, 2021). The principal reason for the network is to forecast potential damage to crops, mainly cereals, from aphids and the viruses that they transmit, based on the timing and size of aphid migrations (Bell et al., 2015). For example, preceding winter temperatures provide a good predictor for the date of spring aphid migrations (Harrington et al., 2007b; Harrington and Woiwod, 2007), and aphid forecasting models have been developed for cereals as well as hops, beans, potatoes and sugar beet (Woiwod and Harrington, 1994). Any insecticidal applications can then be timed correctly for efficient impact on the pest whilst minimising non-target impacts. Each trap is estimated to be representative of aphid abundance in at least an 80 km radius (Cocu et al., 2005; Rothamsted Research, 2021; Taylor, 1974). Weekly aphid counts are disseminated in electronic bulletins and text alerts to interested parties. These data feed into decision support systems (DSSs) for management of aphid pests. The Northern Irish trap ceased to be part of the RIS network in the late 1980s, but a new trap was erected in 2019. In this case, data are displayed graphically on an aphid-monitoring dashboard that compares current counts with previous years. The chart, summary information and a website link are tweeted out to growers (AFBI, 2022).

In addition to monitoring aphid flight and migration, suction trap catches are also regularly used to assess virus prevalence in aphid vectors and insecticide resistance in aphid populations (Harrington and Woiwod, 2007). Since all samples, including the bycatch, have been stored by Rothamsted since 1974, suction traps have also been used retrospectively to study phenologies and flight activities of other taxa, e.g. biting midges (*Culicoides* spp.) (Fassotte et al., 2008; Sanders et al., 2011), social wasps (Archer, 2001), pollen beetles [*Brassicogethes aeneus* (Fabricius)] (Shortall, 2021) and others (Shortall et al., 2013). However, in recent years, the long-term datasets built up from regular aphid monitoring have attained a unique importance. They have been used to demonstrate the impact of climate change on aphid populations, with aphid spring migrations becoming progressively earlier and increased duration of the flight season, as winters have become warmer (Bell et al., 2015; Harrington et al., 2007a).

4.3.2 How suction traps meet the key challenges of pest monitoring (e.g. selectivity, accuracy, processing and cost)

Suction traps are somewhat different to the other trapping systems in that they are designed to sample background insect populations and are not attractive; therefore, their non-selectivity is an advantage. Nevertheless, the likes of the 12.2 m Rothamsted suction traps are carefully positioned in such a way as to intercept migrating aphids from what has been termed the 'aerial plankton', to give a representative forewarning of the timing and density of aphid migrations. These traps are operated on a regional basis because they are comparatively expensive to build, run and process the aphid samples. However, the outputs from the traps are disseminated to a wide range of growers, so they are cost effective.

4.4 Interception traps (window, Malaise and cross-vane)

Various designs of flight interception trap are used to sample insects. Window traps typically consist of a vertical, clear plastic panel (i.e. acrylic, Perspex or Plexiglass) with a collecting trough or drop tray immediately beneath and a rain-guard above (Leather, 2015). Insects fly against the window and fall or crawl downwards in the drop tray, which may be filled with a suitable preservative. Window traps are particularly good for sampling beetles, as they bounce off the window into the sample tray, and the traps are often used in forestry to sample bark beetles (Ozanne, 2005) and even minute Ptiliidae (Komosiński and Marczak, 2018). They have been used for sampling crop pest beetles during experimentation, e.g. pollen beetles on oilseed rape (Ahmed et al., 2013). As they are double sided, they can be used to assess the direction of insect flight (Riis and Nachman, 2006; Williams et al., 2007). However, for routine sampling, they are bulky and prone to damage in high winds. Alternative versions have been developed using a fine permeable mesh as the window. Masner and Goulet (1981) used a mesh window trap treated with pyrethroid insecticide to sample for small Hymenoptera, which are under-represented in Malaise traps as they show a weaker phototrophic response or have difficulty crawling on the Malaise trap material. A very cheap window trap made with a 2 litre soft-drink bottle has been trialled as a potential citizen science trap for monitoring bark and ambrosia beetles (Steininger et al., 2015).

The Malaise trap is the one of the commonest and most widely used flight interception traps. The trap was originally developed by the Swedish entomologist René Malaise who was inspired by the capture of flying insects inside a tent (Malaise, 1937), and correspondingly the structure of the trap resembles an open-sided ridge tent, constructed from fine mesh material (Fig. 4). The trap is erected like a tent with poles and guy ropes holding the mesh trap in place. A central vertical mesh panel runs the length of the trap. Flying insects are intercepted by this mesh panel and move upwards by positive phototaxis (or negative geotaxis) towards an apex, leading them into a collecting jar containing a preservative, usually ethanol (Sheikh et al., 2016). The roof and end walls of the trap provide some shade and are sloped upwards towards the apex, thus further directing any insects towards the collecting jar. A secondary drop tray may be positioned underneath the central panel in the manner of a window trap, to catch any insects such as beetles which drop down when encountering a barrier. The Malaise trap is a powerful tool for sampling

Figure 4 The Malaise flight interception trap is ideal for biodiversity sampling of flying insects, particularly Diptera and Hymenoptera. As a large, delicate and non-selective trap, its use in crop protection has mainly been for experimental studies. Source: AKM image.

insect biodiversity (Skvarla et al., 2021), and as the catches are retained directly in preservative, the quality of specimens is good. This makes the Malaise trap an ideal technique for collecting Diptera and Hymenoptera (Skvarla et al., 2021), including parasitoids (Darling and Packer, 1988; Fraser et al., 2008), which can have relevance to crop protection. Malaise traps have been often used for experimental purposes and to assess biodiversity in agro-ecosystems (Burgio and Sommaggio, 2007; Chay-Hernández et al., 2006; Letourneau et al., 2012; Macfadyen and Muller, 2013; Volpato et al., 2020). One particular study attracted much attention, with an apparently drastic decline in flying insect biomass over 27 years as derived from Malaise trap samples taken in mixed landscapes in Germany (Hallmann et al., 2017). However, Malaise traps are not generally used for routine pest monitoring as they are large and delicate and produce an extensive and mostly indiscriminate catch.

The cross-vane trap works on a similar principle of flight interception as the window trap (Leather, 2015) but is smaller and usually baited with semiochemical lures or a light source. A basic cross-vane trap consists of two panels (the vanes) perpendicular to each other, with a funnel at the bottom (Hines and Heikkenen, 1977) (Fig. 5). The vanes may be clear plastic or an attractive colour depending on the target species. An upper funnel may serve as a rain-guard or alternative collecting mechanism (Knuff et al., 2019; Wilkening et al., 1981), whilst the

Figure 5 A cross-vane trap used for trapping bark beetles in forestry. The trap is baited with a lure, and beetles impact the vanes and fall into the collecting funnel and jar at the bottom. Source: AKM image.

bottom funnel leads to a collecting jar containing a preservative. Flying insects are intercepted by a clear trap or attracted to the colour, lure or light source whereupon they impact the trap and fall, or crawl, into the collecting funnel. The trap may be suspended from a branch or frame (Hines and Heikkenen, 1977; Wilkening et al., 1981) or mounted on a pole (Murchie et al., 1997).

4.4.1 Case study – experimental cross-vane trap for sampling Dasineura brassicae *and its parasitoids*

Murchie et al. (1997) used a cross-vane trap baited with plant kairomones to trap brassica pod midge (*D. brassicae*) and its parasitoids in an oilseed rape crop. The trap design was modified from that of Wilkening et al. (1981) and consisted of yellow painted cross-vanes, above which was a funnel, leading through a clear Perspex column into a collecting jar containing methanol as a preservative (Fig. 6). Isothiocyanates lures were mounted inside the Perspex column. Isothiocyanates are defence chemicals produced by Brassicaceae plants when damaged. The pest species, *D. brassicae*, was caught significantly more in traps

Figure 6 A cross-vane trap for monitoring brassica pod midge and its parasitoids in oilseed rape. The trap is baited with brassica-derived kairomone lures, with different compounds attracting either pests or their parasitoids. Source: AKM image.

baited with allyl isothiocyanate than the control; whereas *D. brassicae's* main parasitoid, *Platygaster subuliformis* (Kieffer), was caught predominantly in traps baited with 2-phenylethyl isothiocyanate. The advantage of this system is that it can trap both the pest and natural enemy selectively dependent on the plant kairomone lure. Most pest monitoring systems are targeted towards the pest but do not necessarily monitor the presence of natural enemies. Knowledge of the occurrence and prevalence of both would lead to more refined economic threshold parameters and targeted integrated pest management approaches (Zhang and Swinton, 2009).

4.4.2 How interception traps meet the key challenges of pest monitoring (e.g. selectivity, accuracy, processing and cost)

There is a difference in approach between a Malaise or a window trap and a semiochemical-baited cross-vane trap. The former traps are largely non-selective and designed to intercept insects as they move along flight paths. Therefore, these traps are ideal for biodiversity assessment but are not commonly used for in-crop sampling. Whilst cross-vane traps can act as interception traps in their own right, for crop or forestry protection purposes they are usually baited with semiochemical lures. In this set-up, they behave more like a cone or a funnel trap, where the cross-vanes form part of the trap directing insects towards the semiochemical plume rather than acting as the main trapping component. This illustrates how an addition of a lure can dramatically alter the characteristics of the trap design.

4.5 Funnel and cone traps

Funnel and cone traps typically rely on semiochemicals to attract pest species. They form the basis of many commercially available trap designs for pheromone-based on-crop monitoring. They have the advantages of being reusable and more durable than sticky traps as they are not affected by dust and detritus and they do not lose effectiveness as catch numbers build up.

Funnel traps are designed for use with pheromone lures. The term 'funnel trap' is used for two main designs. A common design for capturing pest moth species consists of a pheromone dispenser mounted above a funnel structure leading into a collecting jar (Fig. 7). This design may also be commonly referred to as a 'bucket trap'. The funnel acts as a catch-retaining mechanism, as it is easier to enter the collecting jar through the funnel than to exit. They are normally constructed of three plastic sections that clip together. The upper section is moulded to form a rain-guard and pheromone dispenser receptacle. The middle section is the funnel unit and lower section is the collecting jar. Flying insects that get attracted to the pheromone flutter around the dispenser, fall into the funnel and are captured. Some designs may have cross-vanes to intercept insect flight and aid in capture (Fountain et al., 2017). These traps are used in field crop situations to monitor moth pests such as diamond back moth [*Plutella xylostella* (L.)] and silver Y moth [*Autographa gamma* (L.)].

The Lindgren funnel trap consists of a stack of several wide black funnels, convex side down, loosely hung together to form a column that mimics a tree trunk (Lindgren, 1983) (Fig. 8). The funnels can be compressed into each other to minimise space for transport and storage. Lindgren funnel traps are used predominantly to capture bark beetles in forests, rather than crop pest monitoring. Beetles attracted to the pheromone lures attached to the trap collide with it and fall through the series of funnels to a collection jar at the bottom containing a preservative such as ethylene glycol. A rain-guard sits above the uppermost funnel to prevent flooding.

Cone traps are the inversion of funnel traps and work by insects flying or crawling upwards inside a cone (or funnel), towards an attractant. The apex of the cone or spout of the funnel protrudes upwards into the collecting jar, and insects are retained on the lobster-pot principle that it is easier to enter the trap, being channelled by the cone, than it is to escape back through the small entrance hole. For many insects, a positive phototropic response will retain them in the collecting jar as well, especially if the trap bottom is opaque and

Figure 7 A moulded plastic funnel trap for trapping moths. Source: AKM image.

Figure 8 A Lindgren funnel trap used for trapping beetle pests in forest. The trap is baited with semiochemical lures, and the elongated shape mimics a tree trunk. Beetles that impact the trap fall into the funnels and ultimately the collecting jar at the bottom. Source: Image from Pavuk and Wadsworth (2013) under the Creative Commons Attribution (CC BY) licence. No changes were made to the picture.

the collecting jar transparent. For some cone traps, liquid preservatives, such as ethanol, can be held in the collecting jar or insecticidal kill strips can be used to further retain insects and aid in processing (Hardee et al., 1996). Cone traps can be mounted on crop for flying insects or placed on the ground for crawling pests. Depending on the design, the latter may be referred to as pyramid traps or Leggett traps and are used for trapping pest beetles which may crawl as well

Figure 9 A McPhail trap for monitoring fruit flies (Tephritidae). Flies enter the trap through an inverted funnel entrance in the bowl base attracted by pheromone lures or liquid food baits. Source: Image by Peggy Greb, USDA Agricultural Research Service.

as fly to their host crops (Mulder et al., 2012; Nielsen and Jensen, 1993; Prokopy and Wright, 1998; St Onge et al., 2018).

The McPhail trap is a common trap design that works on the cone trap principle and is used mainly for sampling pest flies or social wasps. In this design, the cone is built into the base of the main receptacle, with insects entering through an invaginated, inwardly tapering entrance in the bottom, attracted either by a pheromone lure or by liquid food baits (Hernández-Ortiz et al., 2021) (Fig. 9). Commercial traps are made of moulded plastic in a similar way as the funnel/bucket traps. They comprise two detachable parts: the receptacle bowl with the inverted funnel entrance, typically coloured yellow; and a transparent domed-cylinder upper section forming a cover to the bowl and inside which a pheromone lure can be suspended. A liquid bait is poured into the base of the trap and can contain a preservative to preserve the insect catch; alternatively, if a pheromone lure is used, liquid preservative (e.g. water, propylene glycol and sodium borate) can also be held in the base of the trap (Hernández-Ortiz et al., 2021; Thomas, 2008). In a cropping situation, McPhail traps are used for monitoring pest fruit flies (Tephritidae). The trap is named after M. McPhail who used the traps to study the Mexican fruit fly [*Anastrepha ludens* (Loew)] in the 1930s and 1940s, but the basic concept is older (Steyskal, 1977).

4.5.1 Case study suction traps - boll weevil monitoring using cone traps

One of the most extensive uses of cone traps is for monitoring cotton boll weevil (*Anthonomus grandis* Boheman) using the boll weevil 'grandlure'

Figure 10 A boll weevil cone trap. Weevils are attracted to the colour and pheromone plume from the baited traps. Upon landing, they move upwards into the cone and are retained within the collecting jar at the top. Source: Image by Dale Spurgeon, USDA Agricultural Research Service.

synthetic pheromone (Hardee et al., 1972; Tumlinson et al., 1969). The basic boll weevil trap consists of three parts: a light green or yellow upwardly tapered plastic cylinder, with a mesh funnel above leading to a transparent collecting jar (Fig. 10). Boll weevils are attracted to the colour of the trap and the pheromone plume. On landing on the tapered cylinder, they move upwards to the pheromone, through the overlapping funnel into the collecting jar. The specifics of the trap have been under constant refinement since the early 1970s, but most current traps follow the basic design of Mitchell and Hardee (1974) (Dickerson, 1986; Hardee et al., 1996). Boll weevil cone traps are used extensively in the Cotton Boll Weevil Eradication Program in the USA, a collaborative venture involving the USDA, state governments, researchers and cotton producers, which has seen boll weevil eradicated from most of the USA (Raszick and Bernaola, 2021).

4.5.2 How funnel and cone traps meet the key challenges of pest monitoring (e.g. selectivity, accuracy, processing and cost)

Funnel and cone traps mostly rely on semiochemical attraction. They are therefore highly selective and, depending on the attractant source, have a low bycatch. Accuracy is dependent on the range of the semiochemical plume and, as with all lure-based traps, dependent on weather conditions, the age of the lure and the response of the target species. The trap design of the moulded plastic funnel and McPhail-type traps can be used for a wide variety of pest species, which reduces manufacturing costs, and the traps are robust and reusable.

5 Automated traps

5.1 Time-sorting traps

The earliest form of trap automation was to time-sort trap catches. A light trap fitted with a clockwork time-sorting mechanism was in use as early as 1933 to determine the flight activity of nocturnal Lepidoptera (Williams, 1935). Timing automation serves to provide more information on the time and date of trap catches whilst avoiding the extra effort in changing sampling receptacles. The mechanisms for time-sorting are rotating turntables (Bolliger et al., 2020; Taylor et al., 1982; Williams, 1935; Williams, 1958), rotating sticky surfaces (Riedl and Croft, 1981) or mechanical separation of the catch. For the mechanical separation of the catch, the Johnson–Taylor suction trap used a disc-dropping mechanism to time-sort catches (Johnson, 1950; Taylor, 1951). An electrical time switch connected to a solenoid releases metal discs, fringed with cotton, to drop down the sample tube, stacking the suction trap catch into successive hourly samples. In an extensive study, Lewis and Taylor (1965) used the disc-dropping timing mechanism to derive the diurnal flight periodicity of 400 taxa of insects at multiple sites in the UK. Others have used clockwork mechanisms to time-sort catches. Murchie et al. (2001) used a 24 h quartz clock motor to rotate a funnel mechanism that directed Malaise trap catches into 2-hourly samples. This had the advantage of being cheap to run, as it relied on a single 1.5 V AA battery, which could power the motor for at least 3 months. They used the mechanism to derive the diel flight periodicity of *D. brassicae* on oilseed rape. Rohitha and Stevenson (1987) used a clockwork mechanism to drop and expose sticky trap drums in sequence to provide information on daily aphid catches. For the practical purposes of avoiding staff working during the weekends, the 12.2 m Rothamsted suction traps rotate sample jars using a stepping motor to provide 24 h trap counts over a 4-day period (Bouchery, 1979; Macaulay et al., 1988).

5.2 Automated counting and identification

The next development was traps that could automatically count and identify trapped insects. This is a rapidly evolving area as electronic sensors become cheaper and more sensitive, whilst traps can now be monitored remotely using mobile phone, wireless and cloud technology (Lima et al., 2020). Initially, most designs to count insects entering traps have used some form of optical detection (Sciarretta and Calabrese, 2019). For example, Hobbs and Hodges (1993) illuminated insects passing down a suction trap against a dark background, with sensors to detect the scattered light. Similarly, the passage of insects through infrared beams has also been used for counting insects entering traps, e.g. tobacco budworms and cabbage loopers (Hendricks,

1985), boll weevil (Beerwinkle, 2001), stored product pests (Epsky and Shuman, 2001) and orchard moth pests (Holguin et al., 2010). More recently a sophisticated system using backscattered infrared coupled with a high-speed photodetector has been used to detect insects within targeted volume of air (Rydhmer et al., 2022). Jiang et al. (2008) counted oriental fruit fly [*Bactrocera dorsalis* (Hendel)] entering methyl eugenol-baited traps using interruption of infrared beams as the counting trigger. Fruit fly counts along with temperature, humidity and wind speed logging were done remotely at the trapping unit and periodically sent to a server PC via the mobile phone (Global System for Mobile communication) network. The eventual idea is that real-time trapping data from a network of traps could be disseminated to regional farmers via the internet.

Other approaches have used digital images to count insect samples. Selby et al. (2014) used baited pyramid (cone) traps modified to house an infrared motion-sensing camera set-up, originally designed for wildlife recording, to count plum curculio [*Conotrachelus nenuphar* (Herbst)] in Michigan. The system took photographs of each weevil entering the trap, so the number of photographs acted as a count for trap entry, whilst the photographs could be retained to confirm identity. Photographs were stored on the camera's memory card and collected physically. Others have collected photographs remotely. Doitsidis et al. (2017) monitored olive fruit fly, *Bactrocera oleae* (Gmelin), in Crete using ammonium sulphate-baited McPhail traps fitted with GSM/GPRS-linked digital cameras that uploaded images to the internet using the mobile phone network. Counting of fruit flies was then accomplished using an automated multi-stage image analysis approach, whereby the image is enhanced and filtered and the fruit flies are counted as the percentage black within the processed image. This process relied on the specificity of the trap and was not able to distinguish fruit flies from other insects, although separation on the basis of size would seem to be possible using this approach.

5.3 Automated visual identification

For non-specific traps, identification of the target species is required. There are two main methods by which this can be achieved: direct visual assessment of the target or identification via wingbeat frequency. Visual assessment relies on a digital image of the catch being captured (reviewed by Barbedo, 2020; Preti et al., 2021). Similar to the examples above, Guarnieri et al. (2011) used a modified mobile phone integrated into a commercial pheromone sticky trap to monitor codling moths [*Cydia pomonella* (L.)] in apple orchards. Daily photographs of the sticky plate were sent from the trap to a remote server where they could be identified and counted on-screen, without disturbing the trap. Similarly, Rassati et al. (2016) used modified security cameras that automatically collected and sent images to an internet repository. The cameras were strapped to Lindgren funnel traps and used to monitor longhorn and bark beetles, with the intention to use them at seaport sites where there was a plant health risk. Such approaches rely on human identification of the photographs.

Increasingly, image analysis algorithms and machine learning are being used to automatically identify and count insect catches (Barbedo, 2020; Preti et al., 2021; Sciarretta and Calabrese, 2019; Xie et al., 2015); this is a rapidly developing area. In a review paper, Hoye et al. (2021) argue that computer vision and deep learning, a subset of machine learning, have tremendous potential to automate high-throughput systems for the identification of insects. In another review paper, Gaston and O'Neill (2004) discuss some of the limitations to automation of insect identification, both technical and cultural. Ding and Taylor (2016) provide a useful overview of the image analyses techniques used to identify insects, including analyses of wing venation and structures, colour histogram features and morphometric measurements. However, they point out that whilst some systems will work in a laboratory set-up, automatic identification of field trap catches has significant challenges related to poor-quality images, trap movement, contamination with dirt and leaves, decay and distortion of the specimens. Nevertheless, by using a sliding window-based detection pipeline with a convolutional neural network approach, they were able to identify codling moths from pheromone sticky traps. Similar deep learning approaches have been used for the identification of red turpentine beetle (*Dendroctonus valens* LeConte) from pheromone traps in pine forests in China (Sun et al., 2018), whereby the detector and processor are embedded in the pheromone trap, with the ultimate aim to develop traps as part of an Internet of Things monitoring network. Sun et al. (2018) also give a useful review of the use of image-based pest detection methods, which have been practically applied.

5.4 Automated identification via wingbeat harmonics

Flying insects can be identified by measurement of wingbeat harmonics (Moore et al., 1986). Wingbeat frequency is usually inversely proportional to the body mass and particularly the size of the insect wing. This, plus other characteristics such as wing shape, stroke and number, can result in a unique wingbeat harmonic signal that is phylogenetically clustered (Tercel et al., 2018). Whilst differences in wingbeat frequency can be detected acoustically (for example, compare the different sounds of a mosquito and bumblebee), for trapping purposes, detection is usually by optoacoustic means. Light scattered off beating wings can give a unique signal. Moore and Miller (2002) demonstrated that an optical sensor and data acquisition system could identify alate aphids passing through a halogen lamp light source. They specified that wingbeat waveform derived from a trained artificial neural network approach rather than wingbeat frequency alone provided the most discriminatory signal, as, for instance, wingbeat frequency altered with ambient temperature.

Potamitis et al. (2014) developed a prototype electronic McPhail trap to monitor olive fruit fly *Bactrocera oleae* (Rossi). The key component was an optoelectronic sensor comprising infrared-emitting LED coupled with an array of five phototransistors (Fig. 11). As olive fruit flies flew upwards into the McPhail trap through the ventral entrance, they passed through the infrared beam generating an analogue signal that was sent to a microcontroller, which enabled counting. Trigger events (insects breaking the beam) were stored and sent by text message from a GSM expansion board at predefined intervals to a remote operator. Potamitis et al. (2015) further developed this trap to include GPS, temperature and humidity monitoring. More pertinently, they were able to identify olive fruit fly by their wingbeat harmonic signature, with all processing conducted by the on-board electronics in the trap. The fruit flies only take 30–50 ms to cross the infrared beam. The photodiodes (replacing the earlier phototransistors) detect the minute fluctuations in the infrared beam, and the

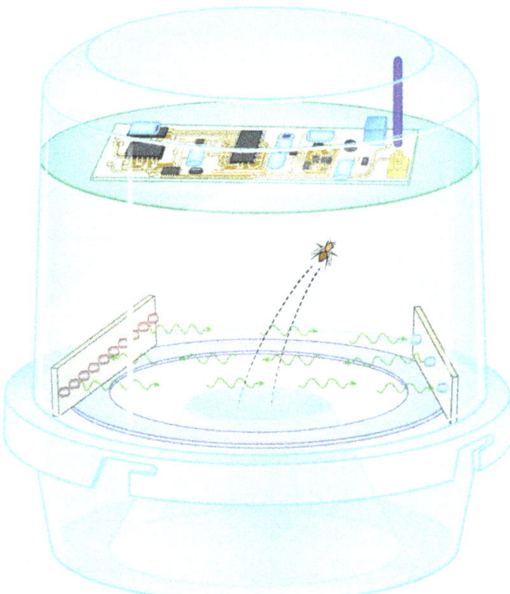

Figure 11 Schematic for an electronic McPhail trap that counts and identifies olive fruit fly. As the flies enter the trap, they pass through an infrared beam, and fluctuations in the beam caused by the flies' wingbeats can generate a unique signal that can be used to identify them. Source: Image from Potamitis et al. (2015) under the Creative Commons Attribution (CC BY) licence. No changes were made to the figure.

signal is processed by Fourier transform, which decomposes the waveform into its constituent frequencies. The wingbeat frequency alone is insufficient to discriminate between species, but the associated wingbeat harmonics can. Potamitis et al. (2015) provide the analogy of musical instruments playing the same note (e.g. frequency), but the human ear can distinguish between say a flute and a violin by other harmonics produced by the instruments. Wingbeat frequency spectra in the trap are compared with those collected in the laboratory, using embedded software. In laboratory trials, the trap was able to discriminate well between *B. oleae* and dissimilar insects such as mosquitoes, non-biting midges and honeybees (>90% accurate) but was less accurate with flies whose wingbeat spectra overlapped.

6 Conclusion

The term precision agriculture is used to describe enhanced temporal and spatial targeting by which crop inputs are applied. To incorporate integrated pest management principles within a precision agriculture framework requires an accurate assessment of pest numbers, preferably before any damage to the crop. Monitoring of pest species occurrence, activity and density is an essential part of integrated pest management. This allows crop protection decisions to be made using economic thresholds, which are a cost/benefit analyses of whether to apply a control measure. In theory, economic thresholds should be flexible and reflect market value of the crop and the cost of any control measure, including machinery costs, labour and fuel. They should also take cognisance of any indirect effects of application, in particular, any effects on natural enemies within the crop that are already exerting a level of pest control. In practice, economic thresholds are often fixed and dated and tend not to be based on published evidence. Furthermore, the method of pest density assessment in relation to the economic threshold must be practicable (Ramsden et al., 2017).

Pest monitoring via trapping falls into three broad categories. Regional-scale monitoring, often for forecasting purposes, utilises specialised traps and experts to process trap catches. The Rothamsted suction trap network is an exemplar of this approach, with aphid survey bulletins freely available to farmers. Another approach involves in-crop monitoring of selected crops as regional sentinels, also with expert trap processing. An overview of pest threat within a region can then be disseminated to growers. Finally, there is on-farm, in-crop monitoring. This is the most accurate method of pest assessment as it relates to the individual crop and therefore can be directly linked to economic spray thresholds. However, to be successful, in-crop monitoring relies on traps being robust, easy to use and easy to process. The trap should be selective, and whilst colour and light act as attractants and are somewhat discriminatory, selectivity has mainly been achieved using semiochemical lures. Since 1959, when the first pheromone was identified, knowledge of insect chemical ecology has expanded to the point where synthetic lures, either pheromones or kairomones, are available for many if not most major pest species. Use of semiochemical lures can reduce bycatch, meaning that most specimens are of the target species, reducing the need for expert identification.

Clean trap catches also make automatic counting more accurate. Whilst trap automation for time-sorting has been used since the 1930s, recent advances in the development and costs of digital cameras and microelectronics mean that automatic counting is now a practical possibility. There are two main ways to count insects in a trap: by image capture of trap catches or by incoming insects disrupting an infrared beam. More ambitious is to both count and identify trapped specimens. Again, the two main approaches are image analyses or identification via wingbeat harmonics derived from reflected infrared beams. Most of the developments in this area have been experimental, and few are adopted commercially, as yet. This is because the technology is still developing but also because the cost of 'smart traps', both purchasing and consumables (e.g. batteries), is considerably more per unit than the basic trap, which is likely to be 10–100 times less expensive, although substantial savings can be made in labour costs. However, a number of entrepreneurial companies are developing traps with an aim to full commercialisation (e.g. FaunaPhotonics www.faunaphotonics.com, DIOPSIS www.diopsis.eu).

The research that is ongoing in trapping technology seeks to improve and refine trap catches whilst crucially minimising processing time. Automated 'smart traps' that count and identify pest species have the potential to substantially increase data collection from crops. This will enable more informed decision-making about insecticide applications. Instead of all-crop spraying or prophylactic applications, control measures will only be targeted at crop areas that need them and only when justified by economic spray thresholds. Such an approach has clear environmental benefits in terms of reduced impact on non-target organisms and potential pollution of waterways and prolongs the life of active ingredients by reducing the conditions for insecticide resistance to develop.

7 Future trends in research

There are many trap designs for monitoring insect pests, and their improvement is a continuous process. Identification and refinement of semiochemical lures will continue as new pheromones and kairomones are identified and synthetic blends improved. New technologies are being utilised to dispense pheromones at a more natural rate, prolonging the life of the lure (Heuskin et al., 2011; Leskey et al., 2021; Muñoz-Pallares et al., 2001). There is also a need to determine the range and accuracy of trapping systems utilising both olfactory and visual responses and relate these to damage on the crop. Whilst this is a basic requirement of an economic threshold approach, it is difficult to quantify trap efficacy in the field, and mathematical approaches coupled with field validation are required.

Trap automation holds great potential. Technologies that are used to count and identify pests on-trap or through remote connections have been developed and will become more refined as technology advances. The use of image and wingbeat analyses using machine learning is at the forefront of these techniques, and there is great scope to advance these methods further. One possible scenario is that traps will connect through an autonomous network, proving real-time data on pest migration to a region or crop. It is also hoped that such monitoring technologies could provide data on the presence and densities of pest natural enemies within the crop, which would provide a holistic monitoring system to feed into DSSs.

Some of the technological developments for insect trapping of human-pathogen vectors could also be utilised in crop protection. Vector biologists have been exploring the use of unmanned aerial vehicles (or drones) to operate traps in field (Faraji et al., 2021; Vacek et al., 2017). For example, traps may be positioned by drones into marshy, inhospitable or dangerous areas. Drones could be used to collect trap contents and deliver physical samples to a collection point for identification or further analyses; or the drone itself is developed into a flying trap that can move easily from location to location (Ye et al., 2020). The Microsoft Premonition project seeks to develop a network of robotic smart traps to monitor mosquitoes to aid in forecasting disease outbreaks. Using wingbeat identification, the traps capture only pathogen-vector species and retain them for detection of any possible pathogens. In addition, these traps are continually monitoring the environmental conditions related to individual mosquito capture, with the potential to build up an immense dataset for modelling mosquito populations (Choney, 2020; Microsoft, 2022). Some approaches looking at disease prevalence do not require capture of the insects per se, but rather their saliva or faeces are collected from traps, which subsequently can be processed for pathogen identification (Hall-Mendelin et al., 2010; Meyer et al., 2019). Such techniques may be of value in monitoring pathogens vectored by phytophagous insects, particularly in cases where insect-feeding damage is not economically important, or surveillance for an invasive plant health pathogen.

If we look to the future, perhaps we have the potential to produce a network of solar-powered, autonomous, interlinked traps sited in individual crops but providing data for a regional dataset. Such traps will be baited with pest and natural-enemy-specific semiochemicals, with automated counting and identification. Real-time data on pests, natural enemies and environmental conditions will be supplied through the mobile phone network to a central server where DSSs linked to market value will calculate the risk to the crop with respect to the economic threshold.

8 Where to look for further information

8.1 Information on the design and types of insect traps

- Southwood, T. R. E. 1978. Ecological Methods: With Particular Reference to the Study of Insect Populations (2nd edn). London: Chapman and Hall.
- Muirhead-Thomson, R. 1991. Trap Responses of Flying Insects - The Influence of Trap Design on Capture Efficiency. London: Academic Press.
- Santos, J. C. and Fernandes, G. W. 2020. Measuring Arthropod Biodiversity: A Handbook of Sampling Methods: Springer Nature.
- Schauff, M. E. (Ed.). 2001. Collecting and Preserving Insects and Mites: Techniques and Tools. Updated and modified version of the USDA Misc. Publication no. 1443 published by the Agricultural Research service in 1986 and edited by George C. Steyskal, William L. Murphy, and Edna M. Hoover. Available at www.ars.usda.gov/ARSUserFiles/80420580/CollectingandPreservingInsectsandMites/collpres.pdf (accessed May 2022).
- Gibb, T. J. and Oseto, C. Y. 2006. Arthropod Collection and Identification: Field and Laboratory Techniques. Amsterdam: Academic Press.

8.2 The principles of pest monitoring

- Pedigo, L. P. and Buntin, G. D. (Eds.). 1994. Handbook of Sampling Methods for Arthropods in Agriculture. Boca Raton, FL: CRC Press.
- Binns, M. R., Nyrop, J. P. and van der Werf, W. 2000. Sampling and Monitoring in Crop Protection (1st edn). New York: CABI Publishing, p. 284.

8.3 Pheromone trapping

- Hardie, J. and Minks, A. K. (Eds.). 1999. Pheromones of Non-Lepidopteran Insects Associated with Agricultural Plants. Wallingford, UK: CABI Publishing.
- Howse, P. E., Stevens, I. D. R. and Jones, O. T. 1998. Insect Pheromones and Their Use in Pest Management. London: Chapman & Hall.

8.4 Automated identification

- Barbedo, J. G. A. 2020. Detecting and classifying pests in crops using proximal images and machine learning: A review. *AI* 1(2): 312-328. doi: 10.3390/ai1020021.
- Lima, M. C. F., Leandro, M. E. D. d. A., Valero, C., Coronel, L. C. P. and Bazzo, C. O. G. 2020. Automatic detection and monitoring of insect pests–A review. *Agriculture* 10(5): 161. doi: 10.3390/agriculture10050161.

- Preti, M., Verheggen, F. and Angeli, S. 2021. Insect pest monitoring with camera-equipped traps: Strengths and limitations. *Journal of Pest Science* 94(2): 203–217. doi: 10.1007/s10340-020-01309-4.
- Potamitis, I., Rigakis, I. and Fysarakis, K. 2015. Insect biometrics: optoacoustic signal processing and its applications to remote monitoring of McPhail type traps. *PLoS ONE* 10: e0140474. doi: 10.1371/journal.pone.0140474.

8.5 Specific trapping systems

8.5.1 Rothamsted suction trap network

- Harrington, R., Hullé, M. and Plantegenest, M. 2007. Monitoring and forecasting. Aphids as Crop Pests. Oxford: CAB International, pp. 515-536.
- Pinto-Zevallos, D. M. and Vänninen, I. 2013. Yellow sticky traps for decision-making in whitefly management: What has been achieved? Crop Protection 47: 74–84. doi: 10.1016/j.cropro.2013.01.009.

9 Acknowledgements

I appreciate the valuable comments on the manuscript by Dr Stephen Jess (AFBI) and the editors, Drs Michelle Fountain and Tom Pope. My work on surveillance and monitoring of insect pests is supported by the Department of Agriculture, Environment & Rural Affairs (DAERA) through both statutory and 'Evidence & Innovation' programme funding.

10 References

AFBI. 2022. Aphid monitoring. Available at: www.afbini.gov.uk/aphid-monitoring (accessed February 2022).

AHDB. 2021. The UK aphid monitoring network. Available at: https://ahdb.org.uk/knowledge-library/the-uk-aphid-monitoring-network (accessed October 2021).

AHDB Potatoes. 2021. Aphid monitoring water trap results. Available at: https://aphmon.fera.co.uk/index.cfm (accessed October 2021).

Ahmed, N., Englund, J. E., Johansson, E. and Ahman, I. 2013. Does insecticide application in a winter oilseed rape field influence the abundance of pollen beetle *Meligethes aeneus* in nearby ornamental flowers and vegetables? *Pest Management Science* 69(11): 1253–1260. doi: 10.1002/ps.3492.

Allison, D. and Pike, K. S. 1988. An inexpensive suction trap and its use in an aphid monitoring network. *Journal of Agricultural Entomology* 5: 103–107.

Archer, M. E. 2001. Changes in abundance of *Vespula germanica* and *V. vulgaris* in England. *Ecological Entomology* 26(1): 1–7.

Barbedo, J. G. A. 2020. Detecting and classifying pests in crops using proximal images and machine learning: A review. *AI* 1(2): 312–328. doi: 10.3390/ai1020021.

Beerwinkle, K. 2001. An automatic capture-detection, time-logging instrumentation system for boll weevil pheromone traps. *Applied Engineering in Agriculture* 17: 893–898.

Belding, M. J., Isard, S. A., Hewings, A. D. and Irwin, M. E. 1991. Photovoltaic-powered suction trap for weakly flying insects. *Journal of Economic Entomology* 84(1): 306–310.

Bell, J. R., Alderson, L., Izera, D., Kruger, T., Parker, S., Pickup, J., Shortall, C. R., Taylor, M. S., Verrier, P. and Harrington, R. 2015. Long-term phenological trends, species accumulation rates, aphid traits and climate: Five decades of change in migrating aphids. *Journal of Animal Ecology* 84(1): 21–34. doi: 10.1111/1365-2656.12282.

Bextine, B., Marshall, P., Mitchell, F. and Brady, J. 2008. Using an orange oil solvent to remove insects from yellow sticky traps: Impacts on the detection of *Xylella fastidiosa* DNA in glassy-winged sharpshooter samples. In: Esser, T. (Ed.), *Proceedings, 2008 Pierce's Disease Research Symposium California Department of Food and Agriculture*. San Diego, California, pp. 12–15.

Blackshaw, R. P. 1983. Some factors influencing variability in water-trap catches of *Tipula* spp. (Diptera: Tipulidae). *Bulletin of Entomological Research* 73(4): 693–699. doi: 10.1017/S0007485300009299.

Blight, M. M. and Smart, L. E. 1999. Influence of visual cues and isothiocyanate lures on capture of the pollen beetle, *Meligethes aeneus* in field traps. *Journal of Chemical Ecology* 25(7): 1501–1516. doi: 10.1023/A:1020876513799.

Bolliger, J., Collet, M., Hohl, M. and Obrist, M. K. 2020. Automated flight-interception traps for interval sampling of insects. *PLoS ONE* 15(7): e0229476. doi: 10.1371/journal.pone.0229476.

Bouchery, Y. 1979. Description d'un collecteur à l'exterérieur, destiné à recueillir automatiquement les captures d'arthropodes d'un piège à succion. *Annales de Zoologie Ecologie Animale* 11: 113–117.

Briscoe, A. D. and Chittka, L. 2001. The evolution of color vision in insects. *Annual Review of Entomology* 46: 471–510.

Brown, G. R. and Matthews, I. M. 2016. A review of extensive variation in the design of pitfall traps and a proposal for a standard pitfall trap design for monitoring ground-active arthropod biodiversity. *Ecology and Evolution* 6(12): 3953–3964. doi: 10.1002/ece3.2176.

Burgio, G. and Sommaggio, D. 2007. Syrphids as landscape bioindicators in Italian agroecosystems. *Agriculture, Ecosystems and Environment* 120(2–4): 416–422. doi: 10.1016/j.agee.2006.10.021.

Chay-Hernández, D. A., Delfín-González, H. and Parra-Tabla, V. 2006. Ichneumonoidea (Hymenoptera) community diversity in an agricultural environment in the state of Yucatan, Mexico. *Environmental Entomology* 35(5): 1286–1297.

Choney, S. 2020. Predicting epidemics like the weather: How Microsoft Premonition can help in the global fight against disease outbreaks. *Microsoft Website*. Available at: https://news.microsoft.com/innovation-stories/microsoft-premonition/ (accessed April 2022).

Clare, G., Suckling, D. M., Bradley, S. J., Walker, J. T. S., Shaw, P. W., Daly, J. M., McLaren, G. F. and Wearing, C. H. 2000. Pheromone trap colour determines catch of non-target insects. *New Zealand Plant Protection* 53: 216–220.

Cocu, N., Conrad, K., Harrington, R. and Rounsevell, M. D. A. 2005. Analysis of spatial patterns at a geographical scale over north-western Europe from point-referenced aphid count data. *Bulletin of Entomological Research* 95(1): 47–56. doi: 10.1079/ber2004338.

Conrad, N., Brandes, M. and Heimbach, U. 2016. Adults of *Psylliodes chrysocephala* in different types of yellow water traps in winter oilseed rape. *IOBC-WPRS Bulletin* 116: 51-52.

Cross, J. V. and Hall, D. R. 2009. Exploitation of the sex pheromone of apple leaf midge *Dasineura mali* Kieffer (Diptera: Cecidomyiidae) for pest monitoring: Part 1. Development of lure and trap. *Crop Protection* 28(2): 139-144. doi: 10.1016/j.cropro.2008.09.008.

Cross, J. V., Hall, D. R., Innocenzi, P. J., Hesketh, H., Jay, C. and Burgess, C. M. 2006a. Exploiting the aggregation pheromone of strawberry blossom weevil *Anthonomus rubi* (Coleoptera: Curculionidae): Part 2. Pest monitoring and control. *Crop Protection* 25(2): 155-166. doi: 10.1016/j.cropro.2005.04.004.

Cross, J. V., Hesketh, H., Jay, C. N., Hall, D. R., Innocenzi, P. J., Farman, D. I. and Burgess, C. M. 2006b. Exploiting the aggregation pheromone of strawberry blossom weevil *Anthonomus rubi* Herbst (Coleoptera : Curculionidae): Part 1. Development of lure and trap. *Crop Protection* 25(2): 144-154. doi: 10.1016/j.cropro.2005.04.002.

Cross, J. V., Hall, D. R., Shaw, P. and Anfora, G. 2009. Exploitation of the sex pheromone of apple leaf midge *Dasineura mali* Kieffer (Diptera: Cecidomyiidae): Part 2. Use of sex pheromone traps for pest monitoring. *Crop Protection* 28(2): 128-133. doi: 10.1016/j.cropro.2008.09.004.

Darling, D. C. and Packer, L. 1988. Effectiveness of malaise traps in collecting Hymenoptera: The influence of trap design, mesh size, and location. *Canadian Entomologist* 120(8-9): 787-796.

Davidson, M. M., Nielsen, M.-C., Butler, R. C., Vellekoop, R., George, S., Gunawardana, D., Muir, C. A. and Teulon, D. A. J. 2015. The effect of adhesives and solvents on the capture and specimen quality of pest thrips on coloured traps. *Crop Protection* 72: 108-111. doi: 10.1016/j.cropro.2015.03.008.

Dicke, M. and Sabelis, M. W. 1988. Infochemical terminology: Based on cost-benefit analysis rather than origin of compounds? *Functional Ecology* 2(2): 131-139. doi: 10.2307/2389687.

Dickerson, W. A. 1986. Grandlure: Use in boll weevil control and eradication programs in the United States. *Florida Entomologist* 69(1): 147-153.

Ding, W. and Taylor, G. 2016. Automatic moth detection from trap images for pest management. *Computers and Electronics in Agriculture* 123: 17-28. doi: 10.1016/j.compag.2016.02.003.

Disney, R. H. L., Erzinçlioğlu, Y. Z., Henshaw, D. J. C., Howse, D., Unwin, D. M., Withers, P. and Woods, A. 1982. Collecting methods and the adequacy of attempted fauna surveys, with reference to the Diptera. *Field Studies* 5: 607-621.

Doitsidis, L., Fouskitakis, G. N., Varikou, K. N., Rigakis, I. I., Chatzichristofis, S. A., Papafilippaki, A. K. and Birouraki, A. E. 2017. Remote monitoring of the *Bactrocera oleae* (Gmelin) (Diptera: Tephritidae) population using an automated McPhail trap. *Computers and Electronics in Agriculture* 137: 69-78. doi: 10.1016/j.compag.2017.03.014.

Domingue, M. J., Andreadis, S. S., Silk, P. J., Ryall, K. L. and Baker, T. C. 2016. Interaction of visual and chemical cues in promoting attraction of *Agrilus planipennis*. *Journal of Chemical Ecology* 42(6): 490-496. doi: 10.1007/s10886-016-0706-y.

Döring, T. F. 2014. How aphids find their host plants, and how they don't. *Annals of Applied Biology* 165(1): 3-26. doi: 10.1111/aab.12142.

Döring, T. F. and Chittka, L. 2007. Visual ecology of aphids–A critical review on the role of colours in host finding. *Arthropod-Plant Interactions* 1(1): 3-16. doi: 10.1007/s11829-006-9000-1.

Döring, T. F., Kirchner, S. M., Kuhne, S. and Saucke, H. 2004. Response of alate aphids to green targets on coloured backgrounds. *Entomologia Experimentalis et Applicata* 113(1): 53–61. doi: 10.1111/j.0013-8703.2004.00208.x.

El-Sayed, A. M. 2022. The pherobase: Database of insect pheromones and semiochemicals. Available at: https://www.pherobase.com/ (accessed April 2022).

Endo, N., Hironaka, M., Honda, Y. and Iwamoto, T. 2022. Combination of UV and green light synergistically enhances the attractiveness of light to green stink bugs *Nezara* spp. *Scientific Reports* 12(1): 12279. doi: 10.1038/s41598-022-16295-z.

Epsky, N. D. and Shuman, D. 2001. Laboratory evaluation of an improved electronic grain probe insect counter. *Journal of Stored Products Research* 37(2): 187–197.

European Commission. 2007. Commission regulation number 1266/2007 on implementation rules for Council Directive 2005/75/EC as regards the control, monitoring, surveillance and restrictions on movements of certain animals of susceptible species in relation to Bluetongue. *Official Journal of the European Union* 283: 37–52.

Faraji, A., Haas-Stapleton, E., Sorensen, B., Scholl, M., Goodman, G., Buettner, J., Schon, S., Lefkow, N., Lewis, C., Fritz, B., Hoffman, C. and Williams, G. 2021. Toys or tools? Utilization of unmanned aerial systems in mosquito and vector control programs. *Journal of Economic Entomology* 114(5): 1896–1909. doi: 10.1093/jee/toab107.

Fassotte, C., Delecolle, J. C., Cors, R., Defrance, T., De Deken, R., Haubruge, E. and Losson, B. 2008. *Culicoides* trapping with Rothamsted suction traps before and during the bluetongue epidemic of 2006 in Belgium. *Preventive Veterinary Medicine* 87(1–2): 74–83. doi: 10.1016/j.prevetmed.2008.06.007.

Fennell, J. T., Fountain, M. T. and Paul, N. D. 2019. Direct effects of protective cladding material on insect pests in crops. *Crop Protection* 121: 147–156. doi: 10.1016/j.cropro.2019.04.003.

Fountain, M. T., Baroffio, C., Borg-Karlson, A.-K., Brain, P., Cross, J. V., Farman, D. I., Hall, D. R., Ralle, B., Rendina, P., Richoz, P., Sigsgaard, L., Storberget, S., Trandem, N. and Wibe, A. 2017. Design and deployment of semiochemical traps for capturing *Anthonomus rubi* Herbst (Coleoptera: Curculionidae) and *Lygus rugulipennis* Poppius (Hetereoptera: Miridae) in soft fruit crops. *Crop Protection* 99: 1–9. doi: 10.1016/j.cropro.2017.05.001.

Fox, R., Randle, Z., Hill, L., Anders, S., Wiffen, L. and Parsons, M. S. 2011. Moths count: Recording moths for conservation in the UK. *Journal of Insect Conservation* 15(1–2): 55–68. doi: 10.1007/s10841-010-9309-z.

Fraser, S. E. M., Dytham, C. and Mayhew, P. J. 2008. The effectiveness and optimal use of Malaise traps for monitoring parasitoid wasps. *Insect Conservation and Diversity* 1(1): 22–31. doi: 10.1111/j.1752-4598.2007.00003.x.

Gaston, K. J. and O'Neill, M. A. 2004. Automated species identification: Why not? *Philosophical Transactions of the Royal Society of London. Series B, Biological Sciences* 359(1444): 655–667. doi: 10.1098/rstb.2003.1442.

Greenway, A. R. 1984. Sex pheromone of the pea moth, *Cydia nigricana* (F.) (Lepidoptera: Olethreutidae). *Journal of Chemical Ecology* 10(7): 973-982. doi: 10.1007/BF00987506.

Guarnieri, A., Maini, S., Molari, G. and Rondelli, V. 2011. Automatic trap for moth detection in integrated pest management. *Bulletin of Insectology* 64: 247–251.

Hall, D. R., Amarawardana, L., Cross, J. V., Francke, W., Boddum, T. and Hillbur, Y. 2012. The chemical ecology of cecidomyiid midges (Diptera: Cecidomyiidae). *Journal of Chemical Ecology* 38(1): 2–22. doi: 10.1007/s10886-011-0053-y.

Hall-Mendelin, S., Ritchie, S. A., Johansen, C. A., Zborowski, P., Cortis, G., Dandridge, S., Hall, R. A. and van den Hurk, A. F. 2010. Exploiting mosquito sugar feeding to detect mosquito-borne pathogens. *Proceedings of the National Academy of Sciences of the United States of America* 107(25): 11255–11259. doi: 10.1073/pnas.1002040107.

Hallmann, C. A., Sorg, M., Jongejans, E., Siepel, H., Hofland, N., Schwan, H., Stenmans, W., Müller, A., Sumser, H., Hörren, T., Goulson, D. and de Kroon, H. 2017. More than 75 percent decline over 27 years in total flying insect biomass in protected areas. *PLoS ONE* 12(10): e0185809. doi: 10.1371/journal.pone.0185809.

Hanula, J. L. 1990. Monitoring adult emergence, ovary maturation, and control of the black vine weevil (Coleoptera: Curculionidae). *Journal of Entomological Science* 25(1): 134–142. doi: 10.18474/0749-8004-25.1.134.

Hardee, D. D., McKibben, G. H., Gueldner, R. C., Mitchell, E. B., Tumlinson, J. H. and Cross, W. H. 1972. Boll weevils in nature respond to grandlure, a synthetic Pheromone123. *Journal of Economic Entomology* 65(1): 97–100. doi: 10.1093/jee/65.1.97.

Hardee, D. D., Weathersbee III, A. A., Gillespie, J. M., Snodgrass, G. L. and Quisumbing, A. R. 1996. Performance of trap designs, lures, and kill strips for the boll weevil (Coleoptera: Curculionidae). *Journal of Economic Entomology* 89(1): 170–174. doi: 10.1093/jee/89.1.170.

Hardie, J., Storer, J. R., Cook, F. J., Campbell, C. A. M., Wadhams, L. J., Lilley, R. and Peace, L. 1996. Sex pheromone and visual trap interactions in mate location strategies and aggregation by host-alternating aphids in the field. *Physiological Entomology* 21(2): 97–106. doi: 10.1111/j.1365-3032.1996.tb00841.x.

Harrington, R., Clark, S. J., Welham, S. J., Verrier, P. J., Denholm, C. H., Hullé, M., Maurice, D., Rounsevell, M. D. and Cocu, N. 2007a. Environmental change and the phenology of European aphids. *Global Change Biology* 13(8): 1550–1564. doi: 10.1111/j.1365-2486.2007.01394.x.

Harrington, R., Hullé, M. and Plantegenest, M. 2007b. Monitoring and forecasting. In: *Aphids as Crop Pests*. Oxford: CAB International, pp. 515–536.

Harrington, R. and Woiwod, I. 2007. Foresight from hindsight: The Rothamsted insect survey. *Outlooks on Pest Management* 18(1): 9–14.

Hendricks, D. E. 1985. Portable electronic detector system used with inverted-cone sex pheromone traps to determine periodicity and moth captures. *Environmental Entomology* 14(3): 199–204. doi: 10.1093/ee/14.3.199.

Hernández-Ortiz, V., Hernández-López, M. and Dzul-Cauich, J. F. 2021. Sampling methods of true fruit flies (Tephritidae). In: Santos, J. C. and Fernandes, G. W. (Eds.), *Measuring Arthropod Biodiversity: A Handbook of Sampling Methods*. Cham, Switzerland: Springer Nature, pp. 205–222.

Heuskin, S., Verheggen, F. J., Haubruge, E., Wathelet, J. P. and Lognay, G. C. 2011. The use of semiochemical slow-release devices in integrated pest management strategies. *Biotechnologie, Agronomie, Société et Environnement* 15: 459–470.

Hines, J. W. and Heikkenen, H. J. 1977. Beetles attracted to severed Virginia pine (*Pinus virginiana* Mill.). *Environmental Entomology* 6(1): 123–127.

Hobbs, S. E. and Hodges, G. 1993. An optical method for automatic classification and recording of a suction trap catch. *Bulletin of Entomological Research* 83(1): 47–51. doi: 10.1017/S0007485300041766.

Hodge, S., Hardie, J. and Powell, G. 2011. Parasitoids aid dispersal of a nonpersistently transmitted plant virus by disturbing the aphid vector. *Agricultural and Forest Entomology* 13(1): 83–88. doi: 10.1111/j.1461-9563.2010.00493.x.

Hohbein, R. R. and Conway, C. J. 2018. Pitfall traps: A review of methods for estimating arthropod abundance. *Wildlife Society Bulletin* 42(4): 597–606. doi: 10.1002/wsb.928.

Holguin, G. A., Lehman, B. L., Hull, L. A., Jones, V. P., Jones, V. P. and Park, J. 2010. Electronic traps for automated monitoring of insect populations. *IFAC Proceedings Volumes* 43(26): 49–54. doi: 10.3182/20101206-3-JP-3009.00008.

Holland, J. M., McHugh, N. M. and Salinari, F. 2021. Field specific monitoring of cereal yellow dwarf virus aphid vectors and factors influencing their immigration within fields. *Pest Management Science* 77(9): 4100–4108. doi: 10.1002/ps.6435.

Hoye, T. T., Arje, J., Bjerge, K., Hansen, O. L. P., Iosifidis, A., Leese, F., Mann, H. M. R., Meissner, K., Melvad, C. and Raitoharju, J. 2021. Deep learning and computer vision will transform entomology. *Proceedings of the National Academy of Sciences of the United States of America* 118(2): e2002545117. doi: 10.1073/pnas.2002545117.

Hu, G., Cheng, X. N., Qi, G. J., Wang, F. Y., Lu, F., Zhang, X. X. and Zhai, B. P. 2011. Rice planting systems, global warming and outbreaks of *Nilaparvata lugens* (Stål). *Bulletin of Entomological Research* 101(2): 187–199. doi: 10.1017/S0007485310000313.

Hu, G., Lu, F., Zhai, B. P., Lu, M. H., Liu, W. C., Zhu, F., Wu, X. W., Chen, G. H. and Zhang, X. X. 2014. Outbreaks of the brown planthopper *Nilaparvata lugens* (Stål) in the Yangtze River Delta: Immigration or local reproduction? *PLoS ONE* 9(2): e88973. doi: 10.1371/journal.pone.0088973.

Hu, G., Lu, M. H., Tuan, H. A., Liu, W. C., Xie, M. C., McInerney, C. E. and Zhai, B. P. 2017. Population dynamics of rice planthoppers, *Nilaparvata lugens* and *Sogatella furcifera* (Hemiptera, Delphacidae) in Central Vietnam and its effects on their spring migration to China. *Bulletin of Entomological Research* 107(3): 369–381. doi: 10.1017/S0007485316001024.

Jacquin-Joly, E. and Groot, A. T. 2018. Pheromones, insects. In: Skinner, M. K. (Ed.), *Encyclopedia of Reproduction* (2nd edn.). Amsterdam: Academic Press, pp. 465–471.

Jiang, J.-A., Tseng, C.-L., Lu, F.-M., Yang, E.-C., Wu, Z.-S., Chen, C.-P., Lin, S.-H., Lin, K.-C. and Liao, C.-S. 2008. A GSM-based remote wireless automatic monitoring system for field information: A case study for ecological monitoring of the oriental fruit fly, *Bactrocera dorsalis* (Hendel). *Computers and Electronics in Agriculture* 62(2): 243–259. doi: 10.1016/j.compag.2008.01.005.

Johnson, C. G. 1950. A suction trap for small airborne insects which automatically segregates the catch into successive hourly samples. *Annals of Applied Biology* 37(1): 80–91.

Katsoyannos, B. I., Papadopoulos, N. T. and Stavridis, D. 2000. Evaluation of trap types and food attractants for *Rhagoletis cerasi* (Diptera: Tephritidae). *Journal of Economic Entomology* 93(3): 1005–1010. doi: 10.1603/0022-0493-93.3.1005.

Kim, K. N., Huang, Q. Y. and Lei, C. L. 2019. Advances in insect phototaxis and application to pest management: A review. *Pest Management Science* 75(12): 3135–3143. doi: 10.1002/ps.5536.

Kirk, W. D. J. 1984. Ecologically selective coloured traps. *Ecological Entomology* 9(1): 35–41. doi: 10.1111/j.1365-2311.1984.tb00696.x.

Kisimoto, R. 1968. Yellow pan water trap for sampling the small brown planthopper, *Laodelphax striatellus* (Fallén), a vector of the rice stripe virus. *Applied Entomology and Zoology* 3(1): 37–48.

Knuff, A. K., Winiger, N., Klein, A. M., Segelbacher, G., Staab, M. and Kotze, D. J. 2019. Optimizing sampling of flying insects using a modified window trap. *Methods in Ecology and Evolution* 10(10): 1820–1825. doi: 10.1111/2041-210X.13258.

Komosiński, K. and Marczak, D. 2018. Four species of Ptiliidae (Coleoptera) new to Poland. *The Coleopterists Bulletin* 72(1): 151–156. doi: 10.1649/0010-065X-72.1.151.

Košťál, V. and Finch, S. 1996. Preference of the cabbage root fly, *Delia radicum* (L.), for coloured traps: Influence of sex and physiological status of the flies, trap background and experimental design. *Physiological Entomology* 21(2): 123–130. doi: 10.1111/j.1365-3032.1996.tb00844.x.

Krüger, K. and Laubscher, K. 2013. *The South African Suction Trap Network: First Aphid Monitoring Results. Suction Traps in Studying Distribution and Occurrence of Insects and Forecasting Pests and Vector Borne Viruses NJF Seminar 468*. Sweden: Kristianstad, pp. 16–19.

Leather, S. R. 2015. *Entomological Classics – The Window (Pane) Flight Intercept Trap. Don't Forget the Roundabouts*. This is a Wordpress blog available at https://simonleather.wordpress.com/2015/11/04/entomological-classics-the-window-pane-flight-intercept-trap/ (accessed April 2022).

Leskey, T. C., Andrews, H., Bady, A., Benvenuto, L., Bernardinelli, I., Blaauw, B., Bortolotti, P. P., Bosco, L., Di Bella, E., Hamilton, G., Kuhar, T., Ludwick, D., Maistrello, L., Malossini, G., Nannini, R., Nixon, L. J., Pasqualini, E., Preti, M., Short, B. D., Spears, L., Tavella, L., Vetek, G. and Wiman, N. 2021. Refining pheromone lures for the invasive *Halyomorpha halys* (Hemiptera: Pentatomidae) through collaborative trials in the United States and Europe. *Journal of Economic Entomology* 114(4): 1666–1673. doi: 10.1093/jee/toab088.

Letourneau, D. K., Bothwell Allen, S. G. and Stireman, J. O. 2012. Perennial habitat fragments, parasitoid diversity and parasitism in ephemeral crops. *Journal of Applied Ecology* 49(6): 1405–1416. doi: 10.1111/1365-2664.12001.

Lewis, T. and Taylor, L. R. 1965. Diurnal periodicity of flight by insects. *Transactions of the Royal Entomological Society of London* 116(15): 393–435.

Lima, M. C. F., Leandro, M. E. D. d. A., Valero, C., Coronel, L. C. P. and Bazzo, C. O. G. 2020. Automatic detection and monitoring of insect pests–A review. *Agriculture* 10(5): 161. doi: 10.3390/agriculture10050161.

Lindgren, B. S. 1983. A multiple funnel trap for scolytid beetles (Coleoptera). *The Canadian Entomologist* 115(3): 299–302. doi: 10.4039/Ent115299-3.

Lloyd, L. L. 1922. The control of the greenhouse whitefly (*Asterochiton vaporariorum*) with notes on its biology. *Annals of Applied Biology* 9(1): 1–32.

Lu, M.-H., Chen, X., Liu, W.-C., Zhu, F., Lim, K.-S., McInerney, C. E. and Hu, G. 2017. Swarms of brown planthopper migrate into the lower Yangtze River Valley under strong western Pacific subtropical highs. *Ecosphere* 8(10): e01967. doi: 10.1002/ecs2.1967.

Lykouressis, P., Perdikis, D. C., Chalkia, C. A. and Vardaki, S. C. 2017. Comparisons between alate aphids caught in yellow water traps and aphid populations on tomato plants. *Entomologia Hellenica* 11. doi: 10.12681/eh.14009.

Ma, M., Wu, Y., Peng, Z., Zhao, X., Zhang, Y., Liao, G. and Zhai, B. 2017. Migration analysis of *Sogatella furcifera* (Hemiptera: Delphacidae) in the Northeastern Hunan Province in June. *Environmental Entomology* 46(4): 757–765. doi: 10.1093/ee/nvx092.

Macaulay, E. D. M., Tatchell, G. M. and Taylor, L. R. 1988. The Rothamsted Insect Survey '12-metre' suction trap. *Bulletin of Entomological Research* 78(1): 121–128. doi: 10.1017/S0007485300016126.

Macfadyen, S. and Muller, W. 2013. Edges in agricultural landscapes: Species interactions and movement of natural enemies. *PLoS ONE* 8(3): e59659. doi: 10.1371/journal.pone.0059659.

Maelzer, D., Zalucki, M. P. and Laughlin, R. 1996. Analysis and interpretation of long term light trap data for *Helicoverpa punctigera* (Lepidoptera; Noctuidae) in Australia: Population changes and forecasting pest pressure. *Bulletin of Entomological Research* 86(5): 547–557. doi: 10.1017/S0007485300039341.

Maelzer, D. A. and Zalucki, M. P. 1999. Analysis of long-term light-trap data for *Helicoverpa* spp. (Lepidoptera: Noctuidae) in Australia: The effect of climate and crop host plants. *Bulletin of Entomological Research* 89(5): 455–463. doi: 10.1017/S0007485399000590.

Malaise, R. 1937. A new insect-trap. *Entomologisk Tidskrift* 58: 148–160.

Marshall, P., Hail, D., Mitchell, F. and Bextine, B. 2010. Impacts of an orange oil solvent and Stickem® on the detection of *Xylella fastidiosa* DNA in glassy-winged sharpshooters, *Homalodisca vitripennis* (Hemiptera: Cicadellidae). *Florida Entomologist* 93(3): 378–384.

Masner, L. and Goulet, H. 1981. A new model of flight interception trap for some hymenopterous insects. *Entomological News* 92: 199–202.

Matsumura, M. 2001. The current status of occurrence and forecasting system of rice planthoppers in Japan. *Journal of Asia-Pacific Entomology* 4(2): 195–199. doi: 10.1016/S1226-8615(08)60123-5.

Metspalu, L., Veromann, E., Kaasik, R., Kovacs, G., Williams, I. H. and Mänd, M. 2015. Comparison of sampling methods for estimating the abundance of *Meligethes aeneus* on oilseed crops. *International Journal of Pest Management* 61(4): 312–319. doi: 10.1080/09670874.2015.1058992.

Meyer, D. B., Ramirez, A. L., van den Hurk, A. F., Kurucz, N. and Ritchie, S. A. 2019. Development and field evaluation of a system to collect mosquito excreta for the detection of arboviruses. *Journal of Medical Entomology* 56(4): 1116–1121. doi: 10.1093/jme/tjz031.

Microsoft. 2022. Microsoft premonition. *Microsoft Website*. Available at: https://innovation.microsoft.com/en-us/exploring-premonition (accessed April 2022).

Mitchell, E. B. and Hardee, D. D. 1974. In-field traps: A new concept in survey and suppression of low populations of boll weevils. *Journal of Economic Entomology* 67(4): 506–508.

Mohammed, M. E. A., El-Shafie, H. A. F. and Bin Refdan, A.-H. M. 2018. Design of an automated solar-powered light trap for monitoring and mass trapping of major date palm pests. *Ecology, Environment and Conservation* 24: 177–185.

Moore, A., Miller, J. R., Tabashnik, B. E. and Gage, S. H. 1986. Automated identification of flying insects by analysis of wingbeat frequencies. *Journal of Economic Entomology* 79(6): 1703–1706.

Moore, A. and Miller, R. H. 2002. Automated identification of optically sensed aphid (Homoptera: Aphidae) wingbeat waveforms. *Annals of the Entomological Society of America* 95(1): 1–8.

Mulder, P. G., Jr., Harris, M. K. and Grantham, R. A. 2012. Biology and management of the pecan weevil (Coleoptera: Curculionidae). *Journal of Integrated Pest Management* 3(1): A1–A9. doi: 10.1603/IPM10027.

Muñoz-Pallares, J., Corma, A., Primo, J. and Primo-Yufera, E. 2001. Zeolites as pheromone dispensers. *Journal of Agricultural and Food Chemistry* 49(10): 4801–4807. doi: 10.1021/jf010223o.

Murchie, A. K., Burn, D. J., Kirk, W. D. J. and Williams, I. H. 2001. A novel mechanism for time-sorting insect catches, and its use to derive the diel flight periodicity of brassica pod midge *Dasineura brassicae* (Diptera : Cecidomyiidae). *Bulletin of Entomological Research* 91(3): 199-204.

Murchie, A. K., Hall, C. E., Gordon, A. W. and Clawson, S. 2018. Black border increases *Stomoxys calcitrans* catch on white sticky traps. *Insects* 9(1): 13. doi: 10.3390/insects9010013.

Murchie, A. K., Smart, L. E. and Williams, I. H. 1997. Responses of *Dasineura brassicae* and its parasitoids *Platygaster subuliformis* and *Omphale clypealis* to field traps baited with organic isothiocyanates. *Journal of Chemical Ecology* 23(4): 917-926.

Murphy, W. L. 1985. Procedure for the removal of insect specimens from sticky-trap material. *Annals of the Entomological Society of America* 78(6): 881.

Nielsen, B. S. and Jensen, T. S. 1993. Spring dispersal of *Sitona lineatus*: The use of aggregation pheromone traps for monitoring. *Entomologia Experimentalis et Applicata* 66(1): 21-30.

Northing, P., Walters, K., Barker, I., Foster, G., Harrington, R., Taylor, M., Tones, S. and Morgan, D. 2004. Use of Internet for provision of user specific support for decisions on the control of aphid-borne virus. In: *Aphids in a New Millennium: Proceedings of the Sixth International Symposium on Aphids*, September 2001, Rennes, France: Institut National de la Recherche Agronomique, pp. 331-336.

Nowinszky, L. 2003. The orientation of insects by light–major theories. In: Nowinszky, L. (Ed.), *The Handbook of Light Trapping*. Szombathely: Savaria University Press, pp. 15-18.

Ozanne, C. M. P. 2005. Sampling methods for forest understory vegetation. In: Leather, S. R. (Ed.), *Insect Sampling in Forest Ecosystems*. Oxford: Blackwell Publishing Science Ltd, pp. 58-76.

Pavuk, D. M. and Wadsworth, A. M. 2013. Longhorned beetle (Coleoptera: Cerambycidae) diversity in a fragmented temperate forest landscape [version 2; peer review: 2 approved]: *F1000Research*,1:25. doi: 10.12688/f1000research.1-25.v2.

Pedigo, L. P. and Buntin, G. D. 1994. *Handbook of Sampling Methods for Arthropods in Agriculture*. Boca Raton, FL: CRC Press, p. 714.

PGRO. 2022. *Technical Update 03. Pea Moth*. Processors & Growers Research Organisation.

Pinto-Zevallos, D. M. and Vänninen, I. 2013. Yellow sticky traps for decision-making in whitefly management: What has been achieved? *Crop Protection* 47: 74-84. doi: 10.1016/j.cropro.2013.01.009.

Potamitis, I., Rigakis, I. and Fysarakis, K. 2014. The electronic McPhail trap. *Sensors (Basel)* 14(12): 22285-22299. doi: 10.3390/s141222285.

Potamitis, I., Rigakis, I. and Fysarakis, K. 2015. Insect biometrics: Optoacoustic signal processing and its applications to remote monitoring of McPhail type traps. *PLoS ONE* 10(11): e0140474. doi: 10.1371/journal.pone.0140474.

Preti, M., Verheggen, F. and Angeli, S. 2021. Insect pest monitoring with camera-equipped traps: Strengths and limitations. *Journal of Pest Science* 94(2): 203-217. doi: 10.1007/s10340-020-01309-4.

Prokopy, R. J. and Owens, E. D. 1983. Visual detection of plants by herbivorous insects. *Annual Review of Entomology* 28(1): 337-364. doi: 10.1146/annurev.en.28.010183.002005.

Prokopy, R. J. and Wright, S. E. 1998. Plum curculio (Coleoptera: Curculionidae) responses to unbaited pyramid and cone traps. *Journal of Economic Entomology* 91(1): 226-234. doi: 10.1093/jee/91.1.226.

Qin, Q., Liang, H., Miao, L., Wang, H., Jiang, L. and Qiao, G. 2013. *Development and Application of Suction Trap Network in China. Suction Traps in Studying Distribution and Occurrence of Insects and Forecasting Pests and Vector Borne Viruses NJF Seminar 468*. Kristianstad, Sweden, p. 20.

Ramsden, M. W., Kendall, S. L., Ellis, S. A. and Berry, P. M. 2017. A review of economic thresholds for invertebrate pests in UK arable crops. *Crop Protection* 96: 30–43. doi: 10.1016/j.cropro.2017.01.009.

Rassati, D., Faccoli, M., Chinellato, F., Hardwick, S., Suckling, D. M. and Battisti, A. 2016. Web-based automatic traps for early detection of alien wood-boring beetles. *Entomologia Experimentalis et Applicata* 160(1): 91–95. doi: 10.1111/eea.12453.

Raszick, T. J. and Bernaola, L. 2021. Boll weevil eradication: A success story of science in the service of policy and industry. *Annals of the Entomological Society of America* 114(6): 702–708. doi: 10.1093/aesa/saab031.

Remund, U. and Boller, E. 1978. Kirschenfliegenfallen für Prognosewesen und biotechnische Bekämpfung im Vormarsch [Cherry fruit fly traps on the rise for biotechnical pest management and prognosis]. *Schweizer Zeitschrift für Obst- und Weinbau* 114: 229–232.

Rhainds, M., Yoo, H. J., Steffey, K. L., Voegtlin, D. J., Sadof, C. S., Yaninek, S. and O'Neil, R. J. 2010. Potential of suction traps as a monitoring tool for *Aphis glycines* (Hemiptera: Aphididae) in soybean fields. *Journal of Economic Entomology* 103(1): 186–189. doi: 10.1603/ec09121.

Riedl, H. and Croft, B. A. 1981. A timing trap for segregating catches of insects by discrete intervals. *The Canadian Entomologist* 113(8): 765–768.

Riis, L. and Nachman, G. 2006. Migration, trapping and local dynamics of whiteflies (Homoptera: Aleyrodidae). *Agricultural and Forest Entomology* 8(3): 233–241.

Rizvi, S. A. H., George, J., Reddy, G. V. P., Zeng, X. and Guerrero, A. 2021. Latest developments in insect sex pheromone research and its application in agricultural pest management. *Insects* 12(6). doi: 10.3390/insects12060484.

Rohitha, B. H. and Stevenson, B. E. 1987. An automatic sticky trap for aphids (Hemiptera: Aphididae) that segregates the catch daily. *Bulletin of Entomological Research* 77(1): 67–71. doi: 10.1017/S0007485300011548.

Rondoni, G., Chierici, E., Marchetti, E., Nasi, S., Ferrari, R. and Conti, E. 2022. Improved captures of the invasive brown marmorated stink bug, *Halyomorpha halys*, using a novel multimodal trap. *Insects* 13(6): 13. doi: 10.3390/insects13060527.

Rothamsted Research. 2021. The insect survey. Available at: www.rothamsted.ac.uk/insect-survey (accessed November 2021).

Rydhmer, K., Bick, E., Still, L., Strand, A., Luciano, R., Helmreich, S., Beck, B. D., Gronne, C., Malmros, L., Poulsen, K., Elbaek, F., Brydegaard, M., Lemmich, J. and Nikolajsen, T. 2022. Automating insect monitoring using unsupervised near-infrared sensors. *Scientific Reports* 12(1): 2603. doi: 10.1038/s41598-022-06439-6.

Sanders, C. J., Shortall, C. R., Gubbins, S., Burgin, L., Gloster, J., Harrington, R., Reynolds, D. R., Mellor, P. S. and Carpenter, S. 2011. Influence of season and meteorological parameters on flight activity of *Culicoides* biting midges. *Journal of Applied Ecology* 48(6): 1355–1364.

Schmidt, N. P., O'Neal, M. E., Anderson, P. F., Lagos, D., Voegtlin, D., Bailey, W., Caragea, P., Cullen, E., DiFonzo, C., Elliott, K., Gratton, C., Johnson, D., Krupke, C. H., McCornack, B., O'Neil, R., Ragsdale, D. W., Tilmon, K. J. and Whitworth, J. 2012. Spatial distribution of *Aphis glycines* (Hemiptera: Aphididae): A summary of the suction trap network. *Journal of Economic Entomology* 105(1): 259–271. doi: 10.1603/ec11126.

Schultz, B. and Saucke, H. 2005. Einfluss verschiedener Saattermine auf den Erbsenwicklerbefall (*Cydia nigricana* Fabr.) in ökologischen Gemüseerbsen. In: Heß, J. and Rahmann, G. (Eds.), *Ende der Nische, Beiträge zur 8 Wissenschaftstagung Ökologischer Landbau*. Kassel University Press GmbH, pp. 105–108.

Sciarretta, A. and Calabrese, P. 2019. Development of automated devices for the monitoring of insect pests. *Current Agriculture Research Journal* 7(1): 19-25. doi: 10.12944/CARJ.7.1.03.

Selby, R. D., Gage, S. H. and Whalon, M. E. 2014. Precise and low-cost monitoring of plum curculio (Coleoptera: Curculionidae) pest activity in pyramid traps with cameras. *Environmental Entomology* 43(2): 421–431. doi: 10.1603/EN13136.

Sheikh, A. H., Thomas, M., Bhandari, R. and Meshram, H. 2016. Malaise trap and insect sampling: Mini review. *Biology Bulletin* 2: 35–40.

Shortall, C., Harrington, R., Taylor, M., Alderson, L., Kruger, T., Izera, D. and Verrier, P. 2013. *Not Just Aphids: Uses of the Rothamsted Insect Survey's Specimen Collection. Suction Traps in Studying Distribution and Occurrence of Insects and Forecasting Pests and Vector Borne Viruses NJF Seminar 468*. Kristianstad, Sweden, pp. 25-27.

Shortall, C. R. 2021. *The Temporal and Spatial Ecology of Brassicogethes (Meligethes) aeneus. PhD*. University of Reading.

Silver, J. B. 2007. *Mosquito Ecology: Field Sampling Methods*. Dordrecht: Springer Verlag.

Skvarla, M. J., Larson, J. L., Fisher, J. R., Dowling, A. P. G. and Reddy, G. V. P. 2021. A review of terrestrial and canopy Malaise traps. *Annals of the Entomological Society of America* 114(1): 27-47. doi: 10.1093/aesa/saaa044.

Smart, L. E., Blight, M. M. and Hick, A. J. 1992. Development of a monitoring system for the Cabbage Seed Weevil and the Pollen Beetle. *IOBC Bulletin* 16: 351–354.

Smart, L. E., Blight, M. M. and Hick, A. J. 1997. Effect of visual cues and a mixture of isothiocyanates on trap capture of cabbage seed weevil, *Ceutorhynchus assimilis*. *Journal of Chemical Ecology* 23(4): 889-902. doi: 10.1023/B:JOEC.0000006378.65158.ca.

Solorzano, J. A., Gilles, J., Bravo, O., Vargas, C., Gomez-Bonilla, Y., Bingham, G. V. and Taylor, D. B. 2015. Biology and trapping of stable flies (Diptera: Muscidae) developing in pineapple residues (*Ananas comosus*) in Costa Rica. *Journal of Insect Science* 15(1): 145. doi: 10.1093/jisesa/iev127.

Spalding, A. 2019. The presidential address part 2. Light pollution and the decline of moths. *British Journal of Entomology and Natural History* 32: 17–34.

St Onge, A., Cárcamo, H. A. and Evenden, M. L. 2018. Evaluation of semiochemical-baited traps for monitoring the pea leaf weevil, *Sitona lineatus* (Coleoptera: Curculionidae) in field pea crops. *Environmental Entomology* 47(1): 93–106. doi: 10.1093/ee/nvx180.

Steininger, M. S., Hulcr, J., Sigut, M. and Lucky, A. 2015. Simple and efficient trap for bark and ambrosia beetles (Coleoptera: Curculionidae) to facilitate invasive species monitoring and citizen involvement. *Journal of Economic Entomology* 108(3): 1115-1123. doi: 10.1093/jee/tov014.

Stern, V. M. 1973. Economic thresholds. *Annual Review of Entomology* 18(1): 259–280. doi: 10.1146/annurev.en.18.010173.001355.

Steyskal, G. C. 1977. History and use of the McPhail trap. *Florida Entomologist* 60(1): 11–16.

Sun, Y., Liu, X., Yuan, M., Ren, L., Wang, J. and Chen, Z. 2018. Automatic in-trap pest detection using deep learning for pheromone-based *Dendroctonus valens* monitoring. *Biosystems Engineering* 176: 140–150. doi: 10.1016/j.biosystemseng.2018.10.012.

Taylor, J., Padgham, D. E. and Perfect, T. J. 1982. A light-trap with upwardly directed illumination and temporal segregation of the catch. *Bulletin of Entomological Research* 72(4): 669–673. doi: 10.1017/S0007485300008695.

Taylor, L. R. 1951. An improved suction trap for insects. *Annals of Applied Biology* 38(3): 582–591.

Taylor, L. R. 1962. The absolute efficiency of insect suction traps. *Annals of Applied Biology* 50(3): 405–421. doi: 10.1111/j.1744-7348.1962.tb06036.x.

Taylor, L. R. 1974. *Monitoring Change in the Distribution and Abundance of Insects*. Report of Rothamsted Experimental Station for 1973 Part 2, pp. 202–239.

Tercel, M. P. T. G., Veronesi, F. and Pope, T. W. 2018. Phylogenetic clustering of wingbeat frequency and flight-associated morphometrics across insect orders. *Physiological Entomology* 43(2): 149–157. doi: 10.1111/phen.12240.

Teulon, D. A. J., Stufkens, M. A. W. and Fletcher, J. D. 2004. Crop infestation by aphids is related to flight activity detected with 7.5 metre high suction traps. *New Zealand Plant Protection* 57: 227–232.

Thomas, D. B. 2008. A safe and effective propylene glycol based capture liquid for fruit fly (Diptera: Tephritidae) traps baited with synthetic lures. *Florida Entomologist* 91(2): 210–213. doi: 10.1653/0015-4040(2008)91[210:ASAEPG]2.0.CO;2.

Thöming, G., Pölitz, B., Kühne, A. and Saucke, H. 2011. Risk assessment of pea moth *Cydia nigricana* infestation in organic green peas based on spatio-temporal distribution and phenology of the host plant. *Agricultural and Forest Entomology* 13(2): 121–130. doi: 10.1111/j.1461-9563.2010.00507.x.

Trumble, J. T., Nakakihara, H. and Zehnder, G. W. 1982. Comparisons of traps and visual searches of foliage for monitoring aphid (Heteroptera: Aphididae) population density in broccoli. *Journal of Economic Entomology* 75(5): 853–856.

Tumlinson, J. H., Hardee, D. D., Gueldner, R. C., Thompson, A. C., Hedin, P. A. and Minyard, J. P. 1969. Sex pheromones produced by male boll weevil: Isolation, identification, and synthesis. *Science* 166(3908): 1010–1012.

Vacek, L., Atter, E., Rizo, P., Nam, B., Kortvelesy, R., Kaufman, D., Das, J. and Kumar, V. 2017. sUAS for deployment and recovery of an environmental sensor probe. In: *International Conference on Unmanned Aircraft Systems (ICUAS)* (Vol. 2017), pp. 1022–1029.

van Tol, R. W. H. M., Tom, J., Roher, M., Schreurs, A. and van Dooremalen, C. 2021. Haze of glue determines preference of western flower thrips (*Frankliniella occidentalis*) for yellow or blue traps. *Scientific Reports* 11(1): 6557. doi: 10.1038/s41598-021-86105-5.

Volpato, A., Ahmed, K. S. D., Williams, C. D., Day, M. F., O'Hanlon, A., Ruas, S., Rotchés-Ribalta, R., Mulkeen, C., Huallacháin, D. Ó. and Gormally, M. J. 2020. Using Malaise traps to assess aculeate Hymenoptera associated with farmland linear habitats across a range of farming intensities. *Insect Conservation and Diversity* 13(3): 229–238. doi: 10.1111/icad.12383.

Wainhouse, D. 1980. A portable suction trap for sampling small insects. *Bulletin of Entomological Research* 70(3): 491–494. doi: 10.1017/S000748530000777X.

Wall, C. 1988. Application of sex attractants for monitoring the pea moth, *Cydia nigricana* (F.) (Lepidoptera: Tortricidae). *Journal of Chemical Ecology* 14(10): 1857–1866.

Wilkening, A. J., Foltz, J. L., Atkinson, T. H. and Connor, M. D. 1981. An omnidirectional flight trap for ascending and descending insects. *The Canadian Entomologist* 113(5): 453–455. doi: 10.4039/Ent113453-5.

Williams, C. B. 1935. The times of activity of certain nocturnal insects, chiefly Lepidoptera, as indicated by a light-trap. *Transactions of the Royal Entomological Society of London* 83(4): 523–556. doi: 10.1111/j.1365-2311.1935.tb02998.x.

Williams, G. 1958. Mechanical time-sorting of pitfall captures. *Journal of Animal Ecology* 27(1): 27–35.

Williams, I. H., Frearson, D., Barari, H. and McCartney, A. 2007. Migration to and dispersal from oilseed rape by the pollen beetle, *Meligethes aeneus*, in relation to wind direction. *Agricultural and Forest Entomology* 9(4): 279–286.

Williams, I. H. and Martin, A. P. 1986. Evidence for a female sex pheromone in the Brassica Pod Midge *Dasineura brassicae*. *Physiological Entomology* 11(3): 353–356.

Witzgall, P., Bengtsson, M., Unelius, C. R. and Löfqvist, J. 1993. Attraction of pea moth *Cydia nigricana* F. *Journal of Chemical Ecology* 19(9): 1917–1928. doi: 10.1007/BF00983796.

Witzgall, P., Kirsch, P. and Cork, A. 2010. Sex pheromones and their impact on pest management. *Journal of Chemical Ecology* 36(1): 80–100. doi: 10.1007/s10886-009-9737-y.

Woiwod, I. and Harrington, R. 1994. Flying in the face of change: The Rothamsted insect survey. In: Leigh, R. A. and Johnston, A. E. (Eds.), *Long-Term Experiments in Agricultural and Ecological Sciences*. Wallingford: CAB International, pp. 321–342.

Woodcock, B. A. 2005. Pitfall trapping in ecological studies. In: Leather, S. R. (Ed.), *Insect Sampling in Forest Ecosystems*. Oxford: Blackwell Publishing Science Ltd, pp. 37–57.

Xie, C., Zhang, J., Li, R., Li, J., Hong, P., Xia, J. and Chen, P. 2015. Automatic classification for field crop insects via multiple-task sparse representation and multiple-kernel learning. *Computers and Electronics in Agriculture* 119: 123–132. doi: 10.1016/j.compag.2015.10.015.

Xu, L., Xie, Y., Na, R. and Li, Q. X. 2020. Mini-review: Recent advances in the identification and application of sex pheromones of gall midges (Diptera: Cecidomyiidae). *Pest Management Science* 76(12): 3905–3910. doi: 10.1002/ps.5949.

Ye, K., Correa, G. J., Guda, T., Teng, H., Ray, A. and Karydis, K. 2020. Development and testing of a novel automated insect capture module for sample collection and transfer 2020. In: *IEEE Publications 16th International Conference on Automation Science and Engineering (CASE)*, pp. 1063–1069.

Zalucki, M. P. and Furlong, M. J. 2005. Forecasting *Helicoverpa* populations in Australia: A comparison of regression based models and a bioclimatic based modelling approach. *Insect Science* 12(1): 45–56. doi: 10.1111/j.1672-9609.2005.00007.x.

Zhang, W. and Swinton, S. M. 2009. Incorporating natural enemies in an economic threshold for dynamically optimal pest management. *Ecological Modelling* 220(9–10): 1315–1324. doi: 10.1016/j.ecolmodel.2009.01.027.

Chapter 2

Advances and challenges in monitoring insect pests of major field crops in the United States

Erin W. Hodgson and Ashley N. Dean, Iowa State University, USA; Anders Huseth, North Carolina State University, USA; and William D. Hutchison, University of Minnesota, USA

1 Introduction

The predominant cropping systems within the United States are dictated by several environmental factors. Although mostly in the temperate zone, the United States is comprised of diverse landscapes and nine climate regions (NCEI, 2022), which influence the crops grown and the seasonality of crop production. The three major crops in the United States are corn (*Zea mays* L., also known worldwide as maize), soybean (*Glycine max* [L.] Merr.) and cotton (*Gossypium* spp. L.). Corn is grown in 41 states, but the Corn Belt, comprised of 13 states in the Midwest, accounts for the majority of US corn production. In 2021, corn was planted on 37.7 million hectares in the United States (USDA-NASS, 2021). Soybeans were planted on 35.3 million hectares in 29 states in 2021, primarily in the same states as corn (USDA-NASS, 2021). Cotton is currently grown in 17 states that are generally distributed along the southern tier states from Virginia to California, a geographical region that is often considered the US Cotton Belt. In 2021, farmers grew approximately 4.5 million hectares of cotton in eight distinct ecological regions in the Cotton Belt (NCEI, 2022; USDA-NASS, 2021).

http://dx.doi.org/10.19103/AS.2022.0113.17

Understanding the presence or absence of important pest species in agriculture is particularly important due to the economic value of field crops in the United States. A major challenge for entomologists and farmers is the wide geographical extent of production of each of these crops. This includes numerous arthropod pests with regional differences in crop injury and economic injury (Cook and Threet, 2020; Musser et al., 2021; Huseth et al., 2021), and the ability to infest multiple major crops in certain regions. Indeed, scientists have long known that complex interactions exist among crops and pests in southern production systems (Stinner et al., 1974; Johnson et al., 1975; Rabb, 1978). Pest surveys, predictive models and traditional pest monitoring (i.e. traps and field scouting) are essential for tracking pests at regional and field scales for each of these major crops.

2 Primary crops in the United States

Insect monitoring is essential for the successful development of Integrated Pest Management (IPM) programs, with the main objectives to understand population density, dispersion and dynamics (Pedigo et al., 2021). Monitoring can occur within a growing season or over multiple seasons with spatial scales from field to continent. Frequent surveillance helps confirm first detections of invasive species in the United States and monitor the arrival of annual migratory species. Additionally, frequent surveillance helps to establish when endemic species become active each year, and assessing biological control activity may be important in some cropping systems to understand the extent of pest control by natural enemies. Moreover, from a practical sense, it is imperative for crop consultants and farmers to have access to useful, yet statistically sound monitoring data to use in tandem with economic (or action) thresholds for decision-making IPM (Pedigo et al., 2021).

Crop fields can be sampled to determine local insect activity and make treatment decisions from estimating pest density or plant injury within a field. However, surveying on a larger scale using private consulting companies or regulatory agencies can be beneficial for monitoring a diverse array of pests on a local, state, regional or national scale. Extensive sampling programs conducted over broad scales largely focus on a few species with moderate levels of precision, whereas field-scale assessments focus on characterizing pest populations with a high degree of precision (Naranjo and Hutchison, 1997; Hodgson et al., 2004; Pedigo et al., 2021). A wide variety of sampling techniques are available to collect monitoring data, usually with known trade-offs that affect the precision (closeness of measured values to each other) and accuracy (closeness of measured values to the true value) of estimates.

Establishing the most appropriate sampling method requires a balance between monitoring goals, required effort, cost, availability of monitoring equipment (e.g. pheromones) and risks associated with inaccurate decision-making. Labor-intensive observation and experimentation are often required to determine appropriate methods and design and implement practical sampling plans (Dent, 2000). Some of the most common methods for estimating insect density in field crops include in situ counts, sweep netting, trapping and extraction of pests from soil. Accurately estimating crop injury can be more difficult, but recent technological advances have improved this area of surveillance (e.g. remote sensing, smartphone applications). Although sampling insect populations have been the foundation of IPM for many decades, recent technological advances offer new techniques, enhanced accuracy and opportunities to link pest populations to host abundance in the landscape.

This chapter will review current insect pest monitoring systems for native and invasive pest species of corn, soybean and cotton in the United States. We present case studies of European corn borer, corn earworm, western bean cutworm and corn rootworm as pests of corn; soybean aphid, stink bugs and Japanese beetle (JB) as pests of soybean; and tobacco thrips, plant bugs and bollworm as pests of cotton. We discuss how new technology has driven recent improvements in monitoring of these pests from the lab, field and areawide levels. Although the primary emphasis will be the value of pest monitoring to support improved decision-making for IPM programs, we also discuss recent trends and tools for insect resistance management (IRM), since the importance of some field crop pests has increased as resistance continues to evolve.

3 Corn (*Zea mays* L.)

3.1 European corn borer

European corn borer (ECB), *Ostrinia nubilalis* (Hübner), has been the most widespread insect pest of corn in the United States since its introduction from Western Europe. This invasive pest is believed to have arrived on 'broom corn' in shipments to Atlantic coast and midwestern states as early as 1914. Today, ECB is present in all states east of the Colorado Rocky Mountain range. Though field (dent) corn (*Zea mays* var. *indentata*) and sweet corn (*Zea mays* convar. *saccharata* var. *rugosa*) are preferred hosts for ECB, the pest readily colonizes over 200 plant species, including other field crops, numerous vegetable crops and wild/commercial hemp (Steffey et al., 1999; Mason et al., 2018). In corn, ECB larvae will feed on all plant parts except the roots, and yield is most likely to be reduced when feeding is initiated in the mid-vegetative stages of corn growth (Hodgson and Rice, 2017). Several reviews describe the ECB life cycle, biology and management in corn (Steffey et al., 1999; Hodgson and Rice, 2017; Mason et al., 2018).

The diversity of crops that serve as ECB hosts undoubtedly assisted with the westward and southern spread during the past century. Another key factor contributing to the success of ECB across diverse climatic regions was the introduction of two ecotypes adapted to different regions of the United States. The multivoltine ecotype was first discovered near Boston, Massachusetts, in 1913, and the univoltine ecotype was found along Lake Erie in 1919 (Coates et al., 2013; Hutchison and Cira, 2017; Mason et al., 2018). Finally, there are two dominant races (or strains) of ECB in the United States determined by the male sex pheromone (11-tetradecenyl-acetate) stereochemistry. The *E*-isomer is most common in the northeastern United States and is multivoltine, and the *Z*-isomer is most common in the midwestern and southern states and may be either univoltine or multivoltine. Also, an *E–Z* hybrid occurs in areas where the two races overlap, mostly in the eastern states (Coates et al., 2013). The complex ecology of ECB (ecotypes, races, broad host range) coupled with variable weather conditions all work to influence infestation timing and the magnitude of damage on corn (Steffey et al., 1999). Consequently, this has led to a diversity of monitoring methods, particularly for adult male moths, in an effort to build effective IPM programs for corn.

During the early years of ECB range expansion, blacklight traps (Photo 1) were the most widely used method of monitoring moth flights (Bartels et al., 1997; Mason et al., 2018). Blacklight traps have the advantage of capturing female and male moths, which is useful in terms of forecasting oviposition

Photo 1 A blacklight trap used to monitor moths. Photo by Adam Sisson, Iowa State University.

phenology during each generation (Mason et al., 2018). Consequently, blacklight traps continue to be used in Minnesota and several Midwest states for ECB and several other economically important lepidopteran pests (Wold and Hutchison, 2021; AEL, 2022). Although blacklight traps have the advantage of attracting both sexes of ECB, the disadvantages of requiring a power source and the plethora of non-target insects captured - which is time consuming to sort - are well known.

Beginning in the 1970s, following the identification of the ECB sex pheromone (Klun and Brindley, 1970), the use of traps baited with synthetic pheromone blends became more popular (Photo 2). Pheromone traps and pheromone lures provide a practical alternative to blacklight traps because they are usually species-specific and do not require a power source (Mason et al., 2018). Although pheromone traps have worked well for the vast majority of insect species in corn, the development of commercial pheromone lures for ECB was complex and continues to have some disadvantages due to quality control issues with pheromone lure production. For example, Bartels et al. (1997) confirmed that the blend of 98*Z*:2*E* (98% *Z*-isomer and 2% *E*-isomer) must be carefully maintained for optimal *Z*-strain moth attraction to traps. Any increase in the *E*-component, at >3% of the lure, will substantially suppress moth captures. Moreover, even when the optimal 98*Z*:2*E* blend was used in

Photo 2 An example of a pheromone lure for attracting moths. This photo shows a rubber septum lure in a sticky trap, but lures may be used with other trap styles. Photo by Adam Sisson, Iowa State University.

pheromone traps, there was a significant 4–15-day delay in reaching peak catch (50% of total moths captured) compared to blacklight traps (Bartels and Hutchison, 1998). Where both *E*- and *Z*-races of ECB occur in the eastern United States, pheromone traps are used to distinguish the strains because the *E*-strain is usually associated with potato and susceptible host crops other than corn (Mason et al., 2018). Both blacklight and pheromone traps are used to provide scouting information for farmers. The biofix occurs when the first spring adults are captured and represents the beginning date for tracking degree days to predict when certain life stages or injury will occur (Hodgson and Rice, 2017).

Historically, the most widely used monitoring method for early-season injury to whorl-stage corn has been larval sampling to estimate the density per plant or percentage of plants infested with one or more larvae (Mason et al., 2018). Worksheet decision guides have been developed over the past 30 years to help farmers and crop consultants assess whether an infestation is above or below the economic injury level (EIL), and subsequently employ the economic threshold (ET) to decide whether to use a foliar insecticide (see worksheets within Hodgson and Rice, 2017). An ET used for many years was 50% of plants infested with larvae. Because EILs fluctuate with market price, control costs and yield projections (e.g. EILs decline when corn price or yield increases), detailed worksheet decision guides should be used for making treatment decisions (Mason et al., 2018).

A sequential sampling plan for ECB egg masses was developed by Calvin et al. (1986) to provide advance warning of the need for insecticidal control of ECB, targeting the tassel/early silk stage. Although the authors note that this can be time consuming and costly, they identified specific leaves near the ear zone where sampling can be focused. However, once Bt (*Bacillus thuringiensis* Berliner) corn became a widely used tool for ECB management by the late 1990s, ECB populations were suppressed throughout much of the Corn Belt and there has been less need for egg-mass sampling in Bt cornfields (Hutchison et al., 2010; Bell et al., 2012; Dively et al., 2018). For farmers that plant either conventional non-Bt or organic corn hybrids, this method will still be useful for assessing the need for insecticidal control or the release of *Trichogramma* spp. parasitic wasps that target ECB eggs (Mason et al., 2018).

In addition to within-season sampling for ECB, the annual autumn survey has been a useful tool to assess larval population changes over multiple years. The annual survey has been conducted in Minnesota and Wisconsin for the past 60-70 years. The time series of larval counts resulting from the annual surveys in five Midwest states proved useful for analyzing the impact of Bt corn adoption on ECB population dynamics from 1996 to 2009 (Fig. 1; Hutchison et al., 2010; Bell et al., 2012). From a pragmatic view, the results are also useful for assisting farmers in making decisions about how much Bt corn to plant the following year (Potter et al., 2022). Such annual surveys can also be used to

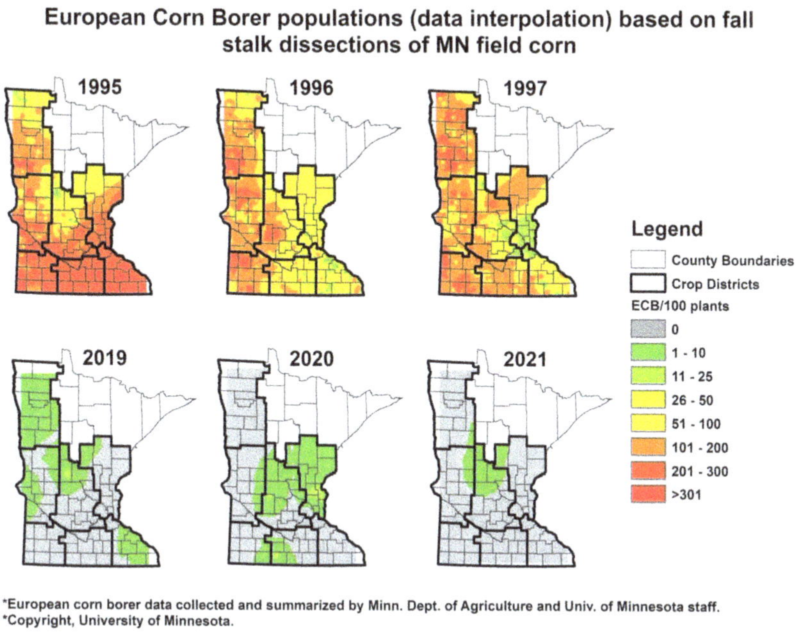

Figure 1 Historical overwintering European corn borer (ECB) populations (1995-7) via the Annual Autumn ECB Survey, comparing a pre-Bt era infestation peak (1995), with the early years of commercialization (1996--7) and recent years (2019-21). Source: Used with permission; Potter et al. (2022), UMN Extension IPM Program.

detect 'unexpected damage' in commercial Bt cornfields and then examine plants for larval infestations. During the past 26 years of commercial Bt corn use, resistance to Bt corn by ECB has been documented under field conditions in two populations: midwestern United States (Crespo et al., 2009) and Nova Scotia, Canada (Smith et al., 2019a).

3.2 Corn earworm

Corn earworm (CEW), *Helicoverpa zea* (Boddie), is native, common throughout the Americas and highly polyphagous. Consequently, the pest has been given several common names depending on the host crop (e.g. tomato fruitworm, bollworm; Hutchison and Cira, 2017). The larvae are the damaging stage, and the highest risk of damage is during the 20-24-day silking period (Steffey et al., 1999). Female moths are highly attracted to fresh silking corn to oviposit; the larvae then enter the ear and begin feeding on the kernels. The larvae are cannibalistic and typically only one completes development in the ear of the corn. The life cycle, biology and management of CEW in field corn have been reviewed by Reay-Jones (2019).

Adult moths are highly mobile and migrate long distances via annual synoptic weather systems that occur in the central and eastern United States. During spring to autumn, low- and high-pressure systems moving west to east create nocturnal low-level jet systems that transport moths from southern to northern states each year at altitudes up to 1500 m. Southern states have been confirmed as the source of moths captured in northern states by identifying pollen grains attached to their bodies (Lingren et al., 1993; Westbrook et al., 1997). Pheromone traps have been used for long-distance migration studies and are useful for IPM purposes. The four-part pheromone lure ([*Z*]-11-hexadecenal, [*Z*]-9-hexadecenal, [*Z*]-7-hexadecenal and hexadecanal; Klun et al., 1980) for CEW is reliable, and either wire-mesh Hartstack (Hartstack et al., 1979; Photo 3) or nylon-mesh traps (Photo 4; Scentry Biologicals Inc, Billings, MT) can be used.

Farmers and crop consultants can access real-time CEW monitoring data during the growing season: Some examples include *PestWatch* for the northeast region (PestWatch, 2017), VegEdge network in Minnesota (Wold and Hutchison, 2021) and Great Lakes and Maritimes Pest Monitoring Network for Canada and the Great Lakes region (Baute, 2021; see Section 6). Crop consultants and farmers can use this data to aid in scouting for CEW in field and sweet corn. Sweet corn is of higher value than field corn, especially in markets

Photo 3 A wire-mesh Hartstack (cone) trap that is used with pheromone lures (Photo 2) to capture corn earworm and other lepidopteran pests. Photo by Anders Huseth, North Carolina State University.

Photo 4 A nylon-mesh cone trap that can be used the same as the Hartstack trap in Photo 3. Photo by Anders Huseth, North Carolina State University.

where farmers sell ears directly to consumers, so ETs vary depending on the growth stage of nearby field corn. When field corn is silking, the threshold for both crops is 10 moths per night, because CEW moths are attracted to any silking corn and the risk to sweet corn is lower. Before and after field corn silks, the threshold in sweet corn is 1 moth per night, because silking sweet corn is at higher risk (Foster, 2016).

Although CEW continues to be the most consistent and damaging pest of sweet corn in the United States, recent studies in the southern United States indicate little or no significant impact of CEW on yields of field corn (Reay-Jones, 2019). Even though corn is the main host for the pest, the primary management concern among field crops is cotton, especially because the use of similar Bt toxins in corn and cotton in the Cotton Belt has implications for resistance development (Reisig et al., 2022). Furthermore, CEW has a high propensity for developing resistance to foliar insecticides in addition to Bt toxins (Jacobson et al., 2009; Dively et al., 2021). A sentinel plot approach using Bt sweet corn has been implemented in 16 US states and 4 Canadian provinces in the past 5 years to monitor for several ear-feeding lepidopterans in corn (Dively et al., 2021). This long-term study found significant increases in CEW resistance to some Bt traits (e.g. Cry1Ab, Cry1A.105 + Cry2Ab2) and demonstrates the effectiveness of sentinel plots to monitor phenotypic resistance, document field-evolved resistance, observe temporal changes in resistance and improve resistance monitoring for Bt crops.

3.3 Western bean cutworm

Western bean cutworm (WBC), *Striacosta albicosta* (Smith), is native and was historically limited to several western and Great Plains states, including Idaho, Colorado and Nebraska (Smith et al., 2019b). However, from 1999 to 2010, a significant eastward range expansion began. Early Midwest detections occurred in Minnesota in 1999 (O'Rourke and Hutchison, 2000), with economic losses documented in Iowa in 2000 (Rice, 2000). From 2010 to 2020, the pest rapidly spread to the northern tier states in the Great Lakes region, the Atlantic coast states and into eastern Canada (Smith et al., 2019b). Despite the 'cutworm' name, this pest does not actually cut vegetative corn at the soil surface. Western bean cutworm is univoltine and moths emerge in June or July, mate and then oviposit on tasseling corn. Female moths typically lay their egg masses on the undersides of leaves near the ear zone, which makes these egg masses easy to detect. Once eggs hatch, larvae will initially feed on leaf tissue but later migrate to the ears to feed on developing kernels. WBC larvae are not cannibalistic and thus multiple larvae per ear can occur and cause substantial damage. Recent reviews describe the life cycle, biology and management of WBC in corn (Michel et al., 2010; Smith et al., 2019b).

As with ECB and CEW, a variety of trap types can be used for monitoring WBC adults, including blacklight traps and pheromone traps. Research has shown that pheromone traps for WBC can replace bulky blacklight traps that require a power source (Mahrt et al., 1987). Pheromone traps for WBC use the female sex pheromone (5:l:5:5 ratio of (*Z*)-5-dodecenyl acetate, (*Z*)-7-dodecenyl acetate, ll-dodencenyl acetate and dodecyl acetate, respectively; Klun et al., 1983), and can be homemade from a milk jug or a purchased wing-style pheromone trap (Photo 5; Scentry Biologicals, Billings, MT). A revised degree-day model for forecasting WBC moth flights via blacklight trap monitoring was developed using 20 years of blacklight trap data from multiple sites in Nebraska (Hanson et al., 2015). Most recently, with the eastward spread of WBC, there has been increased use of pheromone traps for monitoring, and it appears that the degree-day model may need to be revised to take account of this monitoring method (Smith et al., 2019b).

Western bean cutworm sampling in corn has primarily centered on looking for egg masses during the late-vegetative stages. A common ET used for sweet corn is 4% of plants with egg masses during the tasseling stage (Cullen and Jyuotika, 2008). For field corn, the ET is 8% or 20% of plants with egg masses during the tasseling or milk stages, respectively (Seymour et al., 2004; Peairs, 2006; Rice, 2007). The ET has been revised to include percent larval survival

Photo 5 An example of a wing-style trap used in combination with pheromone lures (Photo 2) to monitor western bean cutworm and other lepidopteran pest species. Photo by Adam Sisson, Iowa State University.

in addition to percentage of plants with egg masses (Paula-Moraes et al., 2013). In Nebraska, a rapid, sequential sampling plan for WBC egg masses was developed utilizing ETs for field and sweet corn (Paula-Moraes et al., 2011; AEL, 2022). It was determined that the average sample number to reach a treatment decision was 38–41 plants per field, depending on ET, which is an improvement on the previous recommendation of 100 plants per field (Rice, 2006). Both foliar insecticides and Bt corn hybrids have been used to manage WBC (Hutchison and Cira, 2017). However, like ECB, WBC has developed resistance to the Cry1F trait in Canada (Smith et al., 2017).

3.4 Corn rootworm

Northern (*Diabrotica barberi* Smith & Lawrence) and western (*Diabrotica virgifera* LeConte) corn rootworm (CRW) are the two most economically important soil pests of corn throughout much of North America. Both species are native to North America, and western CRW has invaded Europe (Miller et al., 2005). Most of their univoltine life cycle is subterranean, where the larvae, the most damaging life stage, feed on corn roots in the early stages of crop development. Adults may also cause injury to corn when they feed on aboveground plant parts, but it is usually not economically important; however, adults are generally the target for monitoring efforts as they are easier to sample and a reasonable indicator of the following year's larval pressure. The life cycle, biology, management and challenges of CRW, especially western CRW, have been widely studied and several recent reviews provide details on these topics (Gray et al., 2009; Meinke et al., 2009; Spencer et al., 2009; Gassmann, 2021).

Traditional monitoring for CRW involves scouting cornfields to assess root injury to inform management decisions. The standard approach to assess larval injury is to rate root damage using the widely accepted Iowa State University (ISU) 0-3 Node-Injury Scale (Photo 6; Oleson et al., 2005). This scale improved upon the previous systems (the Iowa 1-6 and the 1-9 root damage rating systems; Hills and Peters, 1971 and Musick and Suttle, 1972, respectively) by providing a linear relationship for the level of larval injury at each rating, and the prediction for yield losses using a nested error component model (Tinsley et al., 2013). Recently, a sequential sampling plan has been developed for more efficient assessments of larval injury and product performance using the ISU 0-3 Node-Injury Scale (Hong et al., 2016).

The most widely accepted scouting technique for adult CRW is to use yellow sticky traps (Photo 7), which rely on the beetles' attraction to the color of the trap and not chemical cues, although some research has evaluated using traps baited with semichemical lures (Whitworth et al., 2002). Northern and western

Photo 6 A corn root with feeding injury from corn rootworm larvae. Using the ISU 0-3 Node-Injury Scale, this root received a score of 3.0. Photo by Erin Hodgson, Iowa State University.

Photo 7 A yellow sticky trap used to monitor corn rootworm adults. Photo by Erin Hodgson, Iowa State University.

CRW are distinct from other leaf beetles and therefore misidentification is not likely for a trained scout. Hein and Tollefson (1985) determined that yellow sticky traps are just as effective in predicting larval injury as the standard adult sampling method, whole-plant adult counts, and suggested a threshold of six or more beetles per trap per day could be used to predict whether economic injury from larvae will occur the following year in continuous corn. Dunbar and Gassmann (2013) revised this threshold to two or more beetles per trap per day, and recently Leaf (2017) suggested adding soil texture, specifically percent clay, as an additional explanatory variable when determining thresholds for predicting larval injury from sticky trap counts. In the model, percent clay content accounted for much of the variation, likely because soil texture impacts survival and activity of corn rootworms (Leaf, 2017). Several sticky trap options exist for monitoring CRW adults (e.g. Trécé Pherocon AM, Trécé Inc., Adair, OK; Scentry Multigard AM, Scentry Biologicals Inc., Billings, MT), as well as different methods of directly counting beetles on plants (Steffey et al., 1982; Sawyer, 1985; Shields et al., 1991; Losey et al., 2003). Due to within-field variability of corn rootworm populations, whole-plant counts, ear counts and sticky traps are equal in precision (~50%) depending on sampling units (2 plants/site at 9 sites/field, 5 ears/site at 9 sites/field and 8 traps/ per field, respectively; Steffey et al., 1982).

Perhaps of equal importance to assessing management practices in the field, using yellow sticky traps annually provides the opportunity to monitor changes in CRW populations over time. A working group of extension, industry and government personnel in 12 US states and 5 Canadian provinces deployed a regional sticky trap network to monitor CRW adults in 2021 (ISU, 2021). Data from this network was captured via Survey123, which is a data collection application that allows data entry in the field and uploads entries with cellular data or Wi-Fi for further analysis (ESRI, 2021a). Data from Survey123 were automatically mapped each day during the trapping season and available for the public to view (ESRI, 2021b). Mass monitoring efforts similar to this may be most important to detect arrival of CRW into areas where it is not currently established. By combining available biological and physiological data, Aragón et al. (2010) modeled the global climatic suitability and invasion risk for western CRW.

Recent advancements in technology, accessibility of georeferenced data and availability of relevant landscape data layers have led to spatial predictions of CRW activity and development of sampling plans. Georeferencing data collection allowed Park and Tollefson (2006) to determine where CRW beetles were aggregated and establish that samples must be obtained at least 84 m apart to be independent. Several studies have used geographical information systems (GIS) and complex analysis procedures to provide information on the interactions between CRW population dynamics and physical factors in the field (e.g. soil type, elevation; Beckler et al., 2005; Davidson et al., 2007). By including soil composition information, Davidson et al. (2007) improved upon the sampling plans developed by Park and Tollefson (2006) and suggested that management zones within a field could be targeted for insecticide applications or planting of CRW resistant hybrids where pressure is greatest. The management zones were areas with high-, medium- and low-yield potential based on soil properties and farmer observation. Davidson et al. (2007) suggests highly productive areas of the field (high clay content and soil moisture, also suggested by Leaf, 2017) likely have more CRW larval feeding, although this may be more useful in fields that are less uniform. Variable within-field applications would decrease input costs and reduce pesticide exposure to pests and non-target organisms in the environment.

Most recently, geospatial analyses are being used to evaluate the role of land use patterns on CRW populations and resistance to corn that expresses Bt toxins. Several studies have linked the continuous cultivation of corn (3 or more years) in a field to CRW populations and resistance development by surveying farmers about field history where data was collected (Gassmann et al., 2011; Dunbar et al., 2016; Reinders et al., 2018; Shrestha et al., 2018; Carrière et al., 2020). However, using a combination of geospatial crop production data and statistical analysis, St. Clair and Gassmann (2021) revealed that fields with confirmed Bt-resistant CRW tend to be located near long-term continuous corn cultivation (within 1.6 km) that enables a combination of persistent populations and continuous resistance selection pressure. This may indicate that, even though field-level management of CRW is common, the intensity of corn production at the landscape level should be considered to delay resistance evolution to future technologies and management tactics.

4 Soybean (*Glycine max* [L.] Merr)

4.1 Soybean aphid

Soybean aphid (SBA), *Aphis glycines* Matsumura, is a well-studied invasive pest in the United States that is native to Asia. In 2000, SBA was confirmed in North America and quickly spread to soybean-growing regions in the north-central states. Soybean aphid overwinters as an egg on *Rhamnus* spp. and disperses to vegetative soybean in June (Ragsdale et al., 2004). During the growing season, asexual clones have multiple, overlapping generations on soybean, and populations can reach exponential growth under ideal conditions for the aphid (McCornack et al., 2004; Tilmon et al., 2011). Soybean aphid uses a piercing-sucking stylet to remove phloem sap from soybean. More than 5000 aphids per plant are possible and economic losses can exceed 40%; however, widespread outbreaks do not occur every year. Therefore, frequent monitoring is critical for making timely management decisions and minimizing unnecessary insecticide applications (Hodgson et al., 2012). Several recent reviews describe the life cycle, biology and management of SBA (Ragsdale et al., 2011; Tilmon et al., 2011; Hodgson et al., 2012; Koch et al., 2018).

Foliar insecticide applications are typically based on whole-plant counts of aphids (Photo 8) until soybeans reach mid-seed fill stage. An ET was developed for SBA (Ragsdale et al., 2007) and recently validated (Koch et al., 2016). Hodgson et al. (2004) recommends a mean sample number of 310 plants for 0.10 precision or 38 plants for 0.25 precision. There is a reasonable expectation to 'break even' or make a profitable decision if mean aphid numbers reach 250 per plant (Ragsdale et al., 2007). Alternatively, a binomial sequential sampling plan for SBA is available for those who want to make rapid treatment decisions (Fig. 2; Hodgson et al., 2004). This efficient method uses a tally threshold of 40 aphids per plant and reduces sampling

Photo 8 Soybean aphids on a small soybean plant. Whole-plant sampling is the standard method for soybean aphids. Photo by Erin Hodgson, Iowa State University.

time when pressure is either very low or high (Hodgson et al., 2007). Like many persistent pests with frequent insecticide exposure, SBA has become resistant to pyrethroids (Hanson et al., 2017; Valmorbida et al., 2022). Dean et al. (2020) provides a decision-making framework for SBA that incorporates anticipated grain market values, input costs and an estimate of pyrethroid-resistant aphid frequency in the population.

Suction traps have been used in the United Kingdom and northwestern United States to better understand aphid migration and management. In 2001, a regional effort to monitor for winged SBA was initiated (Photo 9). At least 31 traps distributed over 10 states have been operational for approximately 5 months every year. First and peak detection of migrating SBA are reported and used by extension personnel to make scouting recommendations. Multiple studies have used data from the suction trapping network, including documentation of over 150 aphid species (Lagos-Kutz et al., 2020).

4.2 Stink bugs

Stink bugs are emerging pests in several US cropping systems, increasing in abundance and frequency in soybean over the last two decades (Koch et al., 2017). Several persistent species include green stink bug, *Chinavia hilaris* (Say); brown stink bug, *Euschistus servus* (Say); one-spotted stink bug, *Euschistus variolarius* (Palisot de Beauvois); and redshouldered stink bug, *Thyanta custator accerra* (McAtee).

Stink bugs typically overwinter as adults in plant debris, human structures and other protected locations; therefore, the composition of surrounding

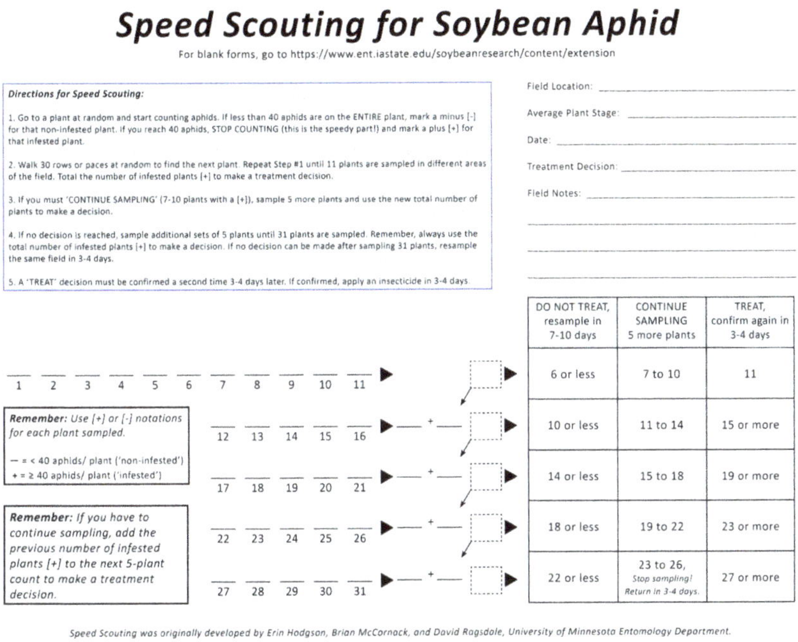

Speed Scouting for Soybean Aphid

For blank forms, go to https://www.ent.iastate.edu/soybeanresearch/content/extension

Directions for Speed Scouting:

1. Go to a plant at random and start counting aphids. If less than 40 aphids are on the ENTIRE plant, mark a minus [-] for that non-infested plant. If you reach 40 aphids, STOP COUNTING (this is the speedy part!) and mark a plus [+] for that infested plant.

2. Walk 30 rows or paces at random to find the next plant. Repeat Step #1 until 11 plants are sampled in different areas of the field. Total the number of infested plants [+] to make a treatment decision.

3. If you must 'CONTINUE SAMPLING' (7-10 plants with a [+]), sample 5 more plants and use the new total number of plants to make a decision.

4. If no decision is reached, sample additional sets of 5 plants until 31 plants are sampled. Remember, always use the total number of infested plants [+] to make a decision. If no decision can be made after sampling 31 plants, resample the same field in 3-4 days.

5. A 'TREAT' decision must be confirmed a second time 3-4 days later. If confirmed, apply an insecticide in 3-4 days.

Field Location: ____________

Average Plant Stage: ____________

Date: ____________

Treatment Decision: ____________

Field Notes: ____________

Remember: *Use [+] or [-] notations for each plant sampled.*

– = < 40 aphids/ plant ('non-infested')
+ = ≥ 40 aphids/ plant ('infested')

Remember: *If you have to continue sampling, add the previous number of infested plants [+] to the next 5-plant count to make a treatment decision.*

Plants sampled	DO NOT TREAT, resample in 7-10 days	CONTINUE SAMPLING 5 more plants	TREAT, confirm again in 3-4 days
1 2 3 4 5 6 7 8 9 10 11	6 or less	7 to 10	11
12 13 14 15 16	10 or less	11 to 14	15 or more
17 18 19 20 21	14 or less	15 to 18	19 or more
22 23 24 25 26	18 or less	19 to 22	23 or more
27 28 29 30 31	22 or less	23 to 26, *Stop sampling! Return in 3-4 days.*	27 or more

Speed Scouting was originally developed by Erin Hodgson, Brian McCornack, and David Ragsdale, University of Minnesota Entomology Department.

Figure 2 The *Speed Scouting* worksheet developed by Hodgson et al. (2004, 2007). Source: Used with permission. Find blank copies here: https://bit.ly/3JDkkB0.

Photo 9 A suction trap used for the soybean aphid trapping network near Ames, Iowa. Photo by Greg VanNostrand, Iowa State University.

landscapes can influence population dynamics in soybean (Koch et al., 2017). The biology of stink bugs contributes to several species becoming frequent agricultural pests (e.g. multivoltine, mobile and polyphagous). Stink bugs typically reach pest status when nymphs and adults attack reproductive structures during plant development (McPherson and McPherson, 2000). Both adults and nymphs have piercing-sucking stylets that is used to egest a digestive enzyme while feeding, causing mechanical and chemical injury to leaves, stems, pods and seeds. Significant feeding can reduce plant quality and quantity by causing direct injury to pods and seeds.

During the past 15 years, an emerging stink bug concern is the invasive brown marmorated stink bug (BMSB), *Halyomorpha halys* (Stål), which is native to Asia and has spread throughout the eastern and midwestern US states (Rice et al., 2014; Koch et al., 2017; Leskey and Nielsen, 2018). Although BMSB has a broad host range, the species does particularly well on soybean, which has resulted in moderate increases in insecticide use in some regions (Koch et al., 2017). As a result of direct feeding on soybean seeds within pods, BMSB can impact soybean yield, seed quality and germination rates. Heavy pod feeding results in delayed maturity of the seeds and stay-green syndrome, which can delay harvest (Koch et al., 2017). Where soybean fields are located adjacent to wooded areas, a common source of BMSB (Leskey et al., 2012; Leskey and Nielsen, 2018), the delayed maturity and stay-green syndrome can be detected along field edges. To date, only selected soybean varieties have been assessed for BMSB injury or tolerance (Koch and Rich, 2015). More research is needed to fully examine the impact of this stink bug on more commonly grown soybean varieties in the United States.

In response to the rapid spread of BMSB in the United States, considerable progress has been made in trap-based monitoring systems for this species (Leskey and Nielsen, 2018), including the identification of the male aggregation pheromone (murgantiol; Khrimian et al., 2014), development of a dual lure (murgantiol + methyl [2*E*,4*E*,6*Z*]-decatrienoate; Weber et al., 2014) and the currently accepted use of a sticky-panel trap design (Photo 10; Leskey and Nielsen, 2018; Acebes-Doria et al., 2020). The sequential sampling plans for in-field monitoring of BMSB have indeed benefited monitoring for other stink bugs, since all major species were included in the plans (Koch et al., 2017). One indication of the effectiveness of the dual lure and sticky trap is that in addition to adult captures, immature nymphs are also captured in high numbers; nymphs approach the traps from nearby infested crops and host plants and scale the 1 m high traps (Acebes-Doria et al., 2020).

Although stink bugs can be present throughout the growing season, concentrating efforts at the initiation of soybean flowering is recommended (Koch et al., 2017). Entomologists who focus on the ecology and management of stink bugs often recommend 'lumping' multiple species together when sampling to make treatment decisions (McPherson and McPherson, 2000). In addition, both nymphs and adults are counted when determining the density within fields. The current ET for stink bugs in soybean is 5 stink bugs/25 sweeps

Photo 10 The black pyramid trap (a) and sticky-panel trap (b) for monitoring brown marmorated stink bug. Photo by Angelita Acebesc-Doria, USDA-ARS.

(1 stink bug/0.3 m of row) for soybean grown for seed and 10 stink bugs/25 sweeps (3 stink bugs/0.3 m of row) for soybean grown for grain (Koch et al., 2017). The species composition is not uniform throughout the country, and the spatial-temporal distribution within a field can be highly variable. Stink bugs can be aggregated around field edges, especially near wooded habitats (Leskey et al., 2012). In soybean, nymphs and adults have a strong startle response and can drop off plants when disturbed; therefore, sweep netting is the most common and convenient sampling technique (Photo 11). Several sampling plans exist for stink bugs in the United States with varying degrees of precision. Pezzini et al. (2019) recently developed an enumerative sequential sampling plan for multiple stink bug species, whereas Aita et al. (2021) published a presence-absence sequential plan for the same stink bug complex, which further reduced the time (cost) of scouting; using this method, the probability of making a correct decision was >95%. Both plans utilize sweep netting with a 25-sweep sample unit.

4.3 Japanese beetle

JB, *Popillia japonica* Newman, is an invasive, polyphagous pest in the United States from Japan. It feeds on more than 300 species of plants, including important turfgrass, horticultural and agricultural species. Much of their univoltine life cycle is spent in the soil, where larvae feed on grass roots.

Photo 11 A sweep net used to sample insect pests in soybean. Photo by Nick Seiter, University of Illinois.

This injury can be economic in turfgrass but is generally not a concern in field crops. The adult beetle feeds on foliage; in soybean, the main concern is skeletonization (defoliation) of the leaves, while silk clipping in corn may interfere with pollination. Several recent reviews provide details on the life cycle, biology and management of JB (Potter and Held, 2002; Shanovich et al., 2019; Hodgson and Dean, 2020).

Currently, JB has infested most areas of North America, east of the Colorado Rocky Mountain range (Shanovich et al., 2019; Hodgson and Dean, 2020). The US Department of Agriculture established a federal quarantine in 1979 to prevent the spread of JB to the western states by aircraft (USDA-APHIS, 2021), and phytosanitary requirements have been outlined in the United States and Canada to prevent introductions of JB via plants or plant materials (Government of Canada, 2019; NPB, 2021). Based on the biology of JB, modeling reveals that areas considered suitable for JB establishment remain unoccupied in North America (Kistner-Thomas, 2019). Additionally, the global distribution, voltinism and severity of JB are expected to change with future climate projections (Kistner-Thomas, 2019). Knowing this information provides an opportunity for monitoring beetles in areas that are not yet infested to eradicate or limit the spread of JB as climates continue to change. Current monitoring efforts have allowed states further west to detect, report and sometimes eradicate isolated infestations (Potter and Held, 2002; CDFA, 2022; CERIS, 2022).

As airports are a common source of introductions, Hamilton et al. (2007) placed traps near the Indianapolis airport to assess how land use impacts trap captures. Their findings suggest removing or reducing agronomic land, specifically fields with preferred hosts such as corn and soybean, within 500 m of the loading terminals could limit JB adults hitchhiking on aircraft to uninfested regions. Hungate et al. (2016) confirmed the utility of using hydrogen isotope composition as a means of determining the natal origin and timing of arrival for JB. By analyzing adults captured in traps placed near the Portland International Airport, it was determined that a local, viable population had not established in Portland, likely due to effective control measures at the Portland and source airports. Hydrogen isotopes may be able to provide information about invasive species on a time scale that enables effective decision-making for eradicating populations that would not be possible through molecular and genetic tools.

Soybean is a favored crop of adult JB, and several studies have reported that adult densities are greatest at the field edge (Morehouse, 2011; Sara et al., 2013). Since soybean yield would be most threatened at field edges, a perimeter treatment could be made to manage JB and limit defoliation while reducing the cost of control and insecticide load in the environment. Other landscape factors that impact JB density, distribution and movement include soil type, soil moisture, distribution of host plants and natural enemies (Potter and Held, 2002). These factors could be further explored using GIS technology to create detailed sampling plans and management recommendations and have already been used to predict future global distributions (Kistner-Thomas, 2019).

JB traps have primarily been used for monitoring in areas where adults are not yet found or to reduce isolated populations, and several trap designs have been evaluated for efficacy and ease of use (Potter and Held, 2002; Alm and Dawson, 2003; Piñero and Dudenhoeffer, 2018; Ebbenga et al., 2022). Past research demonstrates that individual traps (Photo 12) may not be effective at reducing injury to plants for several reasons: Beetles are highly attracted to floral and sex pheromones used in lures, traps fill up quickly, there is a spillover effect to surrounding plants and the odor from decomposing beetles in traps repels JB. However, mass trapping efforts have proven effective for detecting and eradicating beetles in the western states (CDFA, 2022), and mass trapping designs may be effective and economically viable in organic orchards, small fields or other high-value crops (Piñero and Dudenhoeffer, 2018).

In soybean in particular, ETs have not been developed based on beetle counts. Rather, when thresholds are used, management decisions are made for JB based on visual estimates of defoliation across the field (Fig. 3; Ohnesorg and Hunt, 2016; Shanovich et al., 2019). Visual estimates of defoliation are notoriously inaccurate as humans tend to overestimate defoliation, especially at low levels, and regular training is essential to increase accuracy (Johnson et al., 2016). Visual estimates can be done quickly and are effective for a large number of samples or when destructive samples are not desired; however, they may not be reliable when precise measurements are needed. Several technologies have been used in the past to provide measurements of defoliation, such as

Photo 12 Japanese beetle adult trap used with a dual lure (sex pheromone and floral attractant). Photo by Brent Pringnitz, Iowa State University.

computer software (e.g. Image J, National Institutes of Health, Bethesda, MD) and leaf area meters (e.g. LI-COR Biosciences, Lincoln, NE), but these have several drawbacks. They are specialist devices that require maintenance, sensitive to foreign objects, unable to detect non-green colors, time consuming and unable to quantify insect herbivory at the margins of leaves (Johnson et al., 2016; Machado et al., 2016; Getman-Pickering et al., 2020).

Since smartphones are ubiquitous, low-cost and simple to use, it makes sense that they may be used widely in agriculture for a variety of activities given their GPS (global positioning system) and high-quality imaging capabilities (Kwon and Park, 2017). Two mobile applications have been developed to quantify defoliation by insect pests: BioLeaf and LeafByte are available on Android and Apple systems, respectively (Machado et al., 2016; Getman-Pickering et al., 2020). Both of these applications are simple, highly accurate and consistent. Both allow the user to reconstruct leaf edges when defoliation occurs at the leaf margin and can be used in a non-destructive fashion as long as a contrasting background is present. BioLeaf measures the percentage of leaf area that is missing (percent defoliation), while LeafByte can provide absolute measurements of leaf area and herbivory in addition to percentage leaf loss. LeafByte has an additional advantage of automated data storage

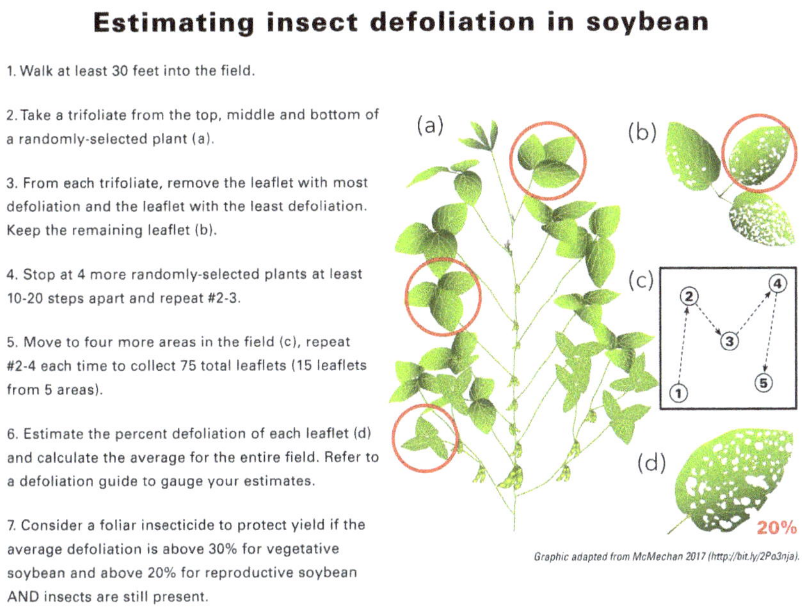

Figure 3 Scouting plan to estimate field-wide defoliation for insect pests. Source: Figure by Erin Hodgson, Iowa State University.

to spreadsheets; this has been useful for research and IPM applications. Two potential limitations of these applications are (a) ruffled leaves, shadows, poor quality photographs and angle of capture impacting the accuracy of estimations and (b) it can be difficult and time consuming to measure herbivory when leaves are mostly consumed or the margins are difficult to recreate. However, even with these limitations, smartphone applications that measure defoliation provide an improvement upon older techniques that are not optimized for insect herbivory and visual estimations that are often inaccurate.

5 Cotton (*Gossypium* spp. L.)

5.1 Tobacco thrips

Thrips are a persistent pest of cotton across the US Cotton Belt. In the Southeastern US production region, tobacco thrips (TT), *Frankliniella fusca* (Hinds), is the most economically important pest of seedling cotton (Cook et al., 2011) and is native to North America. In the southern United States, TT overwinter on winter annual weeds (e.g. dandelion [*Taraxacum officinale*], shepherdspurse [*Capsella bursa-pastoris*] and chickweed [*Stellaria media*]) located in non-crop habitats and in the spring, populations disperse into different cultivated crops (e.g. cotton, soybean, tobacco) (Groves et al., 2001, 2003; Kahn et al., 2005;

Cook et al., 2011). In cotton, adult and immature thrips use piercing-sucking and rasping mouthparts to feed on nearly all parts of cotton plants, causing characteristic silvering associated with thrips injury. The most significant injury occurs on seedlings and developing leaf tissue. Severe infestations can cause a reduction in photosynthetic potential, increase risk for water stress later in the season, loss of apical dominance, yield reductions and plant death (Cook et al., 2011). Following initial colonization, thrips can complete several generations in cotton; however, the plant rapidly exits its stage of susceptibility around four to five true leaves. Cook et al. (2011) provide a review of the life cycle, biology and management of TT.

To minimize early-season thrips injury, cotton farmers have relied on preventative insecticide applications (Cook et al., 2011; Reisig et al., 2012; Stewart et al., 2013; North et al., 2018). Neonicotinoid seed treatments have been the dominant thrips management tool across the US Cotton Belt for nearly two decades and, because of widespread adoption, repeated use has driven the evolution of resistance in the mid-south and southeast cotton production regions (Huseth et al., 2016; Darnell-Crumpton et al., 2018). To understand the drivers of resistance, diet-based neonicotinoid bioassays were linked to the spatiotemporal abundance of neonicotinoid-treated cotton and soybean using publicly available, remotely sensed crop data (Huseth et al., 2018). Results demonstrated that increased use of neonicotinoid-treated seeds in crops, specifically cotton and soybean, was positively related to TT resistance. To counteract the loss of neonicotinoid control, a key component of TT resistance management has been rotation to alternative chemistries (e.g. organophosphates; Marshall et al., 2022) that alleviate selection pressure and exploit fitness costs of resistant individuals (Huseth et al., 2017; D'Ambrosio et al., 2018; Chappell et al., 2019). However, the timing of insecticide applications can be a significant challenge for cotton farmers because these alternative modes of action are often applied during a relatively brief growth stage when foliar sprays are most effective (i.e. one to two true leaves).

To address insecticide application timing challenges, researchers have focused on improved forecasting models for thrips infestations of seedling cotton. The Thrips Infestation Predictor for Cotton (see Section 6), which is an innovative analytical approach, connects statistical models of pest phenology and cotton growth to predict infestation risk based on forecasted weather (Chappell et al., 2020a). The tool has been deployed as an online decision-support resource targeting cotton farmers, county extension agents and independent crop consultants. Since the tool was released in 2017, it has been used to forecast infestation risk before the cotton crop is planted. However, the tool has also been applied to improve timing of foliar applications of insecticides to emerged cotton fields that were planted during periods of high infestation risk. This flexibility allows farmers to make informed post-emergence decisions even if planting windows align with high risk for thrips infestation. This valuable analytical approach to thrips flight prediction has been extended to manage a thrips-transmitted tospovirus, tomato-spotted wilt orthotospovirus (order Bunyavirales, family Tospoviridae, genus *Orthotospovirus*) in peanut (Chappell et al., 2020b) and tobacco (see Section 6).

In the future, the impact of new pest management technologies will likely affect arthropods found in cotton fields across the Cotton Belt. Notably, the deployment of the first thysanopteran and hemipteran active Bt toxin expressed in cotton, Cry51Aa2.834_16 (Bachman et al., 2017; Akbar et al., 2019), will have unique effects on oviposition and feeding deterrence of target pests (Cervantes et al., 2019; Graham et al., 2019; D'Ambrosio et al., 2020; Huseth et al., 2020). However, the use of this novel Bt toxin could have unforeseen effects on the arthropod food web in cotton and, in turn, management decisions by farmers. To address these anticipated changes, the expansion of on-farm pest research projects that focus on pest-crop connections during the release of this technology will be useful to understand changes to the system and adapt cotton IPM practices at broad scales.

5.2 Plant bugs

Since the completion of boll weevil eradication and the release of Bt cotton, two species of plant bugs (Hemiptera: Miridae) have become persistent economic pests of cotton (George et al., 2021): western tarnished plant bug (WTPB; *Lygus hesperus* Knight) and tarnished plant bug (TPB; *Lygus lineolaris* Palisot de Beauvois). These two pests are generally separated, with WTPB being a pest in the western Cotton Belt and TPB in the east. Both species are native, highly polyphagous and have multiple generations on non-crop weeds and multiple host crops each season. In cotton, plant bugs use piercing-sucking stylets to feed on vegetative and reproductive plant tissues, with the majority of economic injury occurring from direct feeding on cotton buds (i.e. squares), flowers and small bolls (Layton, 2000). Direct feeding can also increase the likelihood for stained lint and boll rot due to secondary pathogens (Dorman et al., 2021c). Together, direct and indirect effects of feeding can contribute to significant yield loss to plant bugs in production systems across the US Cotton Belt (George et al., 2021). A few recent reviews describe the life cycle, biology and management of TPB (Layton, 2000; George et al., 2021).

In the arid west, researchers have successfully used a variety of spatiotemporal crop data to document sequential host plant use by WTPB (Carrière et al., 2006, 2012; Sivakoff et al., 2013). Here, geospatial data was used to identify host patches driving source-sink dynamics of WTPB across the cultivated landscape. The limited number of available agricultural hosts within the desert agroecosystem is advantageous when documenting connections between landscape composition, structure, crop phenology and pest management practices.

In the humid subtropical eastern United States, TPB has increased in importance over the past three decades (George et al., 2021). Like WTPB, spatial analysis techniques have been useful to document factors influencing TPB problems at field and landscape scales. Here, on-farm surveys of TPB abundance in cotton were linked with climatic variables and landscape composition to describe variation across a diverse agricultural landscape (D'Ambrosio et al., 2019; Dorman et al., 2020, 2022). Results highlighted the importance of crop composition and climate factors driving populations; however, a limited amount of variance was described by these studies. This outcome is unsurprising due to the large number of crop and non-crop hosts that support TPB populations throughout the growing season (Goethe et al., 2021; Dorman et al., 2022), often resulting in repeated infestation of sampled cotton fields (Dorman et al., 2020). Despite the uncertainty about which landscape factors are most important, the outcomes of these landscape-based studies highlight the importance of large data sets to begin uncovering trends in survey data collected at the scale of commercial cotton production.

At the field scale, spatial analyses have refined understanding of the heterogeneous distribution of hemipteran and thysanopteran pests in cotton. In these studies, Spatial Analysis by Distance IndicEs (SADIE; Perry et al., 1999) has been used to identify either areas of high-density counts or areas of low-density counts using grid-based pest data from cotton (Reay-Jones et al., 2010, 2016, 2019). Outcomes of this work help to inform the design of sampling strategies based on the spatial heterogeneity of pests in cotton and other crops. The recent development of ecoinformatic approaches to observational data sets, typically collected by crop consultants, has provided useful insights into the factors driving the economic importance of pest across a large number of agricultural fields (Rosenheim et al., 2011; Rosenheim and Gratton, 2017). In cotton, researchers have used these large crop scouting data sets to improve knowledge of pest distribution, sampling approaches and management (Rosenheim, 2013, 2021a,b; Meisner et al., 2017).

5.3 Bollworm (corn earworm)

Bollworm, *Helicoverpa zea* (Boddie), is a highly polyphagous pest native to the Americas. As such, this pest has been given several common names, including tomato fruitworm and corn earworm (CEW). To provide continuity throughout the chapter, we will refer to *H. zea* as CEW in the cotton section (Section 5), since CEW was used for this species in the corn section (Section 3). In the southern United States, CEW populations cycle through a predictable sequence of hosts, typically completing development on weeds, corn and then cotton or soybeans (Jackson et al., 2008; Head et al., 2010). In cotton, adult CEW colonize and oviposit when the crop begins blooming (Culin, 1995; Fitt, 1989; Jackson et al., 2008). Larvae feed on reproductive structures, including squares and bolls, causing severe yield reductions when not managed. Corn earworm populations are monitored using a combination of visual counts for larvae and eggs, which is a process that is time consuming and labor intensive (Del Pozo-Valdivia et al., 2021). As a result, early detection tools, like pheromone/blacklight traps, have been instrumental to provide early warnings of annual infestations throughout the Cotton Belt.

The spatial distribution of cotton across the landscape and connectivity with other crop and non-crop plant hosts for pests is an overarching factor driving infestation risk, especially for highly mobile polyphagous pests like CEW (Luttrell et al., 2015). A recent analysis of crop production patterns in the United States detected strong signals of consolidation of cotton and, as a result, this long-term change may favor pests and increase farmer reliance on insecticides (Crossley et al., 2021). Despite the perceived negative effects of increased connectivity of cotton production, the uniform adoption of cotton varieties that express pyramids of Bt toxins may also suppress pest populations. This process has been documented for old world bollworm, *Helicoverpa armigera* (Hübner), in both China and Australia cotton production systems (Wu et al., 2008; Baker and Tann, 2017). In the United States, the consolidation of genetically engineered corn that expresses Bt toxins has been connected to areawide suppression of highly polyphagous lepidopteran pests and, in turn, reduced pest herbivory on multiple crops (Hutchison et al., 2010; Dively et al., 2018).

Although the effects of crop consolidation may vary, the ongoing use of Bt in corn and cotton has led to concern about resistance evolution in CEW throughout the southern cotton-corn production region (Reisig et al., 2018; Reisig and Kurtz, 2018; US-EPA, 2020). For lepidopteran pests, like CEW, this widespread transition to a limited number of Bt plant-incorporated protectants accompanied by a reduced area of non-toxic refuge plants has driven field-evolved resistance globally (Tabashnik et al., 2008, 2013, 2014; Huang et al., 2011). The ongoing shift toward areawide deployment of these Bt toxins in major row crops has increased landscape-level selection pressure and is expected to increase the rate of resistance evolution in pests that move within and among these crops (Peck et al., 1999; Storer et al., 2003; Gustafson et al., 2006; Head et al., 2010; Ives et al., 2017). However, the specific strategy to deploy similar Bt toxins among crops can produce durability differences that are driven by complex interactions between CEW, regionally specific farm management practices and composition of crops in the landscape (Pan et al., 2016).

Documenting changing Bt susceptibility in lepidopteran cotton pests, specifically CEW, has been an ongoing challenge due to the effort needed to accurately complete diet-based bioassays on larvae. Several recent studies have successfully documented variable susceptibility to multiple Bt toxins in field collected CEW populations using plant- or diet-based bioassays in the laboratory (Carrière et al., 2019; Yang et al., 2020; Niu et al., 2021,;Yang et al., 2021). Using a different approach, Reisig et al. (2018) connected laboratory Bt assays for CEW resistance with field surveys of cotton injury and insecticide sprays over time. Results highlighted the importance of linking bioassay responses to field injury to provide context for putative field-evolved resistance to Bt toxins (Reisig et al., 2018). In a subsequent study, Dorman et al. (2021a) unified CEW cotton damage studies collected from longitudinal damage data, areawide surveys across hundreds of fields and replicated on-farm variety tests that included multiple toxin combinations in the same fields. A strength of this approach was the ability to compare damage among varieties that expressed known toxins in commercial fields with replicated on-farm tests nested within the same production region. Results showed that the repeated production of Bt crops in the surrounding landscape was positively related to the risk of CEW injury to cotton bolls (Dorman et al., 2021a). Using longitudinal blacklight trap data, we also know that the abundance of CEW relates to non-Bt host crops (i.e. soybean) in some cotton agroecosystems (Dorman et al., 2021b). This is important because the association of crop composition to Cry1Ac Bt bioassays of CEW populations showed that the proportional abundance of Bt corn relative to non-Bt soybean in the local landscape can influence the effectiveness of refuge for Bt crops (Arends et al., 2021).

6 Conclusion and future trends

Each of the major crops in the United States discussed in this chapter is subject to herbivory from a range of insect pests that present a unique set of challenges. In regions where pests colonize multiple major crops (i.e. the Cotton Belt, where corn, cotton and soybeans are widespread), the interconnected cropping systems can complicate management and expedite development of resistance in polyphagous pest species. With increasing resistance development in many major insect pests in the United States, it is especially important to effectively monitor pests and understand associations with other insects, crops and landscape features. The notion of scouting crops for pests has long been implemented in the United States, but recent research provides new insights on how to effectively and efficiently sample pests, and technological advances provide new monitoring tools and scouting methods for farmers, consultants and researchers.

Although geospatial data and mapping has been around for decades, accessible, open-source GIS tools only recently became popular. With more powerful computers, integrated GPS and easy-to-use software, obtaining and analyzing georeferenced data has become more accessible. These tools have been used for countless applications in applied entomology and pest management: Modeling spatial distributions of pests within fields and larger landscapes, understanding interactions between landscape factors and pests and revealing likely causes of resistance to management tactics (i.e. insecticides and Bt crops) are a few topics discussed in this chapter. Geospatial data is the foundation for much of the recent entomological research that aims to increase accuracy, efficiency and profitability of scouting pests and forecast pest populations and risk of infestations.

As insect development is largely based on external temperatures, degree-day models have been developed to help forecast pest issues. Forecasting pest activity can be done easily with local and nationwide networks that report climatological data online, and these forecasts are most useful when farmers can use them as a decision-making tool (e.g. thrips in cotton). Similarly, climatological data has been combined with insect biology and geospatial data to model suitable climates for pests under current and future projected climate conditions (Aragón et al., 2010; Kistner-Thomas, 2019). This type of analysis is critical in understanding areas that are at risk of infestations, and these areas can use trapping or other surveillance measures to intercept, eradicate or limit the spread of pest species.

Using traps is a keystone for monitoring new, invasive pests or those that are migratory. This chapter discussed how traps are used for several major pests and how trapping networks across regions can provide valuable scouting information. When using traps, it is important to establish a sampling plan and threshold for making management decisions based on trapping information. When traps are not appropriate for monitoring certain pests, scouting plans exist for making treatment decisions based on pest density or plant injury. Research and technology have made scouting for crop pests more efficient and profitable in recent years: For example, Speed Scouting is a way to make efficient, profitable treatment decisions for soybean aphid (Hodgson et al., 2007), and using a smartphone application (e.g. BioLeaf or LeafByte) can increase accuracy of plant defoliation assessments in the field (Machado et al., 2016; Getman-Pickering et al., 2020). Based on the recent advances discussed here as well as others not covered, it is likely that monitoring pests in the future will largely rely on technological advances and connected networks for monitoring activity, infestation risk and resistance development.

7 Where to look for further information

- Pest Tracker: http://pest.ceris.purdue.edu/pests.php.
- Insect Forecast: http://www.insectforecast.com/.

- Corn ipmPIPE: https://corn.ipmpipe.org/insects/.
- VegEdge: https://vegedge.umn.edu/degree-day-models-select-insect-pests-midwest-region.
- EDDMapS: https://www.eddmaps.org/.
- Ag Pest Monitor: https://agpestmonitor.org/.
- myFields: https://www.myfields.info/.
- Cotton Thrips: https://products.climate.ncsu.edu/ag/cottontip/.
- Pest Watch: http://www.pestwatch.psu.edu/.
- Great Lakes and Maritime Pest Monitoring Network: https://fieldcropnews.com/category/great-lakes-and-maritimes-pest-monitoring-network/.
- Corn rootworm adult monitoring network: https://www.arcgis.com/apps/MapSeries/index.html?appid=008cd878003f44fca4d8a6b5f0fe7b1c.
- Thrips Infestation Predictor for Cotton: https://products.climate.ncsu.edu/ag/cottontip/.
- Tobacco Thrips Flight and TSWC Intensity Predictor: https://products.climate.ncsu.edu/ag/tobacco-tswv/.

8 References

Acebes-Doria, A. L., Agnello, A. M., Alston, D. G., Andrews, H., Beers, E. H., Bergh, J. C., Bessin, R., Blaauw, B. R., Buntin, G. D., Burkness, E. C., Chen, S., Cottrell, T. E., Daane, K. M., Fann, L. E., Fleischer, S. J., Guédot, C., Gut, L. J., Hamilton, G. C., Hilton, R., Hoelmer, K. A., Hutchison, W. D., Jentsch, P., Krawczyk, G., Kuhar, T. P., Lee, J. C., Milnes, J. M., Nielsen, A. L., Patel, D. K., Short, B. D., Sial, A. A., Spears, L. R., Tatman, K., Toews, M. D., Walgenbach, J. D., Welty, C., Wiman, N. G., Van Zoeren, J. and Leskey, T. C. 2020. Season-long monitoring of the brown marmorated stink bug (Hemiptera: Pentatomidae) throughout the United States using commercially available traps and lures. *J. Econ. Entomol.* 113(1): 159–171.

Agroecosystems Entomology Lab (AEL) 2022. Western bean cutworm central. https://entomology.unl.edu/agroecosystems/western-bean-cutworm-central (accessed 14 February 2022).

Aita, R. C., Pezzini, D. T., Burkness, E. C., DiFonzo, C. D., Finke, D. L., Hunt, T. E., Knodel, J. J., Krupke, C. H., McCornack, B., Michel, A. P., Philips, C. R., Varenhorst, A. J., Wright, R. J., Marchi-Werle, L., Hutchison, W. D. and Koch, R. L. 2021. Presence-absence sampling for herbivorous stink bugs in soybean in the north central region of the United States. *J. Econ. Entomol.* 114: 1362–1372.

Akbar, W., Gowda, A., Ahrens, J. E., Stelzer, J. W., Brown, R. S., Bollman, S. L., Greenplate, J. T., Gore, J., Catchot, A. L., Lorenz, G., Stewart, S. D., Kerns, D. L., Greene, J. K., Toews, M. D., Herbert, D. A., Reisig, D. D., Sword, G. A., Ellsworth, P. C., Godfrey, L. D. and Clark, T. L. 2019. First transgenic trait for control of plant bugs and thrips in cotton. *Pest Manag. Sci.* 75(3): 867–877.

Alm, S. R. and Dawson, C. G. 2003. Evaluation of two prototype traps and existing trap designs for captures of Japanese beetles (Coleoptera: Scarabaeidae). *J. Econ. Entomol.* 96(2): 453–455.

Aragón, P., Baselga, A. and Lobo, J. M. 2010. Global estimation of invasion risk zones for the western corn rootworm *Diabrotica virgifera*: Integrating distribution models and physiological thresholds to assess climatic favourability. *J. Appl. Ecol.* 47(5): 1026–1035.

Arends, B., Reisig, D. D., Gundry, S., Huseth, A. S., Reay-Jones, F. P. F., Greene, J. K. and Kennedy, G. G. 2021. Effectiveness of the natural resistance management refuge for Bt-cotton is dominated by local abundance of soybean and maize. *Sci. Rep.* 11(1): 17601.

Bachman, P. M., Ahmad, A., Ahrens, J. E., Akbar, W., Baum, J. A., Brown, S., Clark, T. L., Fridley, J. M., Gowda, A., Greenplate, J. T., Jensen, P. D., Mueller, G. M., Odegaard, M. L., Tan, J., Uffman, J. P. and Levine, S. L. 2017. Characterization of the activity spectrum of MON 88702 and the plant-incorporated protectant Cry51Aa2.834_16. *PLoS ONE* 12(1): e0169409. doi: 10.1371/journal.pone.0169409.

Baker, G. H. and Tann, C. R. 2017. Broad-scale suppression of cotton bollworm, *Helicoverpa armigera* (Lepidoptera: Noctuidae), associated with Bt cotton crops in Northern New South Wales, Australia. *Bull. Entomol. Res.* 107(2): 188–199.

Bartels, D. W. and Hutchison, W. D. 1998. Comparison of pheromone trap designs for monitoring Z-strain European corn borer (Lepidoptera: Crambidae). *J. Econ. Entomol.* 91(6): 1349–1354.

Bartels, D. W., Hutchison, W. D. and Udayagiri, S. 1997. Pheromone trap monitoring of Z-strain European corn borer (Lepidoptera: Pyralidae): Optimum pheromone blend, comparison with light traps, and trap number requirements. *J. Econ. Entomol.* 90(2): 449–457.

Baute, T. 2021. https://fieldcropnews.com/category/great-lakes-and-maritimes-pest-monitoring-network/ (accessed 25 February 2022). Great Lakes and Maritimes Pest Monitoring Network.

Beckler, A. A., French, B. W. and Chandler, L. D. 2005. Using GIS in areawide pest management: A case study in South Dakota. *Trans. GIS* 9(2): 109–127.

Bell, J. R., Burkness, E. C., Milne, A. E., Onstad, D. W., Abrahamson, M., Hamilton, K. L. and Hutchison, W. D. 2012. Putting the brakes on a cycle: Bottom-up effects damp cycle amplitude. *Ecol. Lett.* 15(4): 310–318.

California Department of Food and Agriculture (CDFA) 2022. Japanese beetle (JB). https://www.cdfa.ca.gov/plant/JB/ (accessed 5 January 2022).

Calvin, D. D., Knapp, M. C., Xingquan, K., Poston, F. L. and Welch, S. M. 1986. Using a decision model to optimize European corn borer (Lepidoptera: Pyralidae) egg-mass sampling. *Environ. Entomol.* 15(6): 1212–1219.

Carrière, Y., Brown, Z., Aglasan, S., Dutilleul, P., Carroll, M., Head, G., Tabashnik, B. E., Jørgensen, P. S. and Carroll, S. P. 2020. Crop rotation mitigates impacts of corn rootworm resistance to transgenic Bt corn. *Proc. Natl. Acad. Sci. U. S. A.* 117(31): 18385–18392.

Carrière, Y., Ellsworth, P. C., Dutilleul, P., Ellers-Kirk, C., Barkley, V. and Antilla, L. 2006. A GIS-based approach for areawide pest management: The scales of *Lygus hesperus* movements to cotton from alfalfa, weeds, and cotton. *Entomol. Exp. Appl.* 118(3): 203–210.

Carrière, Y., Goodell, P. B., Ellers-Kirk, C., Larocque, G., Dutilleul, P., Naranjo, S. E. and Ellsworth, P. C. 2012. Effects of local and landscape factors on population dynamics of a cotton pest. *PLoS ONE* 7(6): e39862. doi: 10.1371/journal.pone.0039862.

Carrière, Y., DeGain, B., Unnithan, G. C., Harpold, V. S., Li, X. and Tabashnik, B. E. 2019. Seasonal declines in Cry1Ac and Cry2Ab concentration in maturing cotton favor faster evolution of resistance to pyramided Bt cotton in *Helicoverpa zea* (Lepidoptera: Noctuidae). *J. Econ. Entomol.* 112(6): 2907–2914.

Center for Environmental and Research Information Systems (CERIS) 2022. Survey Status of Japanese beetle – *Popillia japonica* (2020). http://pest.ceris.purdue.edu/map?code=INBPAZA (accessed 5 January 2022).

Cervantes, F. A., Backus, E. A., Godfrey, L., Rojas, M. G., Akbar, W. and Clark, T. L. 2019. Quantitative differences in feeding behavior of *Lygus lineolaris* (Hemiptera: Miridae) on transgenic and nontransgenic cotton. *J. Econ. Entomol.* 112(4): 1920–1925.

Chappell, T. M., Ward, R. V., DePolt, K. T., Roberts, P. M., Greene, J. K. and Kennedy, G. G. 2020a. Cotton thrips infestation predictor: A practical tool for predicting tobacco thrips (*Frankliniella fusca*) infestation of cotton seedlings in the south-eastern United States. *Pest Manag. Sci.* 76(12): 4018–4028.

Chappell, T. M., Codod, C. B., Williams, B. W., Kemerait, R. C., Culbreath, A. K. and Kennedy, G. G. 2020b. Adding epidemiologically important meteorological data to peanut Rx, the risk assessment framework for spotted wilt of peanut. *Phytopathology* 110(6): 1199–1207.

Chappell, T. M., Huseth, A. S. and Kennedy, G. G. 2019. Stability of neonicotinoid sensitivity in *Frankliniella fusca* populations found in agroecosystems of the southeastern USA. *Pest Manag. Sci.* 75(6): 1539–1545.

Coates, B. S., Johnson, H., Kim, K. S., Hellmich, R. L., Abel, C. A., Mason, C. and Sappington, T. W. 2013. Frequency of hybridization between *Ostrinia nubilalis* E- and Z-pheromone races in regions of sympatry within the United States. *Ecol. Evol.* 3(8): 2459–2470.

Cook, D., Herbert, A., Akin, D. S. and Reed, J. 2011. Biology, crop injury, and management of thrips (Thysanoptera: Thripidae) infesting cotton seedlings in the United States. *J. Integr. Pest Manag.* 2(2): 1–9.

Cook, D. R. and Threet, M. 2020. Cotton insect losses. https://www.biochemistry.msstate.edu/resources/2020loss.php (accessed 5 December 2021). Mississippi State University.

Crespo, A. L. B., Spencer, T. A., Alves, A. P., Hellmich, R. L., Blankenship, E. E., Magalhães, L. C. and Siegfried, B. D. 2009. On-plant survival and inheritance of resistance to Cry1Ab toxin from *Bacillus thuringiensis* in a field-derived strain of European corn borer, *Ostrinia nubilalis*. *Pest Manag. Sci.* 65(10): 1071–1081.

Crossley, M. S., Burke, K. D., Schoville, S. D. and Radeloff, V. C. 2021. Recent collapse of crop belts and declining diversity of US agriculture since 1840. *Glob. Change. Biol.* 27(1): 151–164.

Culin, J. D. 1995. Local dispersal of male *Helicoverpa zea*. *Entomol. Exp. Appl.* 74(2): 165–176.

Cullen, E. and Jyuotika, J. 2008. Western bean cutworm: A pest of sweet corn. A3856. http://corn.agronomy.wisc.edu/management/pdfs/a3856.pdf (accessed 8 April 2022). University of Wisconsin, Extension, Cooperative Extension.

D'Ambrosio, D. A., Huseth, A. S. and Kennedy, G. G. 2018. Evaluation of alternative mode of action insecticides in managing neonicotinoid-resistant *Frankliniella fusca* in cotton. *Crop Prot.* 113: 56–63.

D'Ambrosio, D. A., Kennedy, G. G. and Huseth, A. S. 2020. Feeding behavior of *Frankliniella fusca* on seedling cotton expressing Cry51Aa2. 834_16 Bt toxin. *Pest Manag. Sci.* 76(8): 2781–2786.

D'Ambrosio, D. A., Peele, W., Hubers, A. and Huseth, A. S. 2019. Seasonal dispersal of *Lygus lineolaris* (Hemiptera: Miridae) from weedy hosts into differently fragmented cotton landscapes in North Carolina. *Crop Prot.* 125. doi: 10.1016/j.cropro.2019.104898.

Darnell-Crumpton, C., Catchot, A. L., Cook, D. R., Gore, J., Dodds, D. M., Morsello, S. C. and Musser, F. R. 2018. Neonicotinoid insecticide resistance in tobacco thrips (Thysanoptera: Thripidae) of Mississippi. *J. Econ. Entomol.* 111(6): 2824–2830.

Davidson, S. A., Peairs, F. B. and Khosla, R. 2007. Insect pest densities across site-specific management zones of irrigated corn in northeastern Colorado. *J. Econ. Entomol.* 100(3): 781–789.

Dean, A. N., Niemi, J. B., Tyndall, J. C., Hodgson, E. W. and O'Neal, M. E. 2020. Developing a decision-making framework for insect pest management: A case study using *Aphis glycines* (Hemiptera: Aphididae). *Pest Manag. Sci.* 77(2): 886–894. doi: 10.1002/ps.6093.

Del Pozo-Valdivia, A. I., Reisig, D. D., Braswell, L., Greene, J. K., Roberts, P. and Taylor, S. V. 2021. Economic injury levels for Bt-resistant *Helicoverpa zea* (Lepidoptera: Noctuidae) in cotton. *J. Econ. Entomol.* 114(2): 747–756.

Dent, D. 2000. *Insect Pest Management* (2nd edition). CABI Publishing, Wallingford, United Kingdom.

Dively, G. P., Kuhar, T. P., Taylor, S., Doughty, H. B., Holmstrom, K., Gilrein, D., Nault, B. A., Ingerson-Mahar, J., Whalen, J., Reisig, D., Frank, D. L., Fleischer, S. J., Owens, D., Welty, C., Reay-Jones, F. P. F., Porter, P., Smith, J. L., Saquez, J., Murray, S., Wallingford, A., Byker, H., Jensen, B., Burkness, E., Hutchison, W. D. and Hamby, K. A. 2021. Sweet corn sentinel monitoring for Lepidopteran field-evolved resistance to Bt toxins. *J. Econ. Entomol.* 114(1): 307–319.

Dively, G. P., Venugopal, P. D., Bean, D., Whalen, J., Holmstrom, K., Kuhar, T. P., Doughty, H. B., Patton, T., Cissel, W. and Hutchison, W. D. 2018. Regional pest suppression associated with widespread Bt maize adoption benefits vegetable growers. *Proc. Natl. Acad. Sci. U. S. A.* 115(13): 3320–3325.

Dorman, S. J., Hopperstad, K. A., Reich, B. J., Kennedy, G. and Huseth, A. S. 2021a. Soybeans as a non-Bt refuge for *Helicoverpa zea* in maize-cotton agroecosystems. *Agric. Ecosyst. Environ.* 322. doi: 10.1016/j.agee.2021.107642.

Dorman, S. J., Hopperstad, K. A., Reich, B. J., Majumder, S., Kennedy, G., Reisig, D. D., Greene, J. K., Reay-Jones, F. P. F., Collins, G., Bacheler, J. S. and Huseth, A. S. 2021b. Landscape-level variation in Bt crops predict *Helicoverpa zea* (Lepidoptera: Noctuidae) resistance in cotton agroecosystems. *Pest Manag. Sci.* 77(12): 5454–5462.

Dorman, S. J., Opoku, J., Mehl, H. L. and Taylor, S. V. 2021c. Fusarium hardlock associated with *Lygus lineolaris* (Hemiptera: Miridae) injury in southeastern cotton. *Plant Dis.* 105(1): 53–59.

Dorman, S. J., Schürch, R., Huseth, A. S. and Taylor, S. V. 2020. Landscape and climatic effects driving spatiotemporal abundance of *Lygus lineolaris* (Hemiptera: Miridae) in cotton agroecosystems. *Agric. Ecosyst. Environ.* 295. doi: 10.1016/j.agee.2020.106910.

Dorman, S. J., Taylor, S. V., Malone, S., Roberts, P. M., Greene, J. K., Reisig, D. D., Smith, R. H., Jacobson, A. L., Reay-Jones, F. P. F., Paula-Moraes, S. and Huseth, A. S. 2022. Sampling optimization and crop interface effects on *Lygus lineolaris* populations in southeastern USA cotton. *Insects* 13(1). doi: 10.3390/insects13010088.

Dunbar, M. W. and Gassmann, A. J. 2013. Abundance and distribution of western and northern corn rootworm (*Diabrotica* spp.) and prevalence of rotation resistance in eastern Iowa. *J. Econ. Entomol.* 106(1): 168–180.

Dunbar, M. W., O'Neal, M. E. and Gassmann, A. J. 2016. Effects of field history on corn root injury and adult abundance of northern and western corn rootworm (Coleoptera: Chrysomelidae). *J. Econ. Entomol.* 109(5): 2096–2104.

Ebbenga, D. N., Burkness, E. C. and Hutchison, W. D. 2022. Optimizing the use of semiochemical-based traps for efficient monitoring of *Popillia japonica* (Coleoptera: Scarabaeidae): Validation of a volumetric approach. *J. Econ. Entomol.* 115(3): 869-876 (in press 3/2022).

Environmental Systems Research Institute (ESRI) 2021a. ArcGIS Survey123. https://survey123.arcgis.com/ (accessed 21 December 2021).

Environmental Systems Research Institute (ESRI) 2021b. Corn rootworm adult monitoring network. https://www.arcgis.com/apps/MapSeries/index.html?appid=008cd878003f44fca4d8a6b5f0fe7b1c (accessed 21 December 2021).

Fitt, G. P. 1989. The ecology of *Heliothis* species in relation to agroecosystems. *Annu. Rev. Entomol.* 34(1): 17-53.

Foster, R. E. 2016. Vegetable insects: Corn earworm. Purdue extension entomology, E-31-W. https://extension.entm.purdue.edu/publications/E-31/E-31.html (accessed 8 April 2022).

Gassmann, A. J. 2021. Resistance to Bt maize by western corn rootworm: Effects of pest biology, the pest-crop interaction, and the agricultural landscape on resistance. *Insects* 12(2). doi: 10.3390/insects12020136.

Gassmann, A. J., Petzold-Maxwell, J. L., Keweshan, R. S. and Dunbar, M. W. 2011. Field-evolved resistance to Bt maize by western corn rootworm. *PLoS ONE* 6(7): e22629. doi: 10.1371/journal.pone.0022629.

George, J., Glover, J. P., Gore, J., Crow, W. D. and Reddy, G. V. P. 2021. Biology, ecology, and pest management of the tarnished plant bug, *Lygus lineolaris* (Palisot de Beauvois) in southern row crops. *Insects* 12(9). doi: 10.3390/insects12090807.

Getman-Pickering, Z. L., Campbell, A., Aflitto, N., Grele, A., Davis, J. K. and Ugine, T. A. 2020. LeafByte: A mobile application that measures leaf area and herbivory quickly and accurately. *Methods Ecol. Evol.* 11(2): 215–221.

Goethe, J. K., Dorman, S. J. and Huseth, A. S. 2021. Local and landscape scale drivers of *Euschistus servus* and *Lygus lineolaris* in North Carolina small grain agroecosystems. *Agric. For. Entomol.* 23(4): 441–451.

Government of Canada 2019. D-96-15: Phytosanitary requirements to prevent the spread of Japanese beetle. Popillia Japonica, in Canada and the United States. https://inspection.canada.ca/plant-health/invasive-species/directives/horticulture/d-96-15/eng/1323854808025/1323854908041 (accessed 5 January 2022).

Graham, S. H., Musser, F. M., Jacobson, A. L., Chitturi, A., Catchot, B. and Stewart, S. D. 2019. Behavioral responses of thrips (Thysanoptera: Thripidae) and tarnished plant bug (Hemiptera: Miridae) to a new Bt toxin, Cry51Aa2. 834_16 In cotton. *J. Econ. Entomol.* 112(4): 1695–1704.

Gray, M. E., Sappington, T. W., Miller, N. J., Moeser, J. and Bohn, M. O. 2009. Adaptation and invasiveness of western corn rootworm: Intensifying research on a worsening pest. *Annu. Rev. Entomol.* 54: 303–321.

Groves, R. L., Walgenbach, J. F., Moyer, J. W. and Kennedy, G. G. 2001. Overwintering of *Frankliniella fusca* (Thysanoptera: Thripidae) on winter annual weeds infected with tomato spotted wilt virus and patterns of virus movement between susceptible weed hosts. *Phytopathology* 91(9): 891–899.

Groves, R. L., Walgenbach, J. F., Moyer, J. W. and Kennedy, G. G. 2003. Seasonal dispersal patterns of *Frankliniella fusca* (Thysanoptera: Thripidae) and tomato spotted wilt virus occurrence in central and eastern North Carolina. *J. Econ. Entomol.* 96(1): 1–11.

Gustafson, D. I., Head, G. P. and Caprio, M. A. 2006. Modeling the impact of alternative hosts on *Helicoverpa zea* adaptation to Bollgard cotton. *J. Econ. Entomol.* 99(6): 2116–2124.

Hamilton, R. M., Foster, R. E., Gibb, T. J., Sadof, C. S., Holland, J. D. and Engel, B. A. 2007. Distribution and dynamics of Japanese beetles along the Indianapolis airport perimeter and the influence of land use on trap catch. *Environ. Entomol.* 36(2): 287–296.

Hanson, A. A., Menger-Anderson, J., Silverstein, C., Potter, B. D., MacRae, I. V., Hodgson, E. W. and Koch, R. L. 2017. Evidence for soybean aphid (Hemiptera: Aphididae) resistance to pyrethroid insecticides in the upper midwestern United States. *J. Econ. Entomol.* 110(5): 2235–2246.

Hanson, A. A., Moon, R. D., Wright, R. J., Hunt, T. E. and Hutchison, W. D. 2015. Degree-day prediction models for the flight phenology of western bean cutworm (Lepidoptera: Noctuidae) assessed with the concordance correlation coefficient. *J. Econ. Entomol.* 108(4): 1728–1738.

Hartstack, A. W., Witz, J. A. and Buck, D. R. 1979. Moth traps for the tobacco budworm12. *J. Econ. Entomol.* 72(4): 519–522.

Head, G., Jackson, R. E., Adamczyk, J., Bradley, J. R., Van Duyn, J., Gore, J., Hardee, D. D., Leonard, B. R., Luttrell, R., Ruberson, J., Mullins, J. W., Orth, R. G., Sivasupramaniam, S. and Voth, R. 2010. Spatial and temporal variability in host use by *Helicoverpa zea* as measured by analyses of stable carbon isotope ratios and gossypol residues. *J. Appl. Ecol.* 47(3): 583–592.

Hein, G. L. and Tollefson, J. J. 1985. Use of the Pherocon AM trap as a scouting tool for predicting damage by corn rootworm (Coleoptera: Chrysomelidae) larvae. *J. Econ. Entomol.* 78(1): 200–203.

Hills, T. M. and Peters, D. C. 1971. A method of evaluating postplanting insecticide treatments for control of western corn rootworm larvae. *J. Econ. Entomol.* 64(3): 764–765.

Hodgson, E. W., Burkness, E. C., Hutchison, W. D. and Ragsdale, D. W. 2004. Enumerative and binomial sequential sampling plans for soybean aphid (Homoptera: Aphididae) in soybean. *J. Econ. Entomol.* 97(6): 2127–2136.

Hodgson, E. W., McCornack, B. P., Koch, K. A., Ragsdale, D. W., Johnson, K. D., O'Neal, M. E., Cullen, E. M., Kraiss, H. J., DiFonzo, C. D., Jewett, M. and Behnken, L. M. 2007. Field validation of Speed Scouting for soybean aphid. *Crop Manag.* doi: 10.1094/CM-2007-0511-01-RS.

Hodgson, E. W., McCornack, B. P., Tilmon, K. and Knodel, J. J. 2012. Management recommendations for soybean aphid (Hemiptera: Aphididae) in the United States. *J. Integr. Pest Manag.* 3(1): E1–E10. doi: 10.1603/IPM11019.

Hodgson, E. and Dean, A. 2020. *Popillia japonica (Japanese Beetle). Invasive Species Compendium*. CABI Publishing, Wallingford, UK.

Hodgson, E. W. and Rice, M. E. 2017. Ecology and management of European corn borer in Iowa field corn. *Iowa State Univ. Extension Outreach Crops* 3139R.

Hong, B., Nowatzki, T. M., Sult, T. S., Owens, E. D. and Pilcher, C. D. 2016. Sequential sampling plan for assessing corn rootworm (Coleoptera: Chrysomelidae) larval injury to Bt maize. *Crop Prot.* 82: 36–44.

Huang, F., Andow, D. A. and Buschman, L. L. 2011. Success of the high-dose/refuge resistance management strategy after 15 years of Bt crop use in North America. *Entomol. Exp. Appl.* 140(1): 1–16.

Hungate, B. A., Kearns, D. N., Ogle, K., Caron, M., Marks, J. C. and Rogg, H. W. 2016. Hydrogen isotopes as a sentinel of biological invasion by the Japanese beetle, *Popillia japonica* (Newman). *PLoS ONE* 11(3): e0149599. doi: 10.1371/journal.pone.0149599.

Huseth, A. S., Chappell, T. M., Langdon, K., Morsello, S. C., Martin, S., Greene, J. K., Herbert, A., Jacobson, A. L., Reay-Jones, F. P. F., Reed, T., Reisig, D. D., Roberts, P. M., Smith, R. and Kennedy, G. G. 2016. *Frankliniella fusca* resistance to neonicotinoid insecticides: An emerging challenge for cotton pest management in the eastern United States. *Pest Manag. Sci.* 72(10): 1934–1945.

Huseth, A. S., Chappell, T. M., Chitturi, A., Jacobson, A. L. and Kennedy, G. G. 2018. Insecticide resistance signals negative consequences of widespread neonicotinoid use on multiple field crops in the US cotton belt. *Environ. Sci. Technol.* 52(4): 2314–2322.

Huseth, A. S., D'Ambrosio, D. A. and Kennedy, G. G. 2017. Responses of neonicotinoid resistant and susceptible *Frankliniella fusca* life stages to multiple insecticide groups in cotton. *Pest Manag. Sci.* 73(10): 2118–2130.

Huseth, A. S., D'Ambrosio, D. A., Yorke, B. T., Head, G. P. and Kennedy, G. G. 2020. Novel mechanism of thrips suppression by Cry51Aa2. 834_16 Bt toxin expressed in cotton. *Pest Manag. Sci.* 76(4): 1492–1499.

Huseth, A. S., Koch, R. L., Reisig, D., Davis, J. A., Paula-Moraes, S. V. and Hodgson, E. W. 2021. Current distribution and population persistence of five lepidopteran pests in US soybean. *J. Integr. Pest Manag.* 12(1). doi: 10.1093/jipm/pmab004.

Hutchison, W. D., Burkness, E. C., Mitchell, P. D., Moon, R. D., Leslie, T. W., Fleischer, S. J., Abrahamson, M., Hamilton, K. L., Steffey, K. L., Gray, M. E., Hellmich, R. L., Kaster, L. V., Hunt, T. E., Wright, R. J., Pecinovsky, K., Rabaey, T. L., Flood, B. R. and Raun, E. S. 2010. Areawide suppression of European corn borer with Bt maize reaps savings to non-Bt maize growers. *Science* 330(6001): 222–225.

Hutchison, W. D. and Cira, T. M. 2017. Economically important maize insect pests. In: Watson, D. (Ed.). *Achieving Sustainable Cultivation of Maize* (vol. 2): Cultivation techniques, pest and disease control,pp. 263-292. Burleigh Dodds Science Publishing, Cambridge, UK. ISBN: 978-1-78676-012-8.

ISU 2021. Iowa State University, corn rootworm IPM. https://cornrootworm.extension.iastate.edu/adult-trapping-network (accessed 21 December 2021). Adult Trapping Network.

Ives, A. R., Paull, C., Hulthen, A., Downes, S., Andow, D. A., Haygood, R., Zalucki, M. P. and Schellhorn, N. A. 2017. Spatio-temporal variation in landscape composition may speed resistance evolution of pests to Bt crops. *PLoS ONE* 12(1): e0169167. doi: 10.1371/journal.pone.0169167.

Jackson, R. E., Bradley, J. R., Van Duyn, J., Leonard, B. R., Allen, K. C., Luttrell, R., Ruberson, J., Adamczyk, J., Gore, J., Hardee, D. D., Voth, R., Sivasupramaniam, S., Mullins, J. W.

and Head, G. 2008. Regional assessment of *Helicoverpa zea* populations on cotton and non-cotton crop hosts. *Entomol. Exp. Appl.* 126(2): 89–106.

Jacobson, A., Foster, R., Krupke, C., Hutchison, W., Pittendrigh, B. and Weinzierl, R. 2009. Resistance to pyrethroid insecticides in *Helicoverpa zea* (Lepidoptera: Noctuidae) in Indiana and Illinois. *J. Econ. Entomol.* 102(6): 2289–2295.

Johnson, M. W., Stinner, R. E. and Rabb, R. L. 1975. Ovipositional response of *Heliothis zea* (Boddie) to its major hosts in North Carolina. *Environ. Entomol.* 4(2): 291–297.

Johnson, M. T. J., Bertrand, J. A. and Turcotte, M. M. 2016. Precision and accuracy in quantifying herbivory. *Ecol. Entomol.* 41(1): 112–121.

Kahn, N. D., Walgenbach, J. F. and Kennedy, G. G. 2005. Summer weeds as hosts for *Frankliniella occidentalis* and *Frankliniella fusca* (Thysanoptera: Thripidae) and as reservoirs for tomato spotted wilt Tospovirus in North Carolina. *J. Econ. Entomol.* 98(6): 1810–1815.

Khrimian, A., Zhang, A., Weber, D. C., Ho, H. Y., Aldrich, J. R., Vermillion, K. E., Siegler, M. A., Shirali, S., Guzman, F. and Leskey, T. C. 2014. Discovery of the aggregation pheromone of the brown marmorated stink bug (Halyomorpha halys) through the creation of stereoisomeric libraries of 1Bisabolen-3-ols. *J. Nat. Prod.* 77(7): 1708–1717.

Kistner-Thomas, E. J. 2019. The potential global distribution and voltinism of the Japanese beetle (Coleoptera: Scarabaeidae) under current and future climates. *J. Insect Sci.* 19(2). doi: 10.1093/jisesa/iez023.

Klun, J. A., Blickenstaff, C. C., Schwarz, M., Leonhardt, B. A. and Plimmer, J. R. 1983. Western bean cutworm, *Loxagrotis albicosta* (Lepidoptera: Noctuidae): Female sex pheromone identification. *Environ. Entomol.* 12(3): 714–717.

Klun, J. A. and Brindley, T. A. 1970. Cis-11-tetradecenyl acetate, a sex stimulant of the European corn borer. *J. Econ. Entomol.* 63(3): 779–780.

Klun, J. A., Plimmer, J. R., Bierl-Leonhardt, B. A., Sparks, A. N., Primiani, M., Chapman, O. L., Lee, G. H. and Lepone, G. 1980. Sex pheromone chemistry of female corn earworm moth, *Heliothis zea*. *J. Chem. Ecol.* 6(1): 165–175.

Koch, R. L., Hodgson, E. W., Knodel, J. J., Varenhorst, A. J. and Potter, B. D. 2018. Management of insecticide-resistant soybean aphids in the upper Midwest of the United States. *J. Integr. Pest Manag.* 9(1). doi: 10.1093/jipm/pmy014.

Koch, R. L., Pezzini, D. T., Michel, A. P. and Hunt, T. E. 2017. Identification, biology, impacts, and management of stink bugs (Hemiptera: Heteroptera: Pentatomidae) of soybean and corn in the midwestern United States. *J. Integr. Pest Manag.* 8(1). doi: 10.1093/jipm/pmx004.

Koch, R. L., Potter, B. D., Glogoza, P. A., Hodgson, E. W., Krupke, C. H., Tooker, J. F., DiFonzo, C. D., Michel, A. P., Tilmon, K. J., Prochaska, T. J., Knodel, J. J., Wright, R. J., Hunt, T. E., Jensen, B., Varenhorst, A. J., McCornack, B. P., Estes, K. A. and Spencer, J. L. 2016. Biology and economics of recommendations for insecticide-based management of soybean aphid. *Plant Health Prog.* 17(4): 265–269.

Koch, R. L. and Rich, W. A. 2015. Stink bug (Hemiptera: Heteroptera: Pentatomidae) feeding and phenology on early-maturing soybean in Minnesota. *J. Econ. Entomol.* 108(5): 2335–2343.

Kwon, O. and Park, T. 2017. Applications of smartphone cameras in agriculture, environment, and food: A review. *J. Bio.Syst. Eng.* 42: 330–338.

Lagos-Kutz, D., Voegtlin, D. J., Onstad, D., Hogg, D., Ragsdale, D., Tilmon, K., Hodgson, E., DiFonzo, C., Groves, R., Krupke, C., LaForest, J., Seiter, N. J., Duerr, E., Bradford,

B. and Hartman, G. L. 2020. The soybean aphid suction trap network: Sampling the aerobiologcal "soup." *Am. Entomol.* 66(1): 48–55. doi: 10.1093/ae/tmaa009.

Layton, M. B. 2000. Biology and damage of the tarnished plant bug, Lygus lineolaris, in cotton. *Southwest. Entomol. Suppl.* 23: 7–20.

Leaf, T. M. 2017. Utility of scouting with sticky traps for integrated pest management of corn rootworm. Ph.D. dissertation. University of Minnesota, Minneapolis.

Leskey, T. C., Lee, D. H., Short, B. D. and Wright, S. E. 2012. Impact of insecticides on the invasive *Halyomorpha halys* (Hemiptera: Pentatomidae): Analysis of insecticide lethality. *J. Econ. Entomol.* 105(5): 1726–1735.

Leskey, T. C. and Nielsen, A. L. 2018. Impact of the invasive brown marmorated stink bug in North America and Europe: History, biology, ecology, and management. *Annu. Rev. Entomol.* 63: 599–618.

Lingren, P. D., Bryant, V. M., Jr., Raulston, J. R., Pendleton, M., Westbrook, J. and Jones, G. D. 1993. Adult feeding host range and migratory activities of corn earworm, cabbage looper, and celery looper (Lepidoptera: Noctuidae) moths as evidenced by attached pollen. *J. Econ. Entomol.* 86(5): 1429–1439.

Losey, J. E., Allee, L. L., Zbarsky, V., Waldron, J. K. and Shields, E. J. 2003. Transect sampling to enhance efficiency of corn rootworm (Coleoptera: Chrysomelidae) monitoring in corn. *J. Econ. Entomol.* 96(5): 1420–1425.

Luttrell, R. G., Teague, T. G. and Brewer, M. J. 2015. Cotton insect pest management. *Cotton* 57: 509–546.

Machado, B. B., Orue, J. P. M., Arruda, M. S., Santos, C. V., Sarath, D. S., Goncalves, W. N., Silva, G. G., Pistori, H., Roel, A. R. and Rodrigues, J. F. Jr. 2016. BioLeaf: A professional mobile application to measure foliar damage caused by insect herbivory. *Comput. Electron. Agric.* 129: 44–55.

Mahrt, G. G., Stoltz, R. L., Blickenstaff, C. C. and Holtzer, T. O. 1987. Comparisons between blacklight and pheromone traps for monitoring the western bean cutworm (Lepidoptera: Noctuidae) in south central Idaho. *J. Econ. Entomol.* 80(1): 242–247.

Marshall, M., Greene, J., Plumblee, M. T., Reay-Jones, F., Mueller, J., Anco, D., Moore, K., Powell, B., Heaton, C., Crouch, J. and Beer, B. 2022. *South Carolina Pest Management Handbook* (vol. 2022). Clemson University, Extension, Clemson, SC. 335pp.

Mason, C. E., Rice, M. E., DiFonzo, C. D. and Porter, R. P. 2018. *European Corn Borer: Ecology, Management & Association with Other Corn Pests. NCR Extension*. Iowa State University, Ames, IA, NCR-0327. 81pp.

McCornack, B. P., Ragsdale, D. W. and Venette, R. C. 2004. Demography of soybean aphid (Homoptera: Aphididae) at summer temperatures. *J. Econ. Entomol.* 97(3): 854–861.

McPherson, J. E. and McPherson, R. M. 2000. *Stink Bugs of Economic Importance in America North of Mexico*. CRC Press, LLC, Boca Raton, FL.

Meinke, L. J., Sappington, T. W., Onstad, D. W., Guillemaud, T., Miller, N. J., Komáromi, J., Levay, N., Furlan, L., Kiss, J. and Toth, F. 2009. Western corn rootworm (*Diabrotica virgifera* LeConte) population dynamics. *Agric. For. Entomol.* 11(1): 29–46.

Meisner, M. H., Zaviezo, T. and Rosenheim, J. A. 2017. Landscape crop composition effects on cotton yield, *Lygus hesperus* densities and pesticide use. *Pest Manag. Sci.* 73(1): 232–239.

Michel, A. P., Krupke, C. H., Baute, T. S. and DiFonzo, C. D. 2010. Ecology and management of the western bean cutworm (Lepidoptera: Noctuidae) in corn and dry beans. *J. Integr. Pest Manag.* 1(1): A1–A10. doi: 10.1603/IPM10003.

Miller, N., Estoup, A., Toepfer, S., Bourguet, D., Lapchin, L., Derridj, S., Kim, K. S., Reynaud, P., Furlan, L. and Guillemaud, T. 2005. Multiple transatlantic introductions of the western corn rootworm. *Science* 310(5750): 992. doi: 10.1126/science.1115871.
Morehouse, A. T. 2011. Sampling the spatial distribution of Japanese beetles in Illinois soybean fields and their impact on production. M.S. Thesis. University of Illinois.
Musick, G. J. and Suttle, P. J. 1972. Ohio northern corn rootworm research, part I: Chemical control. Annual Report. Ohio Agricultural. Research and Development Center, Wooster, OH.
Musser, F. R., Catchot, A. L. Jr., Conley, S. P., Davis, J. A., DiFonzo, C., Graham, S. H., Greene, J. K., Koch, R. L., Owens, D., Reisig, D. D., Roberts, P., Royer, T., Seiter, N. J., Stewart, S. D., Taylor, S., Thrash, B., Tilmon, K., Villanueva, R. T. and Way, M. O. 2021. 2020 soybean insect losses in the United States. *Midsouth Entomologist. Way* 13: 1–25.
Naranjo, S. E. and Hutchison, W. D. 1997. Validation of arthropod sampling plans using a resampling approach: Software and analysis. *Am. Entomol.* 43(1): 48–57.
National Centers for Environmental Information (NCEI) 2022. Geographical reference maps. https://www.ncei.noaa.gov/monitoring-references/maps/us-climate-regions (accessed 30 March 2022).
National Plant Board (NPB) 2021. Japanese beetle harmonization plan. https://nationalplantboard.org/japanese-beetle-harmonization-plan/ (accessed 5 January 2022).
Niu, Y., Oyediran, I., Yu, W., Lin, S., Dimase, M., Brown, S., Reay-Jones, F. P. F., Cook, D., Reisig, D., Thrash, B., Ni, X., Paula-Moraes, S. V., Zhang, Y., Chen, J. S., Wen, Z. and Huang, F. 2021. Populations of *Helicoverpa zea* (Boddie) in the southeastern United States are commonly resistant to Cry1Ab, but still susceptible to Vip3Aa20 expressed in MIR 162 corn. *Toxins* 13(1). doi: 10.3390/toxins13010063.
North, J. H., Gore, J., Catchot, A. L., Stewart, S. D., Lorenz, G. M., Musser, F. R., Cook, D. R., Kerns, D. L. and Dodds, D. M. 2018. Value of neonicotinoid insecticide seed treatments in mid-south cotton (*Gossypium hirsutum* [Malvales: Malvaceae]) production systems. *J. Econ. Entomol.* 111(1): 10–15.
O'Rourke, P. K. and Hutchison, W. D. 2000. First report of the western bean cutworm (Lepidoptera: Noctuidae) in Minnesota corn. *J. Agric. Urban Entomol.* 17: 213–217.
Ohnesorg, W. and Hunt, T. 2016. Soybean defoliation worksheet. https://cropwatch.unl.edu/2016/soybean-defoliation-worksheet (accessed 5 January 2022).
Oleson, J. D., Park, Y. L., Nowatzki, T. M. and Tollefson, J. J. 2005. Node-injury scale to evaluate root injury by corn rootworms (Coleoptera: Chrysomelidae). *J. Econ. Entomol.* 98(1): 1–8.
Pan, Z., Onstad, D., Crain, P., Crespo, A., Hutchison, W., Buntin, D., Porter, P., Catchot, A., Cook, D., Pilcher, C., Flexner, L. and Higgins, L. 2016. Evolution of resistance by *Helicoverpa zea* (Lepidoptera: Noctuidae) infesting insecticidal crops in the southern United States. *J. Econ. Entomol.* 109(2): 821–831.
Park, Y.-L. and Tollefson, J. J. 2006. Development and economic evaluation of spatial sampling plans for corn rootworm *Diabrotica* spp. (Col., Chrysomelidae) adults. *J. Appl. Entomol.* 130(6-7): 337–342.
Paula-Moraes, S., Burkness, E. C., Hunt, T. E., Wright, R. J., Hein, G. L. and Hutchison, W. D. 2011. Cost-effective binomial sequential sampling of western bean cutworm, *Striacosta albicosta* (Lepidoptera: Noctuidae), egg masses in corn. *J. Econ. Entomol.* 104(6): 1900–1908.

Paula-Moraes, S., Hunt, T. E., Wright, R. J., Hein, G. L. and Blankenship, E. E. 2013. Western bean cutworm survival and the development of economic injury levels and economic thresholds in field corn. *J. Econ. Entomol.* 106(3): 1274–1285.

Peairs, F. B. 2006. Western bean cutworm: Characteristics and management in corn and dry beans. https://extension.colostate.edu/docs/pubs/insect/05538.pdf (accessed 8 April 2022). Colorado State University, Extension.

Peck, S. L., Gould, F. and Ellner, S. P. 1999. Spread of resistance in spatially extended regions of transgenic cotton: Implications for management of *Heliothis virescens* (Lepidoptera: Noctuidae). *J. Econ. Entomol.* 92(1): 1–16.

Pedigo, L. P., Rice, M. E. and Krell, R. K. 2021. *Entomology and Pest Management* (7th edition). Waveland Press, Inc, Long Grove, IL. ISBN: 1-4786-3992-X.

Perry, J. N., Winder, L., Holland, J. M. and Alston, R. D. 1999. Red-blue plots for detecting clusters in count data. *Ecol. Lett.* 2(2): 106–113.

PestWatch. 2017. Pestwatch, Penn State University. http://www.pestwatch.psu.edu/ (accessed 28 March 2022).

Pezzini, D. T., DiFonzo, C. D., Finke, D. L., Hunt, T. E., Knodel, J. J., Krupke, C. H., McCornack, B., Michel, A. P., Moon, R. D., Philips, C. R., Varenhorst, A. J., Wright, R. J. and Koch, R. L. 2019. Spatial patterns and sequential sampling plans for stink bugs (Hemiptera: Pentatomidae) in soybean in the North Central Region of the U.S. *J. Econ. Entomol.* 112(4): 1732–1740.

Piñero, J. C. and Dudenhoeffer, A. P. 2018. Mass trapping designs for organic control of the Japanese beetle, *Popillia japonica* (Coleoptera: Scarabaeidae). *Pest Manag. Sci.* 74(7): 1687–1693.

Potter, D. A. and Held, D. W. 2002. Biology and management of the Japanese beetle. *Annu. Rev. Entomol.* 47: 175-205.

Potter, B., Ostlie, K., Hutchison, B., Peltier, A. and Hanson, A. 2022. Reducing 2022 Bt corn acreage to cut production costs? Implications for European corn borer. https://blog-crop-news.extension.umn.edu/2022/01/reducing-bt-trait-acres-in-2022.html (accessed 31 March 2022).

Rabb, R. L. 1978. A sharp focus on insect populations and pest management from a wide-area view. *B. Entomol. Soc. Am.* 24(1): 55–61.

Ragsdale, D. W., McCornack, B. P., Venette, R. C., Potter, B. D., MacRae, I. V., Hodgson, E. W., O'Neal, M. E., Johnson, K. D., O'Neil, R. J., DiFonzo, C. D., Hunt, T. E., Glogoza, P. A. and Cullen, E. M. 2007. Economic threshold for soybean aphid (Hemiptera: Aphididae). *J. Econ. Entomol.* 100(4): 1258–1267.

Ragsdale, D. W., Landis, D. A., Brodeur, J., Heimpel, G. E. and Desneux, N. 2011. Ecology and management of the soybean aphid in North America. *Annu. Rev. Entomol.* 56: 375–399.

Ragsdale, D. W., Voegtlin, D. J. and O'Neil, R. J. 2004. Soybean aphid biology in North America. *Ann. Entomol. Soc. Am.* 97(2): 204–208.

Reay-Jones, F. P. F. 2019. Pest status and management of corn earworm (Lepidoptera: Noctuidae) in field corn in the United States. *J. Integr. Pest Manag.* 10(1). doi: 10.1093/jipm/pmz017.

Reay-Jones, F. P. F., Toews, M. D., Greene, J. K. and Reeves, R. B. 2010. Spatial dynamics of stink bugs (Hemiptera: Pentatomidae) and associated boll injury in southeastern cotton fields. *Environ. Entomol.* 39(3): 956–969.

Reay-Jones, F. P. F., Greene, J. K. and Bauer, P. J. 2016. Stability of spatial distributions of stink bugs, Boll Injury, and NDVI in Cotton. *Environ. Entomol.* 45(5): 1243–1254.

Reay-Jones, F. P. F., Greene, J. K. and Bauer, P. J. 2019. Spatial distributions of thrips (Thysanoptera: Thripidae) in cotton. *J. Insect Sci.* 19(6). doi: 10.1093/jisesa/iez103.

Reinders, J. D., Hitt, B. D., Stroup, W. W., French, B. W. and Meinke, L. J. 2018. Spatial variation in western corn rootworm (Coleoptera: Chrysomelidae) susceptibility to Cry3 toxins in Nebraska. *PLoS ONE* 13(11): e0208266. doi: 10.1371/journal.pone.0208266.

Reisig, D. D., DiFonzo, C., Dively, G., Farhan, Y., Gore, J. and Smith, J. 2022. Best management practices to delay the evolution of Bt resistance in Lepidopteran pests without high susceptibility to Bt toxins in North America. *J. Econ. Entomol.* 115(1): 10-25.

Reisig, D. D., Herbert, D. A. and Malone, S. 2012. Impact of neonicotinoid seed treatments on thrips (Thysanoptera: Thripidae) and soybean yield in Virginia and North Carolina. *J. Econ. Entomol.* 105(3): 884-889.

Reisig, D. D., Huseth, A. S., Bacheler, J. S., Aghaee, M. A., Braswell, L., Burrack, H. J., Flanders, K., Greene, J. K., Herbert, D. A., Jacobson, A., Paula-Moraes, S. V., Roberts, P. and Taylor, S. V. 2018. Long-term empirical and observational evidence of practical *Helicoverpa zea* resistance to cotton with pyramided Bt toxins. *J. Econ. Entomol.* 111(4): 1824-1833.

Reisig, D. D. and Kurtz, R. 2018. Bt resistance implications for *Helicoverpa zea* (Lepidoptera: Noctuidae) insecticide resistance management in the United States. *Environ. Entomol.* 47(6): 1357-1364.

Rice, M. E. 2000. Western bean cutworm hits northwest Iowa. http://lib.dr.iastate.edu/cropnews/2113 (accessed 17 February 2022).

Rice, M. E. 2006. Eastern movement of the western bean cutworm into Indiana and Ohio. In: Proceedings of the Indiana CCA Conference, Indianapolis, IN.

Rice, M. E. 2007. Economic thresholds for western bean cutworms. https://dr.lib.iastate.edu/server/api/core/bitstreams/727042f4-fdc4-4898-9270-b05e8d833c84/content (accessed 8 April 2022). Iowa State University, Extension, Ames, IA.

Rice, K. B., Bergh, C. J., Bergmann, E. J., Biddinger, D. J., Dieckhoff, C., Dively, G., Fraser, H., Gariepy, T., Hamilton, G., Haye, T., Herbert, A., Hoelmer, K., Hooks, C. R., Jones, A., Krawczyk, G., Kuhar, T., Martinson, H., Mitchell, W., Nielsen, A. L., Pfieffer, D. G., Raupp, M. J., Rodriguez-Soana, C., Shearer, P., Shrewsbury, P., Venugopal, P. D., Whalen, J., Wiman, N. G., Leskey, T. C. and Tooker, J. F. 2014. Biology, ecology, and management of brown marmorated stink bug (Hemiptera: Pentatomidae). *J. Integr. Pest Manag.* 5(3): 1-13.

Rosenheim, J. A. 2013. Costs of *Lygus* herbivory on cotton associated with farmer decision-making: An ecoinformatics approach. *J. Econ. Entomol.* 106(3): 1286-1293.

Rosenheim, J. A. 2021a. Control failures following insecticide applications in commercial agriculture: How often do they occur? a case study of *Lygus hesperus* (Hemiptera: Miridae) control in cotton. *J. Econ. Entomol.* 114(3): 1415-1419.

Rosenheim, J. A. 2021b. Evaluating the quality of ecoinformatics data derived from commercial agriculture: A repeatability analysis of pest density estimates. *J. Econ. Entomol.* 114(4): 1842-1846.

Rosenheim, J. A. and Gratton, C. 2017. Ecoinformatics (big data) for agricultural entomology: Pitfalls, progress, and promise. *Annu. Rev. Entomol.* 62: 399-417.

Rosenheim, J. A., Parsa, S., Forbes, A. A., Krimmel, W. A., Law, Y. H., Segoli, M., Segoli, M., Sivakoff, F. S., Zaviezo, T. and Gross, K. 2011. Ecoinformatics for integrated pest management: Expanding the applied insect ecologist's tool-kit. *J. Econ. Entomol.* 104(2): 331-342.

Sara, S. A., McCallen, E. B. and Switzer, P. V. 2013. The spatial distribution of the Japanese beetle, *Popillia japonica*, in soybean fields. *J. Insect Sci.* 13: 36. doi: 10.1673/031.013.3601.

Sawyer, A. J. 1985. Efficient monitoring of the density of adult northern corn rootworm (Coleoptera: Chrysomelidae) in held corn. *Can. Entomol.* 117(2): 171–183.

Seymour, R. C., Hein, G. L., Wright, R. J. and Campbell, J. B. 2004. Western bean cutworm in corn and dry beans. Neb. Coop. Ext. Guide G1359. https://extensionpublications.unl.edu/assets/pdf/g2013.pdf (accessed 8 April 2022).

Shanovich, H. N., Dean, A. N., Koch, R. L. and Hodgson, E. W. 2019. Biology and management of Japanese beetle (Coleoptera: Scarabaeidae) in corn and soybean. *J. Integr. Pest Manag.* 10(1). doi: 10.1093/jipm/pmz009.

Shields, E. J., Sher, R. B. and Taylor, P. S. 1991. Sequential sampling plan for northern and western corn rootworms (Coleoptera: Chrysomelidae) in New York. *J. Econ. Entomol.* 84(1): 165–169.

Shrestha, R. B., Dunbar, M. W., French, B. W. and Gassmann, A. J. 2018. Effects of field history on resistance to Bt maize by western corn rootworm, *Diabrotica virgifera virgifera* LeConte (Coleoptera: Chrysomelidae). *PLoS ONE* 13(7): e0200156. doi: 10.1371/journal.pone.0200156.

Sivakoff, F. S., Rosenheim, J. A., Dutilleul, P. and Carrière, Y. 2013. Influence of the surrounding landscape on crop colonization by a polyphagous insect pest. *Entomol. Exp. Appl.* 149(1): 11–21.

Smith, J. L., Farhan, Y. and Schaafsma, A. W. 2019a. Practical resistance of *Ostrinia nubilalis* (Lepidoptera: Crambidae) to Cry1F *Bacillus thuringiensis* maize discovered in Nova Scotia, Canada. *Sci. Rep.* 9(1): 18247. doi: 10.1038/s41598-019-54263-2.

Smith, J. L., DiFonzo, C. D., Baute, T. S., Michel, A. P. and Krupke, C. H. 2019b. Ecology and management of the western bean cutworm (Lepidoptera: Noctuidae) in corn and dry beans–revision with focus on the Great Lakes Region. *J. Integr. Pest Manag.* 10(1). doi: 10.1093/jipm/pmz025.

Smith, J. L., Lepping, M. D., Rule, D. M., Farhan, Y. and Schaafsma, A. W. 2017. Evidence for field-evolved resistance of *Striacosta albicosta* (Lepidoptera: Noctuidae) to Cry1F *Bacillus thuringiensis* protein and transgenic corn hybrids in Ontario, Canada. *J. Econ. Entomol.* 110(5): 2217–2228.

Spencer, J. L., Hibbard, B. E., Moeser, J. and Onstad, D. W. 2009. Behaviour and ecology of the western corn rootworm (*Diabrotica virgifera virgifera* LeConte). *Agric. For. Entomol.* 11(1): 9–27.

St. Clair, C. R. and Gassmann, A. J. 2021. Linking land use patterns and pest outbreaks in Bt maize. *Ecol. Appl.* 31(4): e02295. doi: 10.1002/eap.2295.

Steffey, K. L., Rice, M. E., All, J., Andow, D. A., Gray, M. E. and Van Duyn, J. W. (Eds.) 1999. *Handbook of Corn Insects*. Entomological Society of America, 164pp.

Steffey, K. L., Tollefson, J. J. and Hinz, P. N. 1982. Sampling plan for population estimation of northern and western corn rootworm adults in Iowa cornfields. *Environ. Entomol.* 11(2): 287–291.

Stewart, S. D., Akin, D. S., Reed, J., Bacheler, J., Catchot, A., Cook, D., Gore, J., Greene, J., Herbert, A., Jackson, R. E. and Kerns, D. L. 2013. Survey of thrips species infesting cotton across the Southern US Cotton belt. *J. Cotton Sci.* 17: 263–269.

Stinner, R. E., Rabb, R. L. and Bradley, J. R. 1974. Population dynamics of *Heliothis zea* (Boddie) and *H. virescens* (f.) in North Carolina: A simulation model. *Environ. Entomol.* 3(1): 163–168.

Storer, N. P., Peck, S. L., Gould, F., Van Duyn, J. W. and Kennedy, G. G. 2003. Spatial processes in the evolution of resistance in *Helicoverpa zea* (Lepidoptera: Noctuidae) to Bt transgenic corn and cotton in a mixed agroecosystem: A biology-rich stochastic simulation model. *J. Econ. Entomol.* 96(1): 156–172.

Tabashnik, B. E., Brévault, T. and Carrière, Y. 2013. Insect resistance to Bt crops: Lessons from the first billion acres. *Nat. Biotechnol.* 31(6): 510–521.

Tabashnik, B. E., Gassmann, A. J., Crowder, D. W. and Carrière, Y. 2008. Insect resistance to Bt crops: Evidence versus theory. *Nat. Biotechnol.* 26(2): 199–202.

Tabashnik, B. E., Mota-Sanchez, D., Whalon, M. E., Hollingworth, R. M. and Carrière, Y. 2014. Defining terms for proactive management of resistance to Bt crops and pesticides. *J. Econ. Entomol.* 107(2): 496–507.

Tilmon, K. J., Hodgson, E. W., O'Neal, M. E. and Ragsdale, D. W. 2011. Biology of the soybean aphid, *Aphis glycines* (Hemiptera: Aphididae) in the United States. *J. Integr. Pest Manag.* 2: A1–A7.

Tinsley, N. A., Estes, R. E. and Gray, M. E. 2013. Validation of a nested error component model to estimate damage caused by corn rootworm larvae. *J. Appl. Entomol.* 137(3): 161–169.

United States Department of Agriculture – Animal and Plant Health Inspection Service (USDA APHIS). 2021. Japanese Beetle. https://www.aphis.usda.gov/aphis/ourfocus/planthealth/plant-pest-and-disease-programs/pests-and-diseases/japanese-beetle/japanese-beetle (accessed 5 January 2022).

United States Department of Agriculture – National Agricultural Statistics Service (USDA NASS) 2021. *Crop Prod.* (accessed November 2021) https://www.nass.usda.gov/Publications/Todays_Reports/reports/crop1121.pdf (accessed 29 March 2022).

U.S. Environmental Protection Agency (US-EPA) 2020. EPA draft proposal to address resistance risks to lepidopteran pests of Bt following the July 2018 FIFRA scientific advisory panel recommendation (EPA-HQ-OPP-2019-0682-0007). https://www.regulations.gov/document/EPA-HQ-OPP-2019-0682-0007 (accessed 7 December 2021).

Valmorbida, I., Coates, B. S., Hodgson, E. W., Ryan, M. and O'Neal, M. E. 2022. Evidence of enhanced reproductive performance and lack-of-fitness costs among soybean aphids, *Aphis glycines*, with varying levels of pyrethroid resistance. *Pest Manag. Sci.* 78(5): 2000–2010. doi: 10.1002/ps.6820.

Weber, D. C., Leskey, T. C., Walsh, G. C. and Khrimian, A. 2014. Synergy of aggregation pheromone with methyl (E,E,Z)-2,4,6-decatrienoate in attraction of *Halyomorpha halys* (Hemiptera: Pentatomidae). *J. Econ. Entomol.* 107(3): 1061–1068.

Westbrook, J. K., Wolf, W. W., Lingren, P. D., Raulston, J. R., Lopez, J. D. Jr., Matis, J. H., Eyster, R. S., Esquivel, J. F. and Schleider, P. G. 1997. Early-season migratory flights of corn earworm (Lepidoptera: Noctuidae). *Environ. Entomol.* 26(1): 12–20.

Whitworth, R. J., Wilde, G. E., Shufran, R. A. and Milliken, G. A. 2002. Comparison of adult corn rootworm (Coleoptera: Chrysomelidae) sampling methods. *J. Econ. Entomol.* 95(1): 96–105.

Wold, S. and Hutchison, W. D. 2021. https://vegedge.umn.edu/weekly-moth-flights (accessed 9 January 2022). Weekly Moth Flight Network.

Wu, K. M., Lu, Y. H., Feng, H. Q., Jiang, Y. Y. and Zhao, J. Z. 2008. Suppression of cotton bollworm in multiple crops in China in areas with Bt toxin-containing cotton. *Science* 321(5896): 1676–1678.

Yang, F., Head, G. P., Price, P. A., González, J. C. S. and Kerns, D. L. 2020. Inheritance of *Bacillus thuringiensis* Cry2Ab2 protein resistance in *Helicoverpa zea* (Lepidoptera: Noctuidae). *Pest Manag. Sci.* 76(11): 3676–3684.
Yang, F., Kerns, D. L., Little, N. S., González, J. C. S. and Tabashnik, B. E. 2021. Early warning of resistance to Bt toxin Vip3Aa in *Helicoverpa zea*. *Toxins* 13(9). doi: 10.3390/toxins13090618.

Chapter 3

Quantifying captures from insect pest trap networks

Nicholas C. Manoukis, USDA-ARS, USA

1 Introduction

Networks of insect traps are created and maintained around the world for a variety of reasons. Here they are defined as a geographically dispersed set of traps, operated by a single entity, and for which catch data are collated. Among the most common goals of such networks, especially in the case of invasive pest insects, are surveillance, delimitation, and/or monitoring. In some cases, traps may be an important component of eradication programs. The goal of trap networks and the economics of their maintenance must be considered at every stage, since these drive design decisions, service schedules, analysis of captures, and the duration of the network's existence. Quantifying capture probability is extremely helpful for the design, continual optimization, and modification of these networks; there are models to allow this, of which TrapGrid is a recent example (Manoukis et al., 2014). Three examples of networks targeting insect pests include a high-density trapping program to eradicate *Lymantria dispar* in New Zealand, a delimitation network against Japanese beetle (*Popillia japonica*) in Missouri, USA, and an extensive network of traps for surveillance of multiple invasive species of Tephritidae in California, USA (Denning and Goff, 1944; Gray, 2010; Gilbert et al., 2013).

http://dx.doi.org/10.19103/AS.2022.0113.02

Surveillance networks are focused on minimizing the potential establishment of invasive species, which threaten natural ecosystems, agriculture, and economies globally (Paini et al., 2016). Surveillance networks enable early detection of a pest incursion, allowing cost-effective removal when tied to rapid response with effective measures to extirpate them (Simberloff et al., 2013). These sorts of networks were the original targets of TrapGrid development, but additional trap network types and goals are now also addressed by the model.

Delimitation surveys are often implemented after an invasive pest is detected, and may involve traps, especially when a more effective lure is available (FAO/IAEA, 2018; IPPC/FAO, 2018). Trap networks for delimitation have two goals: (1) confirm the presence of a population of the adventive pest and (2) determine the spatial extent of the population (van Havre et al., 2015). For many pests in the United States, a fully-trapped (fixed density trapping over a contiguous area) survey is often used, either an 8 km × 8 km or 14.5 km × 14.5 km grid (APHIS/PPQ, 2003, 2004), with the size determined by an expert assessment of the insect's vagility. In some cases, variable trapping densities are employed with the highest numbers of trap per unit area near the center of an incursion (usually assumed to be near where the first finds are made) and decreasing density in bands toward the edge (Caton et al., 2021a).

Trap networks for monitoring aim to provide information on the size of a pest population that is already established in the area. Often, timely reporting on this information is critical; an example of a monitoring network providing real-time information on pest pressure is the smart-trap network created by researchers in Taiwan targeting the oriental fruit fly *Bactrocera dorsalis* (Jiang et al., 2008). Monitoring is also frequently important as a component of integrated pest management programs (Vargas et al., 2008; Clarke et al., 2011) or provides evidence of 'low prevalence' of a pest, which can be a required first part of approved systems approaches (Quinlan et al., 2020).

Trapping as the only or principal tool of an eradication program is less frequent compared to trapping for surveillance and delimitation, since highly attractive, species-specific lures are required. In an eradication program, traps may be needed to assess progress. When an effective lure is available it may not be employed only in traps as part of an eradication; rather, mating disruption or male annihilation (MAT) might be deployed in eradication programs with such lures. In MAT, male attractants are used to target fruit flies and include a lure mixed with spot applications of insecticide in the affected area, rather than traps (Vargas et al., 2014). This saves cost (trap materials) and time (labor, handling time).

There are many combinations of traps and attractants employed in networks. For instance, the attraction and specificity of some of these might be very low; a set of yellow sticky traps may have visual attraction at a short range but draw in a large range of insect species. Similarly, food-based lures are usually not highly specific. Other lure-trap combinations can be highly attractive to specific insects (Fig. 1).

The most effective lures are those related to or based on sex pheromones, parapheromones, or semiochemicals. For Lepidoptera, sex pheromones for more than 1600 species have been identified, and many are highly effective, e.g. 'disparlure' (2-methyl-7*R*,8*S*-epoxy-octadecane) for spongy moth (*Lymantria dispar*) (Bierl et al., 1970; Jurenka et al., 2003). Generally, these lures are based on sex pheromones produced by the female and can be highly attractive in small concentrations in the air, over large distances, to males.

Tephritid fruit flies are invasive pest insects of major economic importance around the world (White and Elson-Harris, 1992). A class of compounds sometimes called 'male lures' (which include parapheromones and semiochemicals) are particularly important for trapping many of these species (Shelly et al., 2014). Three commonly used products are 'trimedlure' (tert-butyl 4-chloro-2-methylcyclohexane-1-carboxylate; targeting the Mediterranean fruit fly, *Ceratitis capitata*) and the more attractive cuelure (4-(3-Oxobutyl)phenyl acetate) as well as methyl eugenol. Each of the last two attracts a variety of *Bactrocera* and *Zeugodacus* species, (Drew and Hooper, 1981). These lures are generally species-specific and can attract a large portion of the male population with such effectiveness as to eradicate a population via MAT (Steiner and Lee,

Figure 1 Curelure-baited Jackson trap filled with melon fly (Zeugodacus cucurbitae) near Keaau, Hawaii. Though not the most attractive male lure, in an area of high prevalence it is possible to capture large numbers of males within an hour. *Photo by Nadya Kandel.*

1955). Quantifying the attractiveness of these lure/trap combinations via field experiments is critical to modeling trap networks.

The details of trap networks targeting fruit flies vary around the world. In the US state of California, the system is well-documented (Gilbert et al., 2013). There, fruit fly surveillance is the main goal, since California is the largest producer of fresh fruit in the USA and fruit fly establishment would cost billions of dollars in lost revenue in the first year alone (Suckling et al., 2016). Jackson traps, a type of small delta trap with a sticky card insert and lure (sometimes with insecticide) hanging inside are particularly widely used (Fig. 1). These may contain trimedlure (without insecticide) or methyl eugenol plus naled (1,2-dibromo-2,2-dichloroethylphosphate) or cuelure plus naled. In California, male lure-baited traps are supplemented with various protein-/food-based traps including glass McPhail traps with a torula yeast slurry, ChamP traps, Pherocon AM traps (yellow sticky panel) impregnated with ammonium acetate and protein hydrolysate, and multilure traps baited with BioLure (Quilici and Donner, 2012; Gilbert et al., 2013). The main fruit fly species of focus are *Ceratitis capitata*, *Bactrocera dorsalis*, *Bactrocera zonata*, *Anastrepha ludens*, *Bactrocera correcta,* and *Zeugodacus cucurbitae*. Other pest Diptera include *Anastrepha suspensa*, *Bactrocera correcta*, *Rhagoletis indifferens, Rhagoletis pomonella*, and *Zeugodacus cucurbitae*. Other pest insects also targeted by surveillance in California (but with different traps and lures) are Asian citrus psyllid (*Diaphorina citri*), boll weevil (*Anthonomus grandis*), European corn borer (*Ostrinia nubilalis*), European grapevine moth (*Lobesia botrana*), European pine shoot moth (*Rhyacionia buoliana*), Japanese beetle (*Popillia japonica*), and khapra beetle (*Trogoderma granarium*) (Gilbert et al., 2013).

2 TrapGrid and other models to analyze trap networks

2.1 TrapGrid model description

TrapGrid is a public domain, freely available software package that quantifies the probability of insects being captured by a network of traps baited with attractants. It implements a computer model published by Manoukis et al. (2014). Some of the important features of the model implemented by TrapGrid include the ability to set the attractiveness of each trap separately, set arbitrary trap positions and subsequent densities, model the dispersion of insects, and calculate capture probabilities over time. No model can include all the factors that may be relevant in the real-world situation, but it is anticipated that the most important features of many trap/insect systems are included and that running the model can lead to useful operational insight to improve network design.

A critical component of TrapGrid is the function that relates the distance of an insect from a given trap (calculated via the Euclidean distance, i.e. d =

$(x_0 - x_1)^2 + (y_0 - y_1)^2)$ to the probability of its capture in that trap. In TrapGrid that function is the hyperbolic secant [*sech* (*d*, λ)]:

$$f(d,\lambda) = \frac{2}{e^{-\lambda d} + e^{\lambda d}}$$

Where *d* is the distance from the trap (usually in meters) and λ is a parameter that captures the 'attractiveness' of the lure/trap combination. Fig. 2 shows the probability of capture for varying values of attraction as a function of distance.

Attraction in the TrapGrid model is often presented as the reciprocal of λ. One reason for this is that smaller values of λ indicate more attractive lure/trap combinations, which can be counterintuitive. On the other hand, larger values of 1/λ mean a higher probability of individual insects being caught at a given distance. Moreover, 1/λ has real-world meaning and allows easy comparison of trap/lure combinations across insect species.

It is worth noting that other functions have been used to relate distance from a trap to the probability of capture in insects; these include exponential, logistic, or Cauchy distributions, for example (Cunningham and Couey, 1986; Plant and Cunningham, 1991; Meats and Edgerton, 2008). *Sech* has some desirable properties for use in TrapGrid: First, the probability of capture is 1 when the distance from the trap is 0, which is clearly realistic

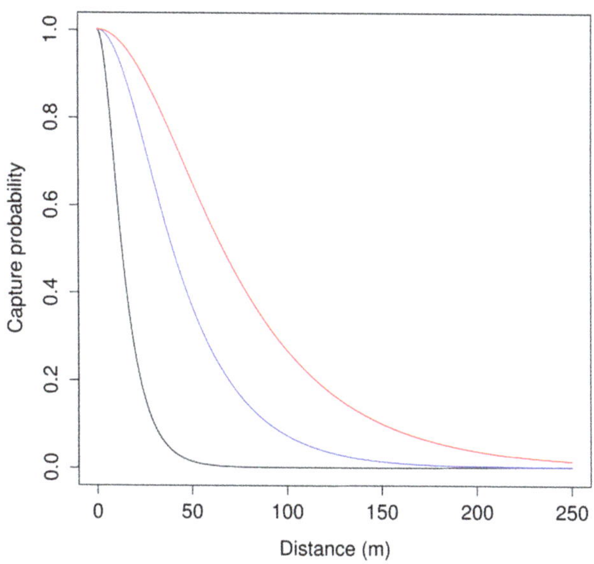

Figure 2 Capture probability as a function of distance from a trap following sech (d, λ). Lines represent traps of varying attraction: 1/λ = 10 (black), 30 (blue), or 50 (red) m. *Figure from Manoukis et al. (2014).*

for a perfectly efficient trap (if the insect is in the trap, it is caught). Second, at a distance of $1/\lambda$, the probability of capture is a constant, about 0.65. This last feature makes it easy to compare the relative attractiveness of different lure/target species combinations. Fig. 3 illustrates capture probability surfaces for two traps, one with a low (Fig. 3a), ($1/\lambda$ = 15 m) and one with high (Fig. 3b) ($1/\lambda$ = 35 m) attraction. Clearly, the 'area under the surface' does not increase linearly with increasing attraction. The volume under Fig. 3b is much larger than might be expected compared with Fig. 3a. However, these numbers give a good 'real-world' idea of how effective a particular lure and trap are over space.

With the capture probability surface defined, it is possible to calculate the 'instantaneous capture probability' for an arbitrary grid. Visually, one can imagine an arena x_m by y_m meters (with area $A = x_m \times y_m$) in size with a height equivalent to the capture probability. If an arena contains a single trap (similar to one of the panels in Fig. 3), the probability of capture of an insect placed randomly in the arena is the area under the curve over the total possible 'volume' of the arena. The average instantaneous capture probability is:

$$\bar{p} = \frac{1}{A} \iint_A \frac{2}{e^{\lambda d_{x,y}} + e^{-\lambda d_{x,y}}} dxdy$$

This same approach can be extended for multiple traps; the sum of the volumes under the curves over the sum of the arena 'volume.' With many traps in close proximity, however, simple addition is problematic since it is possible to result in a larger numerator than the denominator and so an 'instantaneous' capture probability > 1. Biologically, the addition also ignores trap interference (Suckling et al., 2015). It is helpful to consider the inverse of the capture probability, the escape probability q (= $1 - p$). This is because there is only one way to capture zero insects (none of the traps make a capture), but there may be many combinatorial possibilities of one or more captures. Given n traps each a distance $d_{x,y,t}$ from each trap t, we can calculate the average instantaneous escape probability as:

$$\bar{q} = \frac{1}{A} \iint_A \prod_{t=0}^{n} 1 - \frac{2}{e^{\lambda d_{x,y,t}} + e^{-\lambda d_{x,y,t}}} dxdy$$

One way to think about the instantaneous escape probability is that it represents the average probability of escape (=zero captures) if one were to randomly choose a location in the grid - equivalent to throwing a dart at a board. However, real insects in an invading population would have their positions correlated; there would be an incursion point and spread from that point. Furthermore, it is not clear how to move from the instantaneous capture probability to understanding the probability of capture over time as insects move around the area.

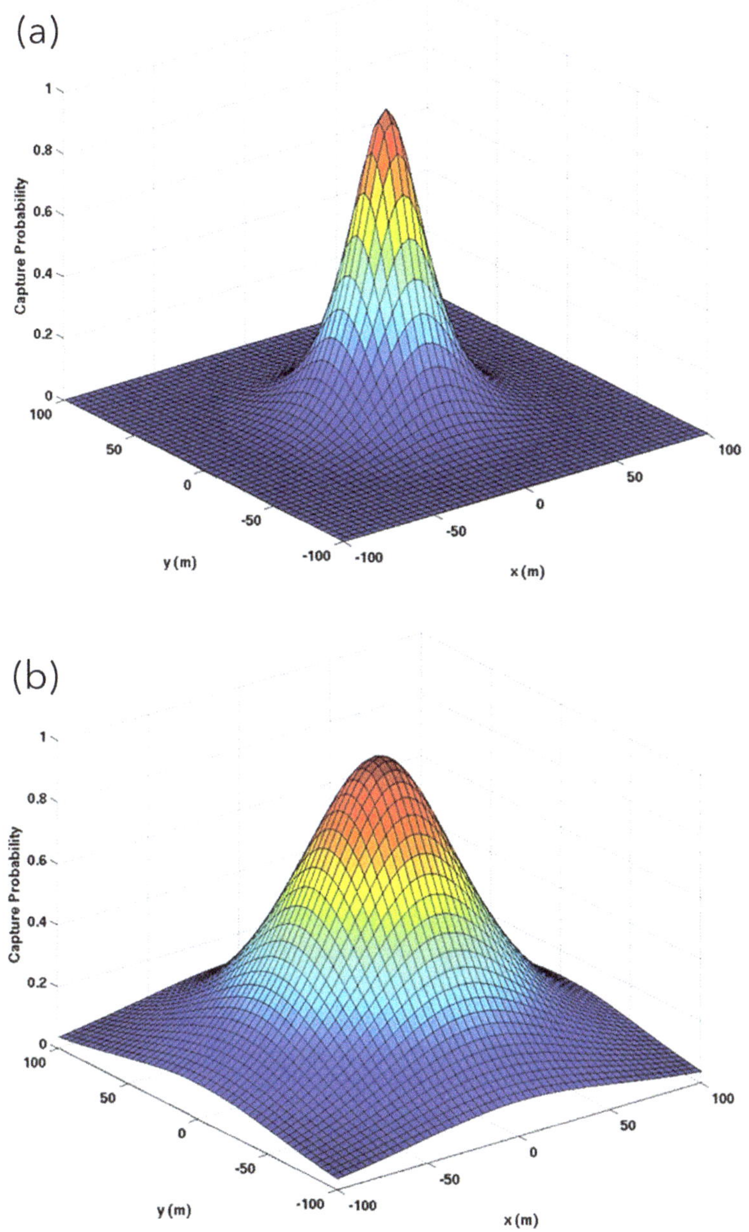

Figure 3 Three-dimensional representation of capture probability over (x,y) for two traps with varying attractiveness. (a) $1/\lambda = 15$ m and (b) $1/\lambda = 35$ m.

One of the simplest ways to model the dispersion of insects over space and time (T) is a diffusion model (Skellam, 1951; Kareiva, 1983):

$$\frac{dN(x,y,T)}{dt} = D\left[\frac{d^2N}{dx^2} + \frac{d^2N}{dy^2}\right]$$

Here N is the population density, time T is in days, D is the diffusion coefficient in m^2/day, and x, y are the positions in meters. The equation above yields a bivariate Gaussian (normal) distribution as a function of diffusion coefficient and T. For any given D, the variance of the bivariate normal will increase with time. The probability density function is:

$$g(x,y) = \frac{1}{2\pi\sigma^2} e^{-\frac{1}{2}\left(\frac{x^2}{\sigma^2} + \frac{y^2}{\sigma^2}\right)}$$

Where σ is the standard deviation (we assume the mean position = 0) and can be calculated as $4\sqrt{(DT)}$. This also assumes standard deviations are equal in x and y dimensions and that there is no correlation in x and y positions. With these, we can calculate the distribution of individual insect positions each time step and then for each insect position calculate the probability of escape. TrapGrid, in its original release, returns the cumulative values of the means for each day over time, thereby estimating the average sensitivity of the grid over time in face of an invasive insect incursion at a given location.

The most recent release of TrapGrid includes an alternate calculation of capture probability: Strictly, the probability of capturing one or more insects. This measure is much more sensitive to the size of the population being modeled but may be useful in low-prevalence, small size trap grids. Full details are given in Manoukis and Hill (2021).

Another important feature of TrapGrid that will not be extensively discussed here is an alternate movement model besides simple diffusion: a random-correlated walk (Kareiva and Shigesada, 1983; Byers, 2001). This is more realistic in the sense that the positions of individual insects are correlated, rather than a new distribution of positions being generated at each time step as in the case of diffusion. It requires two parameters for movement (compared with one, D, in diffusion): mean step size and mean turning angle per time increment. For most insects, these are not available, but work is ongoing to measure these in the field for tephritid fruit flies (Miller et al., 2022).

2.2 Other models

Before discussing applications of TrapGrid, it is worth pausing here to note that TrapGrid and its underlying model are in no way the only attempt to quantify trap network sensitivity. Here a few examples of other methods are provided to give a sense of the range of methods that have already been applied. Further, it is useful to understand other models when interpreting the results of field experiments performed to parameterize them (see Section 3.1).

The concept of trap attraction and insect capture has been represented in a variety of ways. One early idea was the 'Active Space' of Bossert and Wilson (1963), which focused on the distribution of a chemical signal over space modeled via a diffusion process, yielding a gradient in space over a variety of release conditions (still air, puffs of release, and more). Experimental applications of this model proved difficult, but the idea of modeling capture over space was clearly valuable (Nakamura and Kawasaki, 1977; Elkington and Carde, 1984).

Later work simplified the model of distance × capture probability by introducing concepts like the 'sampling range,' 'range of stimulation,' and 'range of attraction' (Wall and Perry, 1987). These were defined, respectively, as the maximum distance from which an insect can reach a source over a given period, the maximum distance at which a behavioral response can be seen in a target insect, and the maximum distance at which insects can be shown to direct their movement toward the source. These are tangible measures that can be measured directly via experimentation, but their link to the application of trapping networks is less direct.

Byers et al. (1989) introduced a direct measure known as the 'effective attraction radius,' which they defined as the size of an unbaited trap that would be needed to have an equivalent capture as the trap with lure. In other words, what is the area within which there is a 100% probability of capture? This measure has a strong direct link to the application of traps, even though it was originally developed to estimate the densities of forest pests. Experimentally, it could be determined via the ratio of passive and baited trap catches. While the original formulation did not include insect movement, this was always possible and indeed was analyzed in subsequent studies (e.g. Byers, 1999).

Not all models of traps and insect captures include a function relating distance to probability of capture. A good example is the 'effective sampling area,' or ESA (Turchin and Odendaal, 1996). ESA is a conversion factor between the density of trappable insects and the number caught over a given period of time, and so tends to be used in situations where the goal is to estimate the population density (Puche and Su, 2004). It can be used to estimate the mean probability of detecting a randomly distributed insect in a trap network (Kean, 2015) and has been applied to questions outside of insects (e.g. crayfish, see Acosta and Perry, 2000). While a trap density for a minimum number of captures can be derived using the concept of the ESA, varying attraction of individual traps given environmental factors within a single grid cannot be as easily gauged or included.

Increasing computational power in the twenty-first century has opened new possibilities in trap network design, including improved models of insect movement, and indeed, its very consideration (Miller et al., 2015). Simulating time, movement, and lure plumes simultaneously is now possible with directly actionable results for agriculture (Adams et al., 2017). TrapGrid is an example of this sort of approach, geared to quantifying capture probability to support management decisions around how many traps to deploy and where. As demonstrated later, TrapGrid can be applied to a variety of situations.

TrapGrid has limitations. Like all models, it does not consider all relevant factors, but only a selected subset and those that it does consider are simplified for mathematical representation and tractability. Some of the biological factors currently not modeled by TrapGrid include the receptiveness of a given insect to the lure being used in the trap network, the efficiency of the trap at capturing an insect that enters, or any consideration of a demographic change in the target insect population. Furthermore, the exiting implementation provides only trap-network-wide results rather than the probability of capture in given traps. Finally, TrapGrid requires an attraction parameter that must be estimated via field studies for given insect-trap/lure combinations, and only a limited number of these have been developed to date.

3 Applications

3.1 Estimating trap attraction

The question of how far an attractant-baited trap can draw its captures has been extensively investigated by entomologists for many different species (see Schlyter, 1992). Perhaps the most common approach has been via Mark-Release-Recapture (MRR) experiments with a centrally located trap and releases at varying distances (e.g. Dodds and Ross, 2002; Shelly et al., 2010; Shelly, 2021). While these experiments clearly provide very useful information, recaptures for a given distance often have high variance and there is difficulty setting sufficient release points at increasing distances since the area under consideration increases rapidly (and with it, increasing variation). As an example of the limitations of this sort of study, it has been noted that estimates of dispersal for given species seem to be highly related to the size of the trapping grid employed in the particular study (Weldon et al., 2014).

There have been novel efforts to improve this situation. A relatively recent study cleverly leveraged the ability to turn blacklight traps on and off to estimate both the range of attraction and spatial interference against *Culicoides* midges (Rigot and Gilbert, 2012). In an example using chemical lures, Jactel et al. (2019) offered a method based on pairs of traps, at varying distances apart, monitoring wild populations of *Monochamus galloprovincialis*

Published by Burleigh Dodds Science Publishing Limited, 2023.

(pine sawyer beetle), and leveraging trap interference to estimate the attractive range of sex pheromone-baited cross vein traps. They reported higher attractiveness at distance than previous MRR- based studies, but the comparison was indirect due to different measures of trap attraction used in the various studies (Section 2.2). The utility and applicability of field results are highest when the experiments are linked at the time of design to mathematical models of trapping (Dufourd et al., 2013).

To date, only two studies have been published with results of field experiments designed specifically to estimate λ (Manoukis and Gayle, 2015; Manoukis et al., 2015), and they illustrate the interplay between field experiments and underlying mathematical models. Both were MRR experiments in a grid of traps, where the proportion of released insects recaptured (*P*) was used to infer the λ value that would result in the observed *P*.

Manoukis et al. (2015) marked and released both Mediterranean fruit fly ('Medfly') and the Oriental fruit fly ('OFF,' *Bactrocera dorsalis*) in three spacings (50 m, 100 m, and 150 m) of 8 × 8 grids of Jackson traps baited with male lures (trimedlure and methyl eugenol, respectively). Experiments were completed in two environments, a macadamia orchard on the east side of Hawaii island and a pahoehoe lava field on the west side. Results are given in Table 1.

Focusing on the first set of results (rows 1–3 in Table 1), a relatively high *P* value is evident for OFF in macadamia in a grid with 150 m spacing in *x* and *y* directions (mean = 0.623). Using TrapGrid (see section 2.1), the relationship between trap attraction and *P* is plotted for an 8 × 8 grid of traps with 150 m spacing. The *P value* of around 0.62 in a 150 m grid translates to a 1/λ over just over 40 m. A second line is plotted for the same relationship, but this time in a grid that has 75 m spacing. From this line, the expected recapture should be between 0.95 and 0.99 in the 75 m grid based on the 1/λ value of ~40 m, from the results in the 150 m grid (Fig. 4). This is an extremely high bar in MRR studies, but the subsequent three rows in the table show that there was a good match to the theoretical expectation. In overall experiments, in both environments, the 1/λ for methyl eugenol was estimated to be around 36 m for receptive OFF males.

Continuing down Table 1, methyl eugenol is confirmed to be a more effective lure for male OFF than trimedlure is for Medfly, and it is more consistent across environments (for Medfly 1/λ approximately 14 m in macadamia, about 7 m in the lava field). While the environment difference in λ between lava field and macadamia orchard for Medfly has some caveats, it does suggest that models of insect traps should include variable attraction for a given lure-target species combination (Manoukis et al., 2015). Further experiments with other tephritid species are needed, and besides the labor and difficulties of MRR experiments, there is no reason similar experiments could not be conducted to estimate 1/λ for other insect species.

Table 1 Results of Mark-Release-Recapture experiment by Manoukis et al. (2015) on Hawaii Island. Three grid spacings, in two environments, are reported for two species Bactrocera dorsalis (OFF) and Ceratitis capitata (Medfly). The proportion of each release cohort responsive to the lure ('Res.') was assessed at the time of the experiment via Y-tube olfactometer assay; this proportion times the estimated number released ('Rel.'), gives the number of available for recapture (denominator of the P calculation, not shown). The number recaptured ('Rec.') is the numerator.

Date	Habitat	Spacing (m)	Species	Rel.	Rec.	Res.	*P*	1/λ
2013-07-08	Macadamia	150	OFF	1256	375	0.58	0.519	34.94
2013-07-10	Macadamia	150	OFF	2011	783	0.60	0.649	41.49
2013-07-12	Macadamia	150	OFF	1987	1256	0.90	0.702	44.58
2013-07-22	Macadamia	75	OFF	1890	1192	0.70	0.904	31.59
2013-07-24	Macadamia	75	OFF	2117	1557	0.88	0.841	28.15
2013-07-26	Macadamia	75	OFF	2062	1824	0.93	0.956	37.75
2013-08-19	Macadamia	150	Medfly	1786	102	0.55	0.104	14.46
2013-08-21	Macadamia	150	Medfly	1716	105	0.55	0.112	15.08
2013-08-23	Macadamia	150	Medfly	1581	126	0.45	0.177	18.9
2013-09-02	Macadamia	75	Medfly	1372	312	0.61	0.374	14.20
2013-09-04	Macadamia	75	Medfly	1956	336	0.57	0.301	12.45
2013-09-06	Macadamia	75	Medfly	2034	275	0.56	0.243	10.97
2013-09-23	Lava	100	OFF	1778	564	0.43	0.746	31.70
2013-09-25	Lava	100	OFF	1907	457	0.28	0.871	38.87
2014-01-15	Lava	100	Medfly	1443	47	0.39	0.084	8.25
2014-01-17	Lava	100	Medfly	1778	38	0.60	0.036	5.65

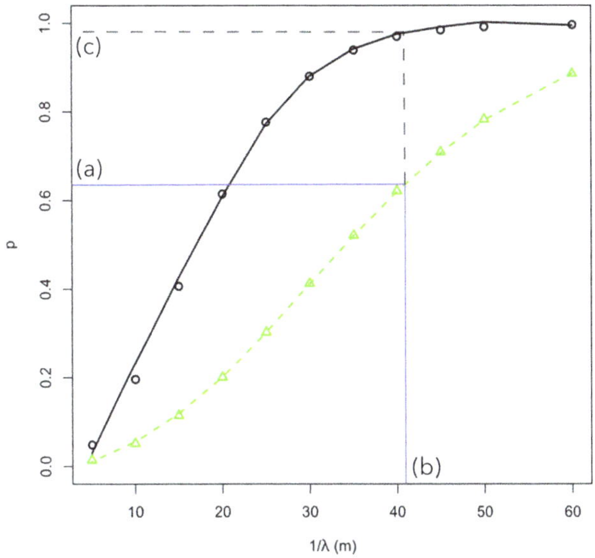

Figure 4 Expected proportion recapture (P) in a 75 m grid (black line, circles) based on attraction (1/λ) inferred for a 150 m grid (green broken line, triangles). The P for the 150 m grid in the first experiments was around 0.62 (a); this corresponds to 1/λ of about 40 m (solid blue line, point b). This attraction would yield a recapture rate of about 0.97 for the 75 m grid (broken line, point c). The observed average P for the 75 m grid in subsequent experiments averaged 0.90; close to expectation *(Table 1)*.

3.2 Large area surveillance

As mentioned in section 1, tephritid fruit flies are some of the most economically important agricultural pests in the world. They are noted for their highly invasive characteristics (Duyck et al., 2004). Some species, such as Medfly and OFF are capable of laying eggs in the fruit of hundreds of hosts, estimated at 419 and 500 plant species and varieties, respectively, with many of commercial interest (Liquido et al., 2020, 2021). After eclosion, the larvae of these fruit flies consume the fruit, making it unsuitable for the market and introducing rotting and spoilage. Third instar larvae leave the fruit to pupate in the soil, finally emerging as adults to begin the life cycle again (Vargas, 1989). Because of their economic impact, many tephritids are considered quarantine pests, and their introduction to new areas can lead to loss of market access in addition to the direct damage (Kruger, 2016).

Fruit fly establishment in production areas thus carries a large potential economic cost. For this reason, surveillance networks are employed around the world (see section 1). An important question arising is: How many traps should be deployed, and in what configuration, to attain an acceptable level of surveillance sensitivity? Many trap networks are designed and deployed without any ability or attempt to quantify capture probability, rather than relying on expert opinion and extrapolation. TrapGrid was originally developed to provide quantitative results addressing this question.

An example can be useful to understand how TrapGrid can be used. The California Department of Food and Agriculture has a well-documented approach to surveillance against fruit flies (Gilbert et al., 2013). For OFF in Southern California, this includes five methyl eugenol-baited Jackson traps per square mile over almost all the greater Los Angeles area plus five torula yeast-baited McPhail traps. A sample grid of the methyl eugenol traps only is shown in Fig. 5a over an area equivalent to Orange County, California. This grid consists of a total of 6738 traps, indicating the effort and cost associated with surveillance trapping in a single county.

Could this or a similar number of traps be deployed more efficiently than a uniform density across a wide area? Incursions of OFF are found several times per year in Southern California, so improvements could have a significant impact on the biosecurity of the State. One very simple approach would be to increase the density of traps in areas with a higher probability of OFF detection based on previous years while decreasing the density in areas where historically not as many OFF have been found. Fig. 5b gives a modified trap network comprising fewer traps (5643) but with variable density: three traps per square mile in areas with fewer detections, and six traps per square mile in areas with higher levels of detection.

TrapGrid can be used to quantify the average capture probability of each of these grids and enable direct comparison. Using the model with simple diffusion and the outbreak locations shown in Fig. 5b, the uniform grid in Fig. 5a has a mean capture probability after 30 days of 0.130 (range: 0.054–0.183). The variable grid in Fig. 5b yields an average capture probability of 0.115 (range: 0.033–0.187). These are comparable sensitivities, with a drop of almost 1100 in the number of traps deployed.

The realism of the results from the TrapGrid model will depend heavily on the quality of the information available on past outbreaks. Additionally, estimates of attraction are critical (Section 3.1). It is also important to have estimates of the diffusion of the pest, a question that has recently been addressed using real-world surveillance data (Caton et al., 2021b). It is also possible that the map of outbreak risk could change over time, so it might be important to redeploy a uniform grid for a few years in line with predicted and observed environmental change to re-assess which areas should be trapped at high and low densities. Despite these issues, this example gives an idea of the potential to improve surveillance efficiency and sensitivity by quantifying capture probability.

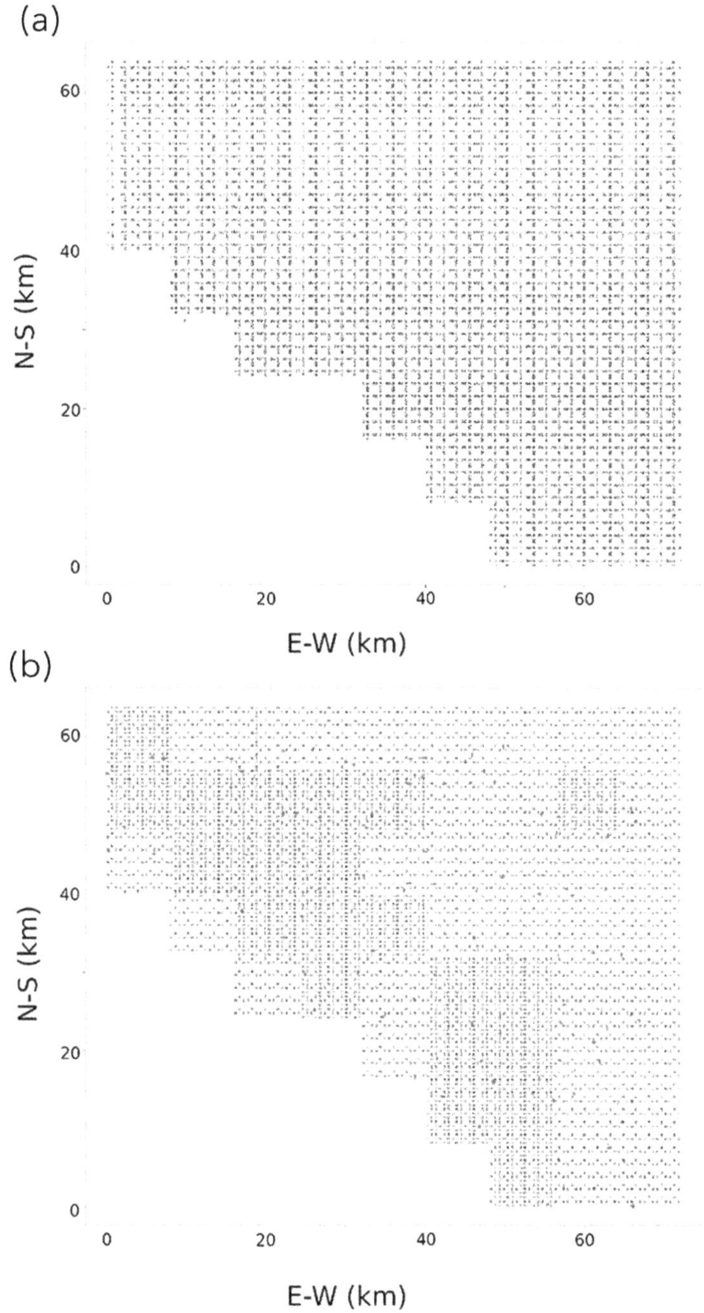

Figure 5 Hypothetical uniform (a) and varying (b) density distribution of traps over Orange County, California. In (b), the red 'x' symbols indicate hypothetical locations of detections that would guide the application of a variable density.

3.3 Other applications

The example in Section 3.2 is at the heart of the original goal for TrapGrid: Landscape-level average trap network sensitivity estimation over time to allow comparison of alternative trap layouts. However, in the time since the model was released, it has been used for other applications. One of these has been to evaluate delimitation grids, as mentioned earlier (Caton et al., 2021a). These same authors employed TrapGrid to vary diffusion coefficients to estimate the dispersal ability of a variety of invasive pests based on surveillance trapping grid data (Caton et al., 2021b).

Low-prevalence situations at a smaller scale, e.g. individual orchards, are also in the process of being examined and quantified via TrapGrid (Manoukis and Hill, 2021). In this context, the goal is to evaluate the potential demographic meaning of a given number of captures across a small trap network - a monitoring situation. Estimating the standing population size of an endemic pest insect can be helpful for systems approaches used to ensure phytosanitary exports and general biosecurity - a low starting population size is regularly an important component of these approaches (Ruesink, 1976; Quinlan et al., 2020).

TrapGrid could also be used to ask questions about how many traps over what area might be needed to create a barrier that traps effectively all insects attempting to cross. Such an application would require a new implementation of the model or significant modification of the existing code since the current version does not track or return information on which traps captured which insects (Manoukis, 2020). Individual-based tracking via a random-correlated walk model would also be critical for this application.

The MRR experiments described in Section 3.1 point to another useful application of TrapGrid: Estimating and comparing trap attraction. In addition to the question of comparing lures between species and in different environments, attraction to wild and colony-reared insects has been examined (Manoukis and Gayle, 2015). Other possible biological questions that could be tested include the effect of lure age, various formulations, the presence or absence of host trees, insect age, etc. Future development of TrapGrid could also include the ability to include habitat maps/geographic information system layers to more realistically represent the environment of the insect and its effect on movement and capture probability.

4 Practicum

In this section, a step-by-step example of the application of TrapGrid is presented as a guide for those interested in using the model. An overview of necessary steps and some starting points for executing each for their own systems are given first.

To illustrate a simple case, let us assume a user wanted to compare the instantaneous capture probability as well as the average cumulative capture probability after 30 days of a single square kilometer with 12 evenly spaced, low attraction traps ($1/\lambda = 8$ m) plus 4 or 16 highly attractive traps ($1/\lambda = 30$ m). The goal of the trapping might be for monitoring, and the user is trying to determine if the additional traps are worth the added cost of materials and servicing, especially since the more attractive traps (pheromone baited) are more costly than the less attractive ones (these are food-based).

The first step is to define the (*x*,*y*) positions of each trap, as well as the total grid. This can easily be done in Excel (Microsoft Corp., Redmond, WA, USA) or similar spreadsheet software. A sample plot from such a spreadsheet is shown in Fig. 6. Positions in the columns for the plot can be changed dynamically to test different layouts.

Two sets of points are saved as 'tab separated value' files: One with the 12 low attraction traps plus 4 high-attraction traps indicated by orange 'X' symbols in Fig. 6, and the other with the 12 low attraction traps plus all 16 ('X' and '+' symbols) high-attraction trap points. The first of these is shown in Box 1 as a sample.

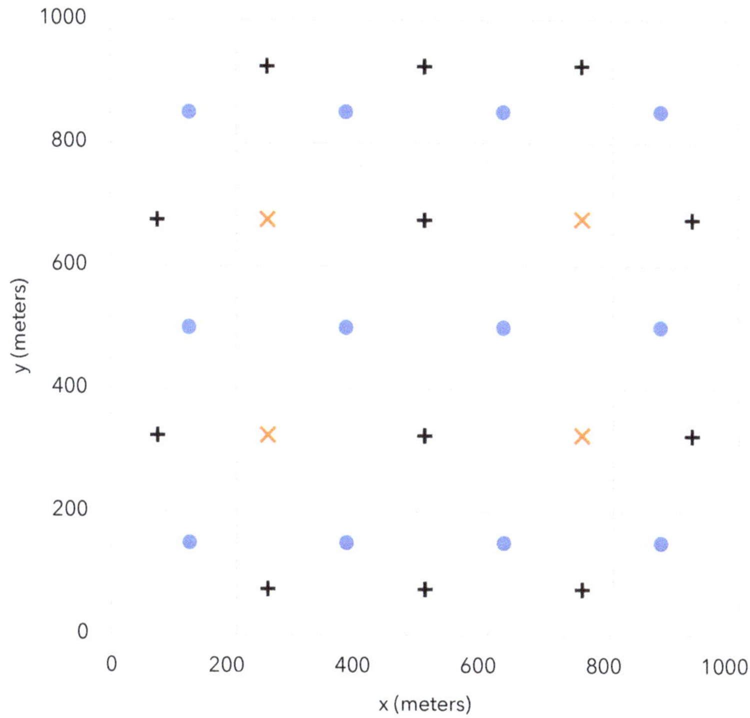

Figure 6 Trap positions in a 1 km^2 area for a test of instantaneous and cumulative average capture probability. Blue dots are low attraction traps ($n = 12$; $1/\lambda = 8$ m). Orange 'X' points are high-attraction traps ($n = 4$; $1/\lambda = 30$ m). Black cross points are additional high-attraction traps ($n = 12$, 16 when combined with 'X' points; $1/\lambda = 30$ m).

Box 1 Sample TrapGrid file, tab-separated-values (*.tsv). First line is grid dimension in x and y. Each subsequent line represents a trap, first value is x, second is y, and third is 1/λ (attraction).

```
1000 1000
125 150 0.125
375 150 0.125
625 150 0.125
875 150 0.125
125 500 0.125
375 500 0.125
625 500 0.125
875 500 0.125
125 850 0.125
375 850 0.125
625 850 0.125
875 850 0.125
250 325 0.033
750 325 0.033
250 675 0.033
750 675 0.033
```

When creating TrapGrid files, it is important to ensure that there are no invisible characters as these may interfere with parsing by the TrapGrid program. For example, an extra tab after the second dimension on line 1 (inserted by default by most spreadsheet programs) will cause a parsing error. So will an additional line break at the end of the file, or additional tabs after an attraction value. It is best to verify these files in a text editor that can display non-printing characters. Use of a word processor such as Microsoft Word is highly discouraged, as this will likely introduce formatting, spellcheck, etc.

With the trap grids defined, the next step includes running the TrapGrid program, freely available online at https://github.com/manoukis/TrapGrid. The program is written in Java and should be executable on any major operating system including Microsoft Windows, Apple MacOS (Apple Corp., Cupertino, CA, USA), or variants of Linux. TrapGrid is run from the command line; help is available by entering the following command in the directory containing the program:

```
java -jar TrapGrid.jar --help
```

To run our first set of simulations, assuming the TrapGrid file (named 'tg16traps .tsv,' contents of which are in Box 1) and the program executable are in the same directory, we use the following command:

```
java -jar TrapGrid.jar -tg tg16traps.tsv -nd 30 -ns 100 -nf 150 -dc
5000 -s 25361 --calculateAvgEscProb > tg16traps-output.txt
```

The first three terms of the command invoke the program by having the system-installed Java Virtual Machine execute the 'jar' file containing the program. After the '-tg' switch, the TrapGrid file is passed to the program. -nd indicates the number of days over which to run the simulations (30), -ns is the total number of simulations to run (100), -nf is the number of flies per simulation (150), -dc is the diffusion coefficient D (5000 m^2/day), and -s is a random seed (optional, but useful for repeatable results; set to 25 361, can be any integer). The next switch '--calculateAvgEscProb' sets the program to calculate the instantaneous escape (capture) probability. Finally, the '>' operator and file name allow the terminal program to redirect the output to a text file for later viewing (file will be named 'tg16traps-output.txt'). The same command as mentioned earlier can be reused but changing the trap file from tg16traps.tsv to a different one with 28 trap lines specified which include the additional high-attraction traps. A different output file is also recommended.

Program execution for these grids took less than a minute on a current commodity portable machine (Intel Core i5-1145G7 @ 2.6 GHz, 16 GB RAM). Larger grids, more simulations, or more traps can increase this execution time significantly, especially if the instantaneous escape probability is calculated. In those instances, more computing resources or time will be required.

The output file will include a block at the top with parameters, program version information, and the instantaneous escape probability. An average cumulative escape probability is then given for each of the 30 days of the simulation - this average is across all 100 simulations. The individual 100 simulation cumulative escape probabilities are then given later. As noted earlier, capture probability P is simply $1 - q$ (where q is the escape probability).

The 16-trap grid has a higher instantaneous escape probability (0.949) compared with the 28-trap grid (0.831) as expected. These values reflect the probability that a point selected at random from anywhere in the 1 km^2 grid yields a capture. Over time, the 28-trap grid also performs better in terms of cumulative capture probability (Table 2). While qualitatively unsurprising, the difference might be - 16 traps have an average cumulative capture probability after 30 days of under 60%, but with the addition of the extra high-attraction traps the value increases to over 93%. Depending on the cost of the traps, goals of the program, and required accuracy in monitoring, the additional traps may be worth the cost.

While the example mentioned earlier is very simple, it illustrates many of the important functions of TrapGrid. One of the additional capabilities worth mentioning is the ability to define population starting points ('outbreaks' in the case of surveillance) via an 'outbreak file.' This is a two-column, tab-delimited file

Table 2 Capture probabilities over time for the 16- and 28-trap grids; initial outbreak positions random for each of the 100 simulations per grid.

Day	16-Trap grid	28-Trap grid
1	0.048	0.156
2	0.090	0.273
3	0.129	0.368
4	0.163	0.445
5	0.195	0.509
6	0.224	0.564
7	0.250	0.610
8	0.275	0.649
9	0.299	0.683
10	0.321	0.713
11	0.342	0.739
12	0.362	0.762
13	0.379	0.782
14	0.396	0.800
15	0.412	0.816
16	0.427	0.831
17	0.441	0.843
18	0.454	0.854
19	0.467	0.865
20	0.479	0.874
21	0.491	0.883
22	0.501	0.891
23	0.512	0.898
24	0.522	0.904
25	0.532	0.910
26	0.541	0.916
27	0.550	0.920
28	0.558	0.925
29	0.565	0.929
30	0.573	0.933

with each row containing an *x* and *y* position, in meters, from which populations are simulated. If this file is specified via the '-ob' switch at the command line, then TrapGrid will only run as many simulations as there are lines in the file. The ability to define population locations is often important for real-world simulations (e.g. MRR experiments with a defined release point in a trap grid).

Finally, a note on transforming latitude/longitude data into *x,y* positions in meters. There are many ways to accomplish this task, but a simple one that is usually sufficient for most grid sizes is to define an origin point and then convert each of the trap positions by using the number of meters per degree latitude and longitude for the study area. Using the WGS84 spheroid this can be approximated for latitude via:

$$M_{d,t}(\phi) = 111132.92 - \left(559.82 * \cos(\phi)\right) + \left(1.175 * \cos(4\phi)\right) + \left(0.0023 * \cos(6\phi)\right)$$

The equivalent equation for longitude is:

$$M_{d,g}(\phi) = ABS\left(111412.84 * \cos(\phi)\right) - \left(93.5 * \cos(3\phi)\right) + \left(0.118 * \cos(5\phi)\right)$$

The difference in each dimension between the origin point and each trap can be calculated in degrees and converted to meters. The origin, trap, or midway point of latitude can be used for φ, results will be very similar for most grid sizes.

5 Conclusion

In this chapter, a high-level overview of the goals of trap networks, some examples and details for tephritid fruit flies, and then a detailed description of the TrapGrid model are given. TrapGrid can be used to quantify the probability of capturing insects instantaneously or over time using a function that relates distance from a given trap to the probability of capture and two models of insect dispersal. Brief descriptions of other modeling approaches to these questions, some of which have seen application outside of research, are discussed followed by ideas for the application of TrapGrid, including a way to determine trap attraction (the parameter λ in the model) and how to compare alternative trap layouts on a landscape scale. Finally, in the practicum, a working example was given comparing two alternative trap layouts in a 1 km^2 area via quantification of capture probability instantaneously and over 30 days. There remain other, undescribed, applications of TrapGrid and similar models to improve insect pest monitoring, surveillance, and control. This chapter serves as a useful starting point for researchers and practitioners interested in potential applications which optimize trapping surveillance and control of pest insects. A version of TrapGrid that includes the features described here is available at https://github.com/manoukis/TrapGrid.

6 Where to look for further information

There are a number of comprehensive sources on trapping and its application. A relatively recent book edited by Shelly et al. (2014) is very helpful for those interested in trapping of fruit flies from biology, to ecology, to programs and application. From a theoretical perspective, the book by Miller et al. (2015) provides a very helpful framework to consider insect trapping in a concise and readable presentation. Muirhead-Thompson's (1991) work is also an excellent resource on trapping across a wide variety of insects.

From the program and regulatory perspective, the regional plant protection organizations as defined by the International Plant Protection Convention of 1997 (https://nappo.org/application/files/5515/8319/2828/IPPC_new_revised_text.pdf) are helpful to follow. Examples include the North American Plant Protection Organization (NAPPO; https://www.nappo.org/) and the Asia and Pacific Plant Protection Commission (APPPC; https://www.fao.org/asiapacific/apppc/en/). A full list of the 10 regional plant protection organizations is available at https://www.ippc.int/en/external-cooperation/regional-plant-protection-organizations/.

The International Atomic Energy Agency and the Food and Agriculture Organization maintain multiple online resources that are useful, especially for those focused on Tephritids. These include the TWD (Tephritid Workers Database) and 'Fruit Fly News'. Trapping comes up regularly in the context of SIT programs. Links to these are available at https://www.iaea.org/resources/nucleus-information-resources and https://nucleus.iaea.org/sites/naipc/twd/Newsletters/Forms/Fruit%20Fly%20News1.aspx.

7 Acknowledgements

Thanks to Dr. Chris Weldon for helpful comments on an earlier version of this chapter. USDA is an equal opportunity provider and employer. This research used resources provided by the SCINet project of the United States Department of Agriculture - Agricultural Research Service and was funded by ARS project number 2040-22430-027-00-D. The USDA is an equal opportunity provider and employer.

8 References

Acosta, C. A. and Perry, S. A. 2000. Effective sampling area: A quantitative method for sampling crayfish populations in freshwater marshes. *Crustaceana*. 73(4): 425–431.

Adams, C. G., Schenker, J. H., McGhee, P. S., Gut, L. J., Brunner, J. F. and Miller, J. R. 2017. Maximizing information yield from pheromone-baited monitoring traps: Estimating plume reach, trapping radius, and absolute density of *Cydia pomonella* (Lepidoptera: Tortricidae) in Michigan apple. *J. Econ. Entomol.* 110(2): 305–318.

APHIS/PPQ. 2003. *Mediterranean Fruit Fly Action Plan: Plant Protection and Quarantine (PPQ).* Animal and Plant Health Inspection Service, United States Department of Agriculture, Washington DC.

APHIS/PPQ. 2004. *New pest response guidelines: Leek moth* Acrolepiopsis assectella *(Zeller): Plant Protection and Quarantine (PPQ).* Animal and Plant Health Inspection Service, United States Department of Agriculture, Washington DC.

Bierl, B. A., Beroza, M. and Collier, C. W. 1970. Potent sex attractant of the gypsy moth: Its isolation, identification, and synthesis. *Science* 170(3953): 87-89.

Bossert, W. H. and Wilson, E. O. 1963. The analysis of olfactory communication among animals. *J. Theor. Biol.* 5(3): 443-469.

Byers, J. A. 1999. Effects of attraction radius and flight paths on catch of scolytid beetles dispersing outward through rings of pheromone traps. *J. Chem. Ecol.* 25(5): 985-1005.

Byers, J. A. 2001. Correlated random walk equations of animal dispersal resolved by simulation. *Ecology* 82(6): 1680-1690.

Byers, J. A., Anderbrant, O. and Löqvist, J. 1989. Effective attraction radius: A method for comparing species attractants and determining densities of flying insects. *J. Chem. Ecol.* 15(2): 749-765.

Caton, B. P., Fang, H., Manoukis, N. C. and Pallipparambil, G. R. 2021a. Simulation-based investigation of the performance of delimiting trapping surveys for insect pests. *J. Econ. Entomol.* 114(6): 2581-2590.

Caton, B. P., Fang, H., Manoukis, N. C. and Pallipparambil, G. R. 2021b. Quantifying insect dispersal distances from trapping detections data to predict delimiting survey radii. *J. Appl. Entomol.* 146: 203-216.

Clarke, A. R., Powell, K. S., Weldon, C. W. and Taylor, P. W. 2011. The ecology of *Bactrocera tryoni* (Diptera: Tephritidae): What do we know to assist pest management? *Ann. Appl. Biol.* 158(1): 26-54.

Cunningham, R. T. and Couey, H. M. 1986. Mediterranean fruit fly (Diptera: Tephritidae): Distance/response curves to trimedlure to measure trapping efficiency. *Environ. Entomol.* 15(1): 71-74.

Denning, J. and Goff, C. 1944. *Eleven Years of Japanese Beetle Control in Missouri - 1934-1944.* Missouri Department of Agriculture.

Dodds, K. J. and Ross, D. W. 2002. Sampling range and range of attraction of *Dendroctonus pseudotsugae* pheromone-baited traps. *Can. Entomol.* 134(3): 343-355.

Drew, R. A. I. and Hooper, G. H. S. 1981. The responses of fruit fly species (Diptera: Tephritidae) in Australia to various attractants. *Aust. J. Entomol.* 20(3): 201-205.

Dufourd, C., Weldon, C., Anguelov, R. and Dumont, Y. 2013. Parameter identification in population models for insects using trap data. *Biomath* 2(2): Article ID: 1312061.

Duyck, P.-F., David, P. and Quilici, S. 2004. A review of relationships between interspecific competition and invasions in fruit flies (Diptera: Tephritidae). *Ecol. Entomol.* 29(5): 511-520.

Elkington, J. D. and Carde, R. T. 1984. Odor dispersion. In: *Insects Chemical Ecology.* Sinauer Associates, Sunderland, MA.

FAO/IAEA. 2018. *Trapping Guidelines for Area-Wide Fruit Fly Programmes* (2nd edn.). International Atomic Energy Agency, Rome, Italy.

Gilbert, A., Bingham, R., Nicolas, M. and Clark, R. 2013. *Insect Trapping Guide* (13th edn.). California Department of Food and Agriculture, Sacramento, California, U.S.A.

Gray, D. R. 2010. Hitchhikers on trade routes: A phenology model estimates the probabilities of gypsy moth introduction and establishment. *Ecol. Appl.* 20(8): 2300–2309.

IPPC/FAO. 2018. *International Standards for Phytosanitary Measures, Publication No. 14 Plant Pest Surveillance*. Food and Agriculture Organization, Rome, Italy.

Jactel, H., Bonifacio, L., van Halder, I., Vétillard, F., Robinet, C. and David, G. 2019. A novel, easy method for estimating pheromone trap attraction range: Application to the pine sawyer beetle *Monochamus galloprovincialis*. *Agric. For. Entomol.* 21(1): 8–14.

Jiang, J.-A., Tseng, C.-L., Lu, F.-M., Yang, E.-C., Wu, Z.-S., Chen, C.-P., Lin, S.-H., Lin, K.-C. and Liao, C.-S. 2008. A GSM-based remote wireless automatic monitoring system for field information: A case study for ecological monitoring of the oriental fruit fly, *Bactrocera dorsalis* (Hendel). *Comput. Electron. Agric.* 62(2): 243–259.

Jurenka, R. A., Subchev, M., Abad, J. L., Choi, M. Y. and Fabriàs, G. 2003. Sex pheromone biosynthetic pathway for disparlure in the gypsy moth, *Lymantria dispar*. *Proc. Natl. Acad. Sci. U. S. A.* 100(3): 809–814.

Kareiva, P. M. 1983. Local movement in herbivorous insects: Applying a passive diffusion model to mark-recapture field experiments. *Oecologia* 57(3): 322–327.

Kareiva, P. M. and Shigesada, N. 1983. Analyzing insect movement as a correlated random walk. *Oecologia* 56(2–3): 234–238.

Kean, J. M. 2015. The effective sampling area of traps: Estimation and application. In: Beresford, R. M., Froud, K., Kean, J. and Worner, S. (Eds.), *Plant Protection Data Toolbox*. New Zealand Plant Protection Society, Chirstchurchm New Zealand, pp. 67–76.

Kruger, H. 2016. Designing local institutions for cooperative pest management to underpin market access: The case of industry-driven fruit fly area-wide management. *Int. J. Commons* 10(1): 176–199.

Liquido, N. J., McQuate, G. T., Birnbaum, A. L., Hanlin, M. A., Nakamichi, K. A., Inskeep, J. R., Ching, A. J. F., Marnell, S. A. and Kurashima, R. S. 2021. A review of recorded host plants of the oriental fruit fly, *Bactrocera dorsalis* (Hendel) (Diptera: Tephritidae) version 4.0. Available at USDA Compendium of Fruit Fly Host Information (CoFFHI).

Liquido, N. J., McQuate, G. T., Hanlin, M. A. and Suiter, K. A. 2020. Host plants of the Mediterranean fruit fly. *Ceratitis capitata* (Wiedemann) Version 4.0. Available at USDA Compendium of Fruit Fly Host Information (CoFFHI).

Manoukis, N. C. 2020. *TrapGrid*. https://github.com/manoukis/TrapGrid. Accessed 2022-12-26.

Manoukis, N. C. and Gayle, S. M. 2015. Attraction of wild-like and colony-reared *Bactrocera cucurbitae* (Diptera:Tephritidae) to cuelure in the field. *J. Appl. Entomol.* 140: 241–249.

Manoukis, N. C., Hall, B. and Geib, S. M. 2014. A computer model of insect traps in a landscape. *Sci. Rep.* 4: 7015.

Manoukis, N. C. and Hill, M. P. 2021. Probability of insect capture in a trap network: low prevalence and detection trapping with trapgrid. *ArXiv:211011432 Q-Bio*.

Manoukis, N. C., Siderhurst, M. and Jang, E. B. 2015. Field estimates of attraction of *Ceratitis capitata* to trimedlure and *Bactrocera dorsalis* to methyl eugenol in varying environments. *Environ. Entomol.* 44(3): 695–703.

Meats, A. and Edgerton, J. E. 2008. Short-and long-range dispersal of the Queensland fruit fly, *Bactrocera tryoni* and its relevance to invasive potential, sterile insect technique and surveillance trapping. *Anim. Prod. Sci.* 48: 1237–1245.

Published by Burleigh Dodds Science Publishing Limited, 2023.

Miller, J. R., Adams, C. G., Weston, P. A. and Schenker, J. H. 2015. *Trapping of Small Organisms Moving Randomly: Principles and Applications to Pest Monitoring and Management*. Springer, Heidelburg, Germany.

Miller, N. D., Yoder, T. J., Manoukis, N. C., Carvalho, L. A. F. N. and Siderhurst, M. S. 2022. Harmonic radar tracking of individual melon flies, *Zeugodacus cucurbitae*, in Hawaii: Determining movement parameters in cage and field settings. *PLoS ONE* 17(11): e0276987.

Muirhead-Thompson, R. C. 1991. *Trap Responses of Flying Insects: The Influence of Trap Design on Capture Efficiency*. Academic Press Ltd., London.

Nakamura, K. and Kawasaki, K. 1977. The active space of the *Spodoptera litura* (F.) sex pheromone and the pheromone component determining this space. *Appl. Entomol. Zool.* 12(2): 162–177.

Paini, D. R., Sheppard, A. W., Cook, D. C., De Barro, P. J., Worner, S. P. and Thomas, M. B. 2016. Global threat to agriculture from invasive species. *Proc. Natl. Acad. Sci. U. S. A.* 113(27): 7575–7579.

Plant, R. E. and Cunningham, R. T. 1991. Analyses of the dispersal of sterile Mediterranean fruit flies (Diptera: Tephritidae) released from a point source. *Environ. Entomol.* 20(6): 1493–1503.

Puche, H. and Su, N. Y. 2004. Estimating population density of the Formosan subterranean termite, *Coptotermes formosanus* (Isoptera: Rhinotermitidae) using the effective sampling area of in-ground monitoring stations. *Bull. Entomol. Res.* 94(1): 47–53.

Quilici, S. and Donner, P. 2012. Analysis of exotic fruit fly trapping networks. *EPPO Bull.* 42(1): 102–108.

Quinlan, M. M., Leach, A., Jeger, M. and Mumford, J. 2020. Pest risk management in trade: The opportunity from using integrated combined measures in a systems approach (ISPM 14). *Outlooks Pest Manag.* 31(3): 106–112.

Rigot, T. and Gilbert, M. 2012. Quantifying the spatial dependence of *Culicoides* midge samples collected by Onderstepoort-type blacklight traps: An experimental approach to infer the range of attraction of light traps. *Med. Vet. Entomol.* 26(2): 152–161.

Ruesink, W. G. 1976. Status of the systems approach to pest management. *Annu. Rev. Entomol.* 21(1): 27–44.

Schlyter, F. 1992. Sampling range, attraction range, and effective attraction radius: Estimates of trap efficiency and communication distance in coleopteran pheromone and host attractant systems1. *J. Appl. Entomol.* 114(1–5): 439–454.

Shelly, T. 2021. Distant-dependent capture probabilities of Mediterranean and oriental fruit flies (Diptera: Tephritidae) in a food-based trap in a Hawaiian mango orchard. *Fla. Entomol.* 104(2): 117–123.

Shelly, T., Epsky, N., Jang, E. B., Reyes-Flores, J. and Vargas, R. (Eds.). 2014. *Trapping and the Detection, Control, and Regulation of Tephritid Fruit Flies*. Springer, New York.

Shelly, T., Nishimoto, J., Diaz, A., Leathers, J., War, M., Shoemaker, R., Al-Zubaidy, M. and Joseph, D. 2010. Capture probability of released males of two *Bactrocera* species (Diptera: Tephritidae) in detection traps in California. *J. Econ. Entomol.* 103(6): 2042–2051.

Simberloff, D., Martin, J. L., Genovesi, P., Maris, V., Wardle, D. A., Aronson, J., Courchamp, F., Galil, B., García-Berthou, E., Pascal, M., Pyšek, P., Sousa, R., Tabacchi, E. and Vilà,

M. 2013. Impacts of biological invasions: What's what and the way forward. *Trends Ecol. Evol.* 28(1): 58–66.

Skellam, J. G. 1951. Random dispersal in theoretical populations. *Biometrika* 38(1-2): 196–218.

Steiner, L. F. and Lee, R. K. S. 1955. Large-area tests of a male-annihilation method for oriental fruit fly control. *J. Econ. Entomol.* 48(3): 311–317.

Suckling, D. M., Kean, J. M., Stringer, L. D., Cáceres-Barrios, C., Hendrichs, J., Reyes-Flores, J. and Dominiak, B. C. 2016. Eradication of tephritid fruit fly pest populations: Outcomes and prospects. *Pest Manag. Sci.* 72(3): 456–465.

Suckling, D. M., Stringer, L. D., Kean, J. M., Lo, P. L., Bell, V., Walker, J. Ts, Twidle, A. M., Jiménez-Pérez, A. and El-Sayed, A. M. 2015. Spatial analysis of mass trapping: How close is close enough? *Pest Manag. Sci.* 71(10): 1452–1461.

Turchin, P. and Odendaal, F. J. 1996. Measuring the effective sampling area of a pheromone trap for monitoring population density of southern pine beetle (Coleoptera: Scolytidae). *Environ. Entomol.* 25(3): 582–588.

van Havre, Z., Whittle, P., Jarrad, F., Low-Choy, S. and Mengersen, K. 2015. Designing surveillance for emergency response. In: *Biosecurity Surveillance: Quantitative Approaches.* CAB International, Wallingford, Oxford, UK, pp. 123–133.

Vargas, R. I. 1989. Mass production of Tephritid fruit flies. In: *World Crop Pests Fruit Flies Their Biol. Nat. Enemies Control.* Elsevier, New York, pp. 141–151.

Vargas, R. I., Leblanc, L., Piñero, J. C. and Hoffman, K. M. 2014. Male annihilation, past, present, and future. In: Shelly, T., Epsky, N., Jang, E. B., Reyes-Flores, J. and Vargas, R. (Eds.), *Trapping and the Detection, Control, and Regulation of Tephritid Fruit Flies.* Springer, Netherlands, pp. 493-511.

Vargas, R. I., Mau, R. F., Jang, E. B., Faust, R. M., Wong, L., Koul, O., Cuperus, G., Elliott, N. 2008. The Hawaii fruit fly areawide pest management programme. In: *Areawide Pest Manag. Theory Implement.* CABI Publishing, Wallingford, UK, pp. 300-325.

Wall, C. and Perry, J. N. 1987. Range of action of moth sex-attractant sources. *Entomol. Exp. Appl.* 44(1): 5–14.

Weldon, C. W., Schutze, M. K. and Karsten, M. 2014. Trapping to monitor tephritid movement: Results, best practice, and assessment of alternatives. In: *Trapping and the Detection, Control, and Regulation of Tephritid Fruit Flies.* Springer, Netherlands, pp. 175–217.

White, I. M. and Elson-Harris, M. M. 1992. *Fruit Flies of Economic Significance: Their Identification and Bionomics.* CAB International, Wallingford, UK.

Chapter 4

Developments in crop insect pest detection techniques

Richard W. Mankin, USDA-ARS, USA

1 Introduction

To meet the challenges agriculture faces now on a planet buffeted by climate change (Skendžić et al., 2021) requires organized efforts across its production and research sectors (Botha et al., 2014; Xia et al., 2022) to minimize food scarcity and reduce environmental and human health impacts (Frye et al., 2012) while conserving dwindling resources (Jordon et al., 2021). Part of this effort involves early detection and identification of targeted pests (Fedor et al., 2009; Lima et al., 2020) to avoid increases in pre- and post-harvest losses and economic harm expected from global warming (Deutsch et al., 2018). Problems of food scarcity are exacerbated because humans have higher metabolic rates than most animals of the same size (Gibbons, 2022) and thus must consume relatively more high-energy food.

An ideal insect-detection system for integrated pest management (IPM) accurately identifies and spatially targets pests early in their life cycles (Thenmozhi and Reddy, 2019; Liu and Wang, 2020) and is affordable relative to the value of

http://dx.doi.org/10.19103/AS.2022.0113.03
Published by Burleigh Dodds Science Publishing Limited, 2023.

the crop being protected, especially when considering the economic needs of the farmer, the impacts of management practices on neighboring and regional farms, and regulatory requirements at local and regional levels (Dara, 2019; Singerman and Rogers, 2020). Because cropland management units have different sizes, the distances over which insect pests are detected by a particular technology can affect their relevance to the data collection needs of farmers and pest managers, as well as government officials who have responsibility for regulatory and conservation oversight in regions adjacent to the croplands. There remains a considerable need for research on conservation in landscape habitats surrounding crops (Karp et al., 2018), especially with respect to edge effects in fragmented habitats (Laurance et al., 2018). A better understanding of predator and parasitoid abundance and behavior in such areas can help resolve uncertainties about best management practices. Fortunately, as discussed in different sections later, growers and pest managers have access to increasingly affordable, complementary insect-detection technologies, i.e. physical- and biochemical-energy detection tools augmented by artificial intelligence (AI) (Partel et al., 2019; Cheng et al., 2022) that makes use of machine learning and deep learning methods (LeCun et al., 2015; Karar et al., 2021; Roch et al., 2021) incorporating convolutional neural networks (Thenmozhi and Reddy, 2019; Spiesman et al., 2021) to quickly recognize patterns in images, acoustic recordings, and other detected signals which enable identification of pest insects. For purposes of this chapter, applications of complementary technologies to insect detection in agriculture first became available with the development of remote visual and radar sensing after the 1940s (Riley, 1989; Reynolds et al. 2005) and accelerated after 1983 when the Global Positioning System was made available for public use (Comparetti, 2011).

Traditional field surveys of pest insects by farmers and scouts, as well as observations of search activities by insect predators and parasitoids that can serve as survey proxies, are being combined with input from recently developed insect behavioral and sensory detection technology, computer-based pest identification, and rapidly improving knowledge of pest-insect host-seeking and feeding behaviors to enhance the effectiveness of pest management activities (Table 1). An example of such combinations is the inclusion of electronic sensors in pheromone traps. Pheromones and other semiochemicals such as kairomones, allomones, and antifeedants (Norin, 2007; Murali-Baskaran et al., 2018) are volatile organic chemicals (VOCs) or contact chemicals which influence behaviors of insects and other animals that sense them. Physiological correlates of such behaviors have been explored by examining neurophysiological signals in the insect central nervous systems antennae, eyes, and mechanoreceptor sensory organs. Of particular interest with respect to insect behavior, electroantennograms (EAGs) (Olsson and Hansson, 2013; Martinez et al., 2014) detect insect antennal electronic responses to

Table 1 Enhancements of traditional methods with complementary technologies for detecting and managing crop insect pests

Traditional sensing methods	Complementary technologies
Vision	Sensors of electromagnetic energy (light, heat, radar, X-ray, etc.) (additional coverage in other chapters of this book)
Audition touch/vibration	Sensors of sound/vibration energy
Taste/smell	Biochemical/molecular/nanobionic/electrophysiological sensors identifying DNA fragments, metabolites, and contact or volatile organic chemicals, electrical penetration graphs (additional coverage in other chapters of this book)
Observation of predator/parasitoid searching behavior activities	Smart traps, mating disruption, pest exclusion, push-pull, refuge plants (additional coverage in other chapters of this book)

volatiles. Electroretinograms (ERGs) (Yinon, 1971; Stowasser et al., 2015) detect retinal electronic responses to light of different wavelengths. Mechanoreceptors (Tuthill and Wilson, 2016) detect substrate vibrations and insect movements.

The combined usage of traditional methods and complementary technology is expanding rapidly. Attractant-baited 'smart' automated traps (Jiang et al., 2008) are now easily fitted with digital red/green/blue ((RGB) wavelength) cameras and piezoelectric or fiber optic acoustic/vibration sensor systems that enable rapid wireless transmission of information to remote data sites. Internet of things (IoT) technology (Potamitis et al., 2017; Rigakis et al., 2021a) connects and analyzes information collected by these sensor networks, enabling rapid processing and delivery of the sensor data back to farmers and other pest managers (Karar et al., 2022). The data they collect can be organized into easily interpreted formats that are superimposed on real-world displays, i.e. augmented reality (Huuskonen and Oksanen, 2018), helping farmers and managers interpret the data collected. Other technologies are used primarily in research at this time but may be adapted for field use in the future.

Typical detection ranges of traditional methods and complementary technology can be subdivided spatially along nano-, micro-, meso-, and macro-scales (Table 2). Nano-scale technology utilizes multiple biochemical tools and nanobionic sensors (Ang and Lew, 2022) operating internally in insect organs, including genetic marker-based analyses which can help distinguish among strains of a pest insect species that have similar morphologies but different host preferences and different levels of pesticide resistance, e.g. Nagoshi et al. (2017). Detection ranges in the micro-scale region include those of visual, auditory, vibrational, and olfactory sense organs which operate from a few millimeters to a few meters distance from the pest insect. In this range, sensors can be used for the detection of insects on individual crop plants or trees.

Several methods used to detect crop insect pests in the micro-scale range also have been applied in the detection of fruit damage during harvesting and later stages of the food supply chain (Mahanti et al., 2022). Such technologies have also been applied to assess the structural characteristics and health of plants (Mankin et al., 2018) and to detect disease vectors like mosquitoes or pollinators through wingbeat detection (Chen et al., 2014; Jakhete et al., 2017).

The meso-scale (Table 2) corresponds to the detection range of several radars, typically 2.5 km for individual insects or up to 10 km or more for insect swarms at dense concentrations (Riley, 1989). Vertical-entomological radars are regularly monitored in the United Kingdom (Chapman et al., 2010; Hu et al., 2016) to determine the timing and sizes of large-scale migrations. Meso-scale technology also includes weather radar, which Stepanian et al. (2020) used to detect airborne mayfly swarms. Drones and small airplanes with digital cameras are deployed to detect insects or insect-initiated damage over meso-scale ranges and, depending on height, can detect 1–10 cm-width objects

Table 2 Spatial scales of insect-detection technologies

Detection range	Traditional methods	Complementary technology
Nano-scale (<1 mm)	Biochemical/molecular	Single nucleotide polymorphisms, nanobionic sensors (additional coverage in other chapters of this book)
Micro-scale (<1 m)	Vision/Audition/Touch	Hand-held, stationary, or drone red/green/blue wavelength cameras (additional coverage in other chapters of this book) acoustic/vibration sensors, electrical penetration graph, or laser vibrometer systems for pest detection within crop plants, trees, and stored products
Meso-scale (<1 km)	Electroantennogram/electroretinogram/single olfactory cell electronics/smell/sound recorders Radar	Multicellular electrophysiological recordings of volatile organic chemicals drone/airplane-cameras multispectral scanning (additional coverage in other chapters of this book)
Macro-scale (>1000 km)	None	Satellite weather/behavior-based dispersal models (additional coverage in other chapters of this book)

on the ground (Huuskonen and Oksanen, 2018; Iost Filho et al., 2020). Lidar (Brydegaard and Jansson, 2019; Tauc et al., 2019) and similar devices using near-infrared light-emitting diodes (LEDs) (Rydhmer et al., 2022) operate at micro- and meso-scale detection ranges. The LED systems can be used as pseudo-acoustic sensors to reflect light from the wings of flying insects and analyze the time-varying reflections from beating wings (Chen et. al., 2014; Rydhmer et al., 2022). These have become useful tools for electronically collecting information from pheromone traps (Table 1).

The macro-scale encompasses landscapes that can be examined by satellites, which resolve objects on the ground with dimensions of 30 m (Huuskonen and Oksanen, 2018; Bright et al., 2020). Insects cannot be individually resolved at this scale, but satellites provide a series of Landsat images of specific regions that can be analyzed to identify areas of crops impacted by insect damage (Bright et al., 2020). Meso- and macro-scale technologies are two of the several methods for the remote detection of insect pests Riley (1989).

Traditional methods of insect monitoring continue as important pest management tools (Tables 1 and 2), but the accuracy of population forecasts is often dependent on dwindling expert knowledge (Zhang et al, 2019; Furuya et al., 2021). Farming operations could benefit from additional data obtained using newer insect-detection, AI, and IoT technologies (Osco et al., 2022) although concerns over ownership (Yu et al., 2021) and security (Alahmadi et al., 2022) of obtained data remain to be clarified. Moreover, traditionally developed knowledge of VOCs, visual, and auditory stimuli that affect feeding and mating behaviors of crop pests has been applied to augment pest control measures (Foster and Harris, 1997; Čokl and Millar, 2009; Davis et al., 2013; Silva et al., 2021). Recent improvements in technologies associated with formulation and dissemination of VOCs, and additionally, technologies associated with molecular biology, nanobionics, electronics, acoustics, and computer technology have benefited understanding of how pest insects detect and use VOCs and vibrational signals, and how these stimuli can be co-opted for pest management. Examples of recent applications of such technology are listed in Table 3. It should be noted for the last entry of Table 3, however, that care must be exercised in the broadcasting of sounds that disrupt mating to ensure that the signals remain below levels of urban noise pollution implicated as one of the potential causes for declines in nonpest invertebrate populations (Goulson, 2019; Raboin, 2021).

This chapter is arranged approximately according to sensory modality (visual, acoustic, vibration, olfaction, etc.) with consideration of how each modality is used in insect pest detection and management activities at different scales of detection.

Table 3 Examples of complementary technology in pest management activities that involve sensing and/or co-opting of stimuli used by pest insects for communication

Technology/Application	References
Enhanced use of volatile organic chemicals for pest insect mating disruption	Miller and Gut (2015); Preti et al. (2021); Higbee and Burks (2021); Gavara et al. (2022)
Seasonal monitoring of regional nonpest species co-attracted to pheromone traps detecting Plusiine pests	Shaw et al. (2021)
Kairomones for biological control	Martini et al. (2014); Murali-Baskaran et al. (2018); Da Silva et al. (2022)
Kairomone-enhanced pheromone lures	Walgenbach et al. (2021)
Floral scent real-time sampling	Kim et al. (2022); Zheng and Zhang (2022)
Electroantennograms and single olfactory cell recordings to determine the temporal patterns of emission and active spaces of natural or synthetically produced volatile organic chemicals	Mayer et al. (1987); Suckling et al. (2007); Mankin et al. (1991); Yang et al. (2022)
Broadcast of sound or vibrations for insect pest mating disruption	Polajnar, et al. (2015); Lujo, et al. (2016); Laumann et al. (2018); Takanashi et al. (2019); Mazzoni et al. (2019); Avosani et al. (2022)

2 Camera systems for pest detection at micro-scale ranges

Moth pheromone traps were initially expected to selectively capture a single species, but experience demonstrated that some pheromones captured insects of multiple, closely related species depending on the geographical region. Camera systems were later incorporated to send images of trapped insects to remote locations where individuals of each caught species were identified and counted by computer analyses or by human identification (Guarnieri et al., 2011). However, it became apparent that fully automated approaches were needed to discriminate among the large numbers of different species captured in multiple regions where trapping was used as a detection tool. In addition, general scouting operations could benefit from the automated identification of insects photographed during surveys. Consequently, collections of multiple datasets with large numbers of images of different insect species and stages, including the Indian Council of Agricultural Research and the National Bureau of Agricultural Insect Resources dataset (Karar et al., 2021) were analyzed for specific features to distinguish among species. Machine learning algorithms were developed (Karar et al., 2021) that identified images of important insect crop pests independently of orientation (Deng et al., 2018: Thenmozhi and Reddy, 2019). Similar collections were developed for other pests worldwide, including VisDrone-VDT2018 with 80 000 drone-collected images of insect pests (Zhu et al., 2019), PlantDiseaseNet, with 79 275 pest images collected under various meteorological conditions (Liu and Wang, 2020), IP102 (Wu et al., 2019), with 75 000 images belonging to 102 pest categories, Xie2 (Xie et al., 2018) with 4000 images of different pest species, and APHID-4K with 4000 images of aphids (Du et al., 2021), who also developed machine learning methods to detect images of aphid clusters. The individual aphids could not be distinguished from each other due to the limited resolution of the camera. Kasinathan et al. (2021) combined several machine learning techniques to classify and detect insects in digital camera images collected in corn, soybean, and wheat fields.

Deep learning methods usually require access to more data than machine learning due to a larger number of neural network layers that must be calibrated. However, they have been used successfully with several of the largest data sets above to classify crop insect pest species. Li et al. (2021), for example, describe the use of deep learning methods to analyze images collected from immatures and adults of ten crop pest species obtained from Bing, Pest24, TPest, and AgriPest.

Several camera-based pest insect detection systems are listed in Table 1. When applying such technology to insect detection, there is a need to recognize the variability of insect morphological characteristics of the same species in different orientations at different life stages. When trapped insects are imaged, discoloration by liquid trapping agents, loss of legs, insect clumping, and debris introduce specific features that cannot be incorporated into the identification process unless such images are also included in the training dataset (Nanni et al., 2022).

Complementary technologies that have been tested but are not yet in common usage include a guarded electrical probe developed by Thomas et al. (2021) to measure electrical impedance. The device was used in experiments to noninvasively identify stalks of maize, *Zea mays* L., that had been partially hollowed out by *Ostrinia nubilalis* Hübner (Lepidoptera: Pyralidae) larvae. In an earlier approach, Labatte et al. (1992) applied X-ray imaging methods to map cavities produced within maize plants by *O. nubilalis* feeding behaviors.

3 Drone/camera systems for pest detection at meso-scale detection ranges

Uses of small drones for IPM in a variety of crop pest insect applications are reviewed by Iost Filho (2020). Drone systems are particularly useful for preparing digital maps of pest insect 'hot spots' that can be precisely targeted for treatment. One example of RGB-camera-based technology to detect crop insect pests (Kalischuk et al., 2019) involved assessments of insect pest damage that caused increases in plant disease severity and plant stress. Testing was conducted in watermelon fields comprising 15–38 ha, using DJI Z3 digital camera operated from a Matrice 100 quadcopter drone (Da-Jiang Innovations Science and Technology Co., Shenzhen, China).

4 Landsat systems for detection of pests at macro-scale detection ranges

Landsat satellite technology has been in operation since 1984, providing the capability for long-term analyses of changes in forest coverage over time (e.g. Senf et al., 2017). Multiple image processing algorithms have been developed to make the data for such analyses easily accessible to researchers (Kennedy et al., 2018). Because crop insect pests are not large enough to be directly detectable at Landsat-scale detection ranges, this technology is not yet widely used by farmers and researchers, although there is potential that future generations of Landsat satellite cameras with higher resolution may enable development of such capabilities.

5 Sound- and vibration-sensors for pest detection

In cropland environments, the weak sounds and vibrations produced by insects in air, soil, or plant tissues are embedded in soundscapes (Pijanowski et al., 2011) and vibroscapes (Šturm et al., 2022) containing varying levels of human-produced background noise, often more audible than the pests targeted for detection. Insects that produce sufficiently audible airborne sounds are easily recorded by currently available, portable passive acoustic monitoring (PAM) systems (Sugai et al., 2020). Acoustic signal analyses and machine learning toolkits have been developed to process the recorded signals, filter out background noise, and identify important features by which different sound-producing animals and insects can be compared. Ulloa et al. (2021) developed a toolkit that can scan large audio datasets and find regions of interest containing specific features that are user-programmable. Wu et al. (2022) developed a kit that contains features and sounds collected from multiple PAM databases to which the user can compare their own acoustic signals. Roch et al. (2021) reviewed supervised and unsupervised forms of machine learning that are most widely used to identify features of relevance to bioacoustics, oceanography, and music.

When targeted insect pests move and feed in soil, plant tissue, or other opaque substrates, the effects of airborne background noise can be reduced by connecting customized piezoelectric attachments or waveguides directly to the substrates (Mankin et al., 2011; Rigakis et al., 2021a). A similar effect occurs when pest-generated vibrations inside a tree are transferred onto a substrate such as an optical fiber that is wrapped around the tree trunk (Ashry et al., 2020). Vibrations produced by insect larvae as they move and feed inside a tree intermittently alter the effective refractive index of fiber wrapped around the tree and, when optical pulses are transmitted at one end of the fiber, the Rayleigh backscattered traces of the pulses show intermittent intensity fluctuations (speckles) that can be monitored and interpreted. Also, the behavior of sapsucking hemipteran insects can be detected by means of electrical penetration graphs, which can be used to guide the development of treatments that reduce feeding (Maluta et al., 2022) or identify mechanisms of plant resistance (Willet et al., 2016; Mercer et al., 2021).

Air- and substrate-borne signals collected at both meso- and micro-scale ranges typically are processed further to reduce background noise (Mankin et al., 2021). As with camera-produced images, machine learning or similar automated analysis tools (Romano et al., 2013; Jordan and Mitchell, 2015; Phung et al., 2017; Dong et al., 2018; Rathore et al., 2019; Bianco et al., 2019; Roch et al., 2021) can be applied to identify and compare salient features of the collected signals, including bursts of impulses from sliding, tapping, and snapping behaviors that distinguish between pest and nonpest insect larvae and also discard background noise (Mankin et al., 2008a,b). The behaviors of adult hemipteran pests that transmit vibrational signals along stems or tree branches can be monitored using vibration sensors assisted by traditional and machine-learning-based signal processing methods (Mankin et al., 2020b). In addition, the capability to disrupt mating communications has been demonstrated in multiple insect species that utilize vibrational signals during mating (Lujo et al., 2016; Zaffaroni Caorsi et al., 2021). The use of sensors that detect airborne or vibrational signals has been enhanced by the concomitant development of inexpensive amplifiers and user-friendly computer interfaces (Mankin et al., 2009a,b, 2010; Potamitis et al., 2009 Potamitis, 2013; Potamitis and Rigakis, 2014).

The instrument used most frequently to record insect vibrational signals in field environments before 2020, the AED-2010 (AEC, Inc. Fair Oaks, CA, USA), ceased production in 2019. Current replacements include Tree Vibes (Insectronics, Crete, Greece), used for detection of insect pests in trees (Rigakis et al., 2021a; Karar et al., 2022) and stored products (Mankin et al., 2021), and the Postharvest Insect Detection System (Custom Engineered Solutions, W Hempstead, NY, USA), used for stored product insect detection Mankin et al. (2020a). Other potential pest insect detection devices for which no scientific reports are yet available are listed in Table 4. Background information about the pseudo-acoustic sensor and signal processing technology used for the FarmSense device is provided in the studies by Batista et al. (2010), Katzanek (2020), and Mercer et al. (2021).

Table 4 Commercially available acoustic-, pseudo-acoustic-, and camera-based pest insect detection systems for field use

Sensor type	Company Name and Location	Website
Accelerometer	Smart Palm, Prince Sultan Univ., Riyadh, Saudi Arabia	https://doi.org/10.3390/agronomy10070987
Acoustic (unknown)	IoTree Pest Detection, Rockville, MD, USA	https://www.agrint.net
Cell phone	Palmear Al Karama, Jordan	https://www.palmear.ai/
Cell phone	Permia Sensor, London, UK	https://permiasensing.com/
Pseudo-acoustic	FarmSense flying insect Detection, Riverside, CA, USA	https://farmsense.io
Unknown	iPalm Digital Platform, Russell IPM, Flintshire, UK	https://russell-iot.com/portfolio/ipalm/
Camera	Trapview, EFOS, Slovenia	https://www.trapview.com/v2/en/
Camera	iSCOUT, Pessi Instruments, Weiz, Austria	https://metos.at/iscout/
Camera	SightTrap, Insects Limited, Westfield, IN, USA	https://www.insectslimited.com/sighttrap
Camera	Spensa Technologies, Burnsville, MN, USA	https://dtn2.flywheelsites.com
Camera	SPOTTA, Cambridge, UK, and Los Angeles, CA, USA	https://www.spotta.com

It should be noted that several of the systems described earlier, including the Smart Palm system for the detection of *Rhynchophorus ferrugineus* (Olivier; Coleoptera: Dryophthoridae) larval vibrations in trees (Koubaa et al., 2020), also employ sensors that detect temperature, humidity, and pH in addition to vibration and sound. Such sensors provide additional, cost-effective information to farmers and pest managers, which can enable 'smart farming' (De Alwis et al., 2022) technologies to play a broader role in agriculture in the future.

6 Case studies: augmenting traditional pest detection and biological control with nano-scale- and micro-scale-sensor technologies

Combinations of traditional pest management methods and complementary technologies can provide broadscale sensing tools which improve understanding of pest behavior over macro- to nano-scale detection ranges (Table 2). An example is the macro-scale modeling of yearly migrations of multiple strains of *Spodoptera frugiperda* (J. E. Smith) (Lepidoptera: Noctuidae) within North America. Early generations are first seen in overwintering areas in southern Texas and Florida. In late winter, first and later generations follow warming temperatures and prevailing winds through spring and summer, causing crop damage in locations as far north as Canada (Westbrook et al., 2019). Two strains of *S. frugiperda* have been identified, one found primarily on grasses, including rice, and one primarily on corn (Nagoshi et al., 2007b). Adults captured by pheromone traps were tested by genetic marker analysis to distinguish between rice and corn strains. The corn strains were subdivided further into four haplotype subgroups using cytochrome C oxidase subunit I markers and single nucleotide polymorphisms (Nagoshi et al., 2007a, 2017), sensing in the nano-scale detection range (Table 2). The combinations of such technologies in the study provided fine-grained information about the migration pathways of each strain for geostatistical dispersal models that Westbrook et al. (2019) developed to predict the magnitudes and times of pest dispersal. It should be noted also that plots of insect population distributions obtained by Spatial Analyses by Distance Indices and other geostatistical analyses can similarly provide important information for targeting crop pest infestations (Mankin et al., 2007).

Traditional biological control efforts also benefit from the complementary use of ERG and EAG electrophysiological tools, as well as three-dimensional printing of specially designed visual and pheromone traps to improve understanding of how insect pests employ visual and chemical cues to orient to their host crops. Combined behavioral and electrophysiological studies of a devastating citrus pest, *Diaphorini citri* (Hemiptera: Liviidae) (Allan et al., 2020), have enabled a better understanding of *D. citri* attraction to differently colored sticky traps, for example. It was determined that traps coated with magnesium oxide and/or barium sulfate were more attractive due to the attraction of *D. citri* to ultraviolet light, which is strongly reflected by such coatings (George et al., 2020). *Diaphorina citri* orientation to light was affected by both wavelength and polarization (Paris et al., 2017). George et al. (2016) discovered also that two degradation products of citrus tree VOCs, formic and acetic acid, play a role in *D. citri* attraction to host trees. Similarly, using EAG methods, Yang et al. (2019) identified four highly active VOCs produced by soybean (*Glycine max* (L) Merrill. and kudzu (*Pueratia montana* (Lour.) Merr. var. Lobata (Willd.) plants that were strongly attractive to the invasive Kudzu bug, *Megacopta cribraria* (Fabricius) (Hemiptera: Platispidae) in olfactometer bioassays. Additional studies of insect pests physiological and behavioral responses may lead to further improvements in trapping systems.

Three-dimensional-printed traps have been developed to improve the detection of the bacterium causing huanglongbing (citrus greening disease, HLB) that causes the death of citrus trees (Singerman and Rogers, 2020). The bacterium is vectored by an invasive citrus pest, *D. citri* Kuyayama (Hemiptera: Liviidae). *Diaphorina citri* were trapped in laboratory, greenhouse, and field environments (Snyder et al., 2022) by three-dimensional traps that contain a liquid preservative which does not quickly degrade the disease pathogen. Such traps not only detect the presence of *D. citri* in citrus groves but also determine whether they are vectors of the bacterium that produces HLB. Management of citrus in areas where HLB is not yet endemic is strongly dependent on how rapidly the disease is spread. The three-dimensional traps can be printed and deployed rapidly to monitor quickly the spread of HLB in a small area and remove infected trees.

For silverleaf whitefly, *Bemisia tabaci* (Biotype B) (Gennadius) (Hemiptera: Aleyrodidae) and other pests that have developed increased resistance to insect growth regulators and neonicotinoids (Dennehy et al., 2010; Perring et al., 2018), there is a need to maintain the presence of predators, parasitoids, and other ecosystem services that would normally reduce pest populations (Bradshaw et al., 2021). Areas with high insecticide usage would benefit from improved habitat manipulation and other biological control techniques to reduce economic losses (Naranjo, 2001). 'Push-pull' strategies have been tested in different ecological contexts, some of which yielded successful results (Oji and Mohamed, 2005; Li et al., 2014). Habitat manipulation and ecosystem engineering (Zhong et al., 2022) that provide or eliminate visual and olfactory cues could facilitate improvements in 'push-pull' effectiveness (Potting et al., 2005), and improved VOC technology and behavioral bioassays have fostered such progress. Recently, mustard plants and oils were found to repel *B. tabaci* (Legaspi et al., 2016) and thus could be used as part of a 'push-pull' strategy welcomed by organic crops growers (Allan 2018; Khan et al., 2014a,b; Tyler-Julian et al., 2018). Plant surface repellency combined with VOC repellency has been successful against *B. tabaci* in studies applying limonene-scented kaolin to tomato crops (Johnston et al., 2022). Natural enemy refuges have been of continued interest for biological control in studies such as those by Corbett and Rosenheim (1996), Meagher et al. (2019), Juliano and Gratton (2020), and Clem and Harmon-Threatt (2021). Increased technological capability to monitor populations of natural enemies in refuges can benefit refuge maintenance efforts. Mechanical barriers such as exclusion screening to keep crop plants free of specific insect pests have been cost-effective where a few pests can vector diseases over large areas (Ebert et al., 2020) and may be used increasingly in areas where the plants would benefit from reductions in solar radiation or increases in relative humidity (Mahmood et al., 2018). Crops within such exclusion barriers benefit from traditional or complementary methods of detecting pest insects that may have evaded barriers to entry.

A recent focus in biological control is the development of a better understanding of the effects of host-associated differentiation (HAD) of pest strains or biotypes on their natural enemies (Thompson et al., 2022). A crop pest may experience HAD when it begins to mate preferentially on a particular host crop, recognizing highly specific tactile, olfactory, and taste cues of the host. Advances in knowledge of chemical cues that affect natural enemy foraging behavior have led to the recognition that, when crop pests evolve through HAD, their natural enemies may be affected in ways that cause them to exhibit HAD as well (Harrison et al., 2022). The adaptation process may be more rapid than previously thought, as recent studies have demonstrated that continuous adaptation to rapid environmental change occurs in *Drosophila melanogaster* Meigen (Diptera: Drosophilidae) (Rudman et al., 2022), which has a relatively short 10-day generation time. Better knowledge of HAD and the associated differences that occur in herbivore-induced plant volatiles (HIPVs) can help researchers select natural enemies that target the insect pest more precisely. Technological improvements in the detection, analysis, and manipulation of HIPVs thus are assisting the process of selecting natural enemies with greater capacity to reduce pest insect populations (Stelinski et al., 2019).

7 Conclusion

Physical energies and processes that can be used to detect pest insects in agricultural settings include electromagnetic energy (light, infrared, radar, etc.), sound and vibration energy, and biochemical reactions and syntheses. All of these are used singly and in combination to complement traditional information from human sensory modalities of vision, audition, touch and vibration, olfaction, and taste. Examples of modern sensor technology are described that enable pest insect detection over a greater range (scale) of distances, often with greater precision than human sensory systems provide. The use of electronic sensors also enables real-time data collection, storage, and analyses by computer systems. Machine learning and deep learning software algorithms have been developed that enable the identification and use of important features of the detected signals to precisely identify pests and distinguish pest species from each other. A goal is to target applications of control treatments to locations where the pest insects are present, thereby reducing treatment costs and reducing harm to nontarget organisms. The costs of purchasing and using the new technology in agricultural environments are decreasing, especially when sensors that detect temperature, humidity, pH, and other environmental conditions important to crops are included together in systems that relay information to farmers and managers. In addition, enhanced knowledge of pest distributions obtained from the use of these sensors enables more effective use of 'push-pull' technologies and other biological control methods.

8 Future trends in research

Deep learning is not the 'Amulet of Yendor,' but much of the upcoming effort to distinguish among insect species by use of data stored as images, sounds, and/or vibrations is likely to apply deep learning or similar methods to assist in crop insect pest detection unless they are quickly supplanted by even more precise methods. If such methods are developed, they are likely to incorporate signal features that entomologists have already identified to be singularly important for pest identification. Exclusion screening or screenhouses may become more important in regions that have become hotter and drier due to climate change, especially if the excluded insects are important pests or vectors of plant disease and have become resistant to pesticides. Out of sight out of mind, soil invertebrates may slowly gain in interest to pest managers, as their significant roles in crop ecosystems become more apparent (Johnson et al., 2007; Inyang et al., 2019; Veen et al., 2019; Helmberger et al., 2022 Mankin, 2022).

Finally, the arms race between crop insect pest resistance and pesticide manufacturer ingenuity is likely to continue, as noted by Sparks et al. (2019), complicated by the effects of climate change that may benefit the pests while harming crops and increasing farmer costs of controlling and monitoring pests with higher metabolic rates (Deutsch et al., 2018; Skendžić et al., 2021). A standoff is the most likely result, although an improved understanding of the bioactivity of commercially developed insecticides and natural products may tip the balance in favor of the farmer.

Table 5 Locations of crop pest detection research centers by region

Region	Institution	References
Europe		
Slovenia	Univ. Ljubljana, Slovenia	Virant-Doberlet et al. (2019)
Greece	Hellenic Mediterranean Univ., Chania, Crete; Univ. West Attica, Athens, Greece	Potamitis et al. (2019); Rigakis et al. (2021b)
Italy	Univ. Trento, Italy	Avosani et al. (2021)
Switzerland	ETH, Zurich	Maeder et al. (2022)
France	CNRS, Paris	Šturm et al. (2022)
Germany, UK	Hoch. Geisenheim Univ. Geisenheim; Univ. York, York; Oxford Univ., Oxford	Görres and Chesmore (2019); Mortimer (2019)
North America		
(Western to	UCLA, Los Angeles, CA	Montgomery et al. (2021)
Eastern centers)	USDA, Parlier, CA; Adair, OK	Higbee and Burks (2021)
	UC Riverside, CA, Univ. BC, Vancouver, Canada, Northern Ariz. Univ., Flagstaff, AZ, Univ.; Canterbury, NZ	Liao et al. (2022)Bedoya et al. (2021)
	U. Missouri, Columbia, MO	Kollasch et al. (2020)
	Univ. Toronto, Toronto, ON; NAFRO, Tsukuba, Japan	Nakano and Mason (2018)
	Carleton, Univ., Ottawa, ON	Low et al. (2021)
	USDA, Gainesville, FL; FAMU, Tallahassee, FL	Inyang et al. (2019)
	Argonne Nat. Lab., Binghamton, NY; Cornell Univ., Ithaca, NY	Zhou et al. (2022)
	Stevens Inst., Hoboken, NY	Sutin et al. (2019)
South America	EMBRAPA, Sao Paulo, Brazil	Laumann et al. (2018)
Asia		
India	Janta Vedic College, Baraut, India	Banga et al. (2020)
Sri Lanka	Rinzen Lab., Rattanapitya, Sri Lanka	Siriwardena et al. (2010)
Australia	Macquarie Univ., Sydney, AU; La Trobe, Univ., Melbourne, AU	Wignall and Taylor (2011); Lubanga et al. (2021)
New Zealand	Massey Univ., Auckland, NZ	Wignall and Herberstein (2022)
Japan	NAFRO, Hiroshima; Kyoto Univ. Kyoto	Kawakita and Ichikawa (2019)
China	Shaanxi Normal Univ. Xi'an; Chinese Acad. Sci, Beijing	Sun et al. (2018); Wu et al. (2022)

While looking forward, it is worthwhile to think back 50 years to the partially fulfilled promise of the Green Revolution (Vandervoet, 2022). One can only hope that scientists, engineers, and governments also can develop improved methods to reverse the ongoing decline of ecosystem services including clean air, potable water, and high-quality soil (Bradshaw et al., 2021) as human populations and consumption increase and the frequencies of occurrence of hotspots of acute food insecurity continue to escalate (WFP and FAO, 2022).

9 Where to look for further information

Centers where arthropod acoustic or vibrational studies of practical interest are conducted frequently include those listed in Table 5, arranged geographically along with references to relevant publications.

Recently, many studies have focused on the development of computer methods to identify digital images of insect pests due to the decreased costs of collecting images. Such studies are discussed in other chapters of this book.

10 References

Alahmadi, A. N., Rehman, S. U., Alhazmi, H. S., Glynn, D. G., Shoaib, H. and Solé, P. (2022). Cyber-security threats and side-channel attacks for digital architecture, *Sensors (Basel)* 22(9), 3520. DOI: 10.3390/s22093520.

Allan, S. A. (2018). Behavioral-based control of insect crop pests. In: Cordoba-Aguilar, A., Gonzalez-Tokman, D. and Gonzalez-Santoyo, I. (Eds.), *Insect Behavior: From Mechanisms to Ecological and Evolutionary Consequences*, Oxford University Press, Oxford, pp. 309-331.

Allan, S. A., George, J., Stelinski, L. L. and Lapointe, S. L. (2020). Attributes of yellow traps affecting attraction of *Diaphorina citri* (Hemiptera: Liviidae), *Insects* 11(7), 452. DOI: 10.3390/insects11070452.

Alwis, S. D., Hou, Z., Zhang, Y., Na, M. H., Ofoghi, B. and Sajjanhar, A. (2022). A survey on smart farming data, applications and techniques, *Comput. Ind.* 138, 103624.

Ang, M. C.-Y. and Lew, T. T. S. (2022). Non-destructive technologies for plant health diagnostics, *Front. Plant Sci.* 13, 884454. DOI: 10.3389/fpls.2022.884454.

Ashry, I., Mao, Y., Al-Feghid, Y., Al-Shawaf, A., Al-Bagshi, M., Al Brahim, S., Ng, T. K. and Ooi, B. S. (2020). Early detection of red palm weevil using distributed optical sensor, *Sci. Rep.* 10(1), 3155. DOI: 10.1038/s41598-020-60171-7.

Avosani, S., Berardo, A., Pugno, N. M., Verrastro, V., Mazzoni, V. and Cornara, D. (2021). Vibrational disruption of feeding behaviors of a vector of plant pathogen, *Entomol. Gen.* 41(5), 481-495.

Avosani, S., Ciolli, M., Verrastro, V. and Mazzoni, V. (2022). Application of vibrational signals to study and manipulate an insect vector: The case of *Philaenus spumarius* (Hemiptera: Aphrophoridae), *Pest Manag. Sci.* 78(10), 4061-4071. DOI: 10.1002/ps.7025.

Banga, K. S., Kotwaliwale, N., Mohapatra, D., Babu, V. B., Giri, S. K. and Bargale, P. C. (2020). Assessment of bruchids density through bioacoustic detection and artificial neural network (ANN) in bulk stored chickpea and green gram, *J. Stored Prod. Res.* 88, 101667.

Batista, G. E., Campana, B., and Keogh, E. (2010). Classification of live moths combining texture, color and shape primitives, 2010 Ninth International Conference on Machine Learning and Applications, pp. 903-906, DOI: 10.1109/ICMLA.210.142.

Bedoya, C. L., Hofstetter, R. W., Nelson, X. J., Hayes, M., Miller, D. R. and Brockerhoff, E. G. (2021). Sound production in bark and ambrosia beetles, *Bioacoustics* 30(1), 58-73. DOI: 10.1080/09524622.2019.1686424.

Bianco, M. J., Gerstoft, P., Traer, J., Ozanich, E., Roch, M. A., Gannot, S. and Deledalle, C. A. (2019). Machine learning in acoustics: Theory and applications, *J. Acoust. Soc. Am.* 146(5), 3590.

Botha, N., Klerkx, L., Small, B. and Turner, J. A. (2014). Lessons on transdisciplinary research in a co-innovation programme in the New Zealand agricultural sector, *Outlook Agric.* 43(3), 219-223.

Bradshaw, C. J. A., Ehrlich, P. R., Beattie, A., Ceballos, G., Christ, E., Diamond, J., Dirzo, R., Ehrlich, A. H., Harte, J., Harte, M. E., Pyke, G., Raven, P. H., Ripple, W. J., Saltré, F., Turnbull, C., Wackernagle, M., and Blumstein, D. T. (2021) *Front. Conservation Sci.* 1, 615419.

Bright, B. C., Hudak, A. T., Meddens, A. J. H., Egan, J. M. and Jorgensen, C. L. (2020). Mapping multiple insect outbreaks across large regions annually using Landsat time series data, *Remote Sens.* 12(10), 1655. DOI: 10.3390/rs12101655.

Brydegaard, M. and Jansson, S. (2019). Advances in entomological laser radar, *J. Eng.* 2019(21), 7542-7545.

Chapman, J. W., Nesbit, R. L., Burgink, L. E., Reynolds, D. R., Smith, A. D., Middleton, D. R. and Hill, J. K. (2010). Flight orientation behaviors promote optimal migration trajectories in high-flying insects, *Science* 327(5966), 682-685.

Chen, Y., Why, A., Batista, G., Mafra-Neto, A. and Keogh, E. (2014). Flying insect classification with inexpensive sensors, *J. Insect Behav.* 27(5), 657-677.

Cheng, X., Doosthosseini, A. and Kunkel, J. (2022). Improve the deep learning midels in forestry based on explanations and expertise, *Front. Plant Sci.* 13, 902105. DOI: 10.3389/fpls.2022.902105.

Clem, C. S. and Harmon-Threatt, A. N. (2021). Field borders provide winter refuge for beneficial predators and parasitoids: A case study on organic farms, *J. Insect Sci.* 21(3), 2. DOI: 10.1093/jisesa/ieab027.

Čokl, A. A. and Millar, J. G. (2009). Manipulation of insect signaling for monitoring and control of pest insects. In: Ishaaya, I. and Horowitz, A. R. (Eds.), *Biorational Control of Arthropod Pests*. Springer Science + Business Media BV, New York, pp. 279-316.

Comparetti, A. (2011). Precision agriculture: Past, present and future. In: *Proceedings of the Agricultural Engineering and Environmental Conference*, Kaunas, Lithuania, September 22-23, 2011, pp. 216-230.

Corbett, A. and Rosenheim, J. A. (1996). Impact of a natural enemy overwintering refuge and its interaction with the surrounding landscape, *Ecol. Entomol.* 21(2), 155-164.

Da Silva, I. T. F. A., Magalhães, D. M., Borges, M., Laumann, R. A., De Brito, C. H., Miranda, J. E. and Blassioli-Moraes, M. C. (2022). Exploitation of herbivoreinduced cotton volatiles by the parasitic wasp *Bracon vulgaris* reveals a dominant chemotactic effect of terpenoids, *Biocontrol* 67(2), 135-148. DOI: 10.1007/s10526-022-10135-9.

Dara, S. K. (2019). The new integrated pest management paradigm for the modern age, *J. Integr. Pest Manag.* 10(1), 12.

Davis, T. S., Crippen, T. L. , Hofstetter, R. W. and Tomberlin, J. K. (2013). Microbial volatile emissions as insect semiochemicals, *J. Chem. Ecol.* 39(7), 840–859.

Deng, L., Wang, Y., Han, Z. and Yu, R. (2018). Research on insect pest image detection and recognition based on bio-inspired methods, *Biosyst. Eng.* 169, 139–148.

Dennehy, T. J., Degain, B. A., Harpold, V. S., Zaborac, M., Morin, S., Fabrick, J. A., Nichols, R. L ., Brown, J. K., Byrne, F. J. and Li, X. C. (2010). Extraordinary resistance to insecticides reveals exotic Q biotype of *Bemisia tabaci* in the New World, *J. Econ. Entomol.* 103(6), 2174–2186.

Deutsch, C. A., Tewksbury, J. J., Tigchelaar, M., Battisti, D. S., Merrill, S. C., Huey, R. B. and Naylor, R. L. (2018). Increase in crop losses to insect pests in a warming climate, *Science* 361(6405), 916-919.

Dong, X., Yan, N. and Wei, Y. (2018). Insect sound recognition based on convolutional neural network. In: *Proceedings of the 2018 IEEE 3rd International Conference on Image, Vision and Computing (ICIVC)*, Chongqing, China, June 27-29, 2018, pp. 855-859.

Du, J., Liu, L., Li, R., Jiao, L., Xie, C. and Wang, R. (2021). Towards densely clustered tiny pest detection in the wild environment, *Neurocomputing* 12, 12. DOI: 10.1016/j.neucom.2021.12.012.

Ebert, T. A., Waldo, L., Stanton, D. and Schumann, A. W. (2020). Screen mesh size for exclusion of *Diaphorina citri* (Hemiptera: Liviidae) in citrus production, *J. Econ. Entomol.* 113(4), 2026–2030.

Fedor, P., Vaňhara, J., Havel, J., Malenovský, I. and Spellerberg, I. (2009). Artificial intelligence in pest insect monitoring, *Syst. Entomol.* 34(2), 398–400.

Foster, S. P. and Harris, M. O. (1997). Behavioral manipulation methods for insect pest-management, *Annu. Rev. Entomol.* 42, 123–146.

Frye, C., Bo, E., Calamandrei, G., Caizà, L., Dessi-Fulgheri, F., Fernández, M., Fusani, L., Kah, O., Kajta, M., Le Page, Y., Patisaul, H. B., Venerosi, A., Wojtowicz, A. K. and Panzica, G. C. (2012). Endocrine disrupters: A review of some sources, effects, and mechanismof actions on behavior and neuroendocrine systems, *J. Neuroendrocrinol.* 24(1), 144–159.

Furuya, D. E. G., Ma, L., Pinheiro, M. M. F., Gomes, F. D. G., Gonçalvez, W. N., Junior, J. M., Rodrigues, D. de C., Blassioli-Moraes, M. C., Michereff, M. F. F., Borges, M., Alaumann, R. A., Ferreira, E. J., Osco, L. P., Ramos, A. P. M., Li, J. and de Jorge, A. C. (2021). Prediction of insect-herbivory-damage and insect-type attack in maize plants using hyperspectral data, *Int. J. Appl. Earth Observ. Geoinf.* 105, 102609.

Gavara, A., Navarro-Llopis, V., Primo, J. and Vacas, S. (2022). Influence of weather conditions on *Lobesia botrana* (Lepidoptera: Tortricidae) mating disruption dispensers' emission rates and efficacy, *Crop Prot.* 155, 105926.

George, J., Paris, T. M., Allan, S. A., Lapointe, S. L. and Stelinski, L. L. (2020). UV reflective properties of magnesium oxide increase attraction and probing behavior of Asian citrus psyllids (Hemiptera:Liviidae), *Sci. Rep.* 10(1), 1890.

George, J., Robbins, P. S., Allessandro, R. T., Stelinski, L. L. and Lapointe, S. L. (2016). Formic and acetic acids in degradation products of plant volatiles elicit olfactory and behavioral responses from an insect vector, *Chem. Senses* 41(4), 325–338.

Gibbons, A. (2022). The calorie counter, *Science* 375(6582), 710-713.

Görres, C. M. and Chesmore, D. (2019). Active sound production of scarab beetle larvae opens up new possibilities for species-specific pest monitoring in soils, *Sci. Rep.* 9(1), 10115.

Goulson, D. (2019). The insect apocalypse and why it matters, *Curr. Biol.* 29(19), R967–R971.

Guarnieri, A., Maini, S., Molari, G. and Rondelli, V. (2011). Automated trap for moth detection in integrated pest management, *Bull. Insectol.* 64(2), 247–251.

Harrison, K., Tarone, A. M., DeWitt, T. and Medina, R. F. (2022). Predicting the occurrence of host-associated differentiation in parasitic arthropods: A quantitative literature review, *Entomol. Exp. Appl.* 170(1), 5–22.

Helmberger, M. S., Miesel, J. R., Tiemann, L. K. and Grieshop, M. J. (2022). Soil invertebrates generate microplastics from polystyrene foam debris, *J. Insect Sci.* 22(1), 1–6.

Higbee, B. S. and Burks, C. S. (2021). Individual and additive effects of insecticide and mating disruption in integrated management of Navel Orangeworm in almonds, *Insects* 12(2), 188. DOI: 10.3390/insects12020188.

Hu, G., Lim, K. S., Horvitz, N., Clark, S. J., Reynolds, D. R., Sapir, N. and Chapman, J. W. (2016). Mass seasonal bioflows of high-flying insect migrants, *Science* 354(6319), 1584–1587.

Huuskonen, J. and Oksanen, T. (2018). Soil sampling with drones and augmented reality in precision agriculture, *Agriculturists Elec. (Comp)* 154, 25–35.

Inyang, E. I., Hix, R. L., Tsolova, V., Rohde, B. B., Dosunmu, O. and Mankin, R. W. (2019). Subterranean acoustic activity patterns of *Vitacea polistiformis* (Lepidoptera; Sesiidae) in relation to abiotic and biotic factors, *Insects* 10(9), 267.

Iost Filho, F. H., Heldens, W. B., Kong, Z. and de Lange, E. S. (2020). Drones: Innovative technology for use in precision pest management, *J. Econ. Entomol.* 113(1), 1–25.

Jakhete, S. S., Allan, S. A. and Mankin, R. W. (2017). Wingbeat frequency-sweep and visual stimuli for trapping male *Aedes aegypti* (Diptera: Culicidae), *J. Med. Entomol.* 54(5), 1415–1419.

Jiang, J.-A., Tseng, C.-L., Lu, F.-M., Yang, E.-C., Wu, Z.-S., Chen, C.-P., Lin, K.-C. and Liao, C.-S. (2008). A GSM-based remote wireless automatic monitoring system for field information: A case study for ecological monitoring of the oriental fruit fly, Bactrocera dorsalis (Hendel), *Agriculturists Elec (Comp)* 62, 243–259.

Johnson, S. N., Crawford, J. W., Gregory, P. J., Grinev, D. V., Mankin, R. W., Masters, G. J. and Murray, P. J. (2007). Noninvasive techniques for investigating and modelling root-feeding insects in managed and natural systems, *Agric. For. Entomol.* 9, 39–46.

Johnston, N., Paris, T., Paret, M. L., Freeman, J. and Martini, X. (2022). Repelling whitefly (*Bemisia tabaci*) using limonene-scented kaolin: A novel pest management strategy, *Crop Prot.* 154, 105905.

Jordan, M. I. and Mitchell, T. M. (2015). Machine learning: Trends, perspectives, and prospects, *Science* 349(6245), 255–260.

Jordon, N., Gutknecht, J., Bybee-Finley, K. A., Hunter, M., Krupnik, T. J., Pittelkow, C. M., Prasad, P. V. V. and Snapp, S. (2021). To meet grand challenges, agricultural scientists must engage in the politics of constructive collective action, *Crop Sci.* 61(1), 24–31.

Juliano, B. and Gratton, C. (2020). Temporal resource (dis)continuity for conservation biological control from field to landscape scales, *Front. Sustain. Food Syst.* 4, 127 DOI: 10.3389/fsufs.2020.00127.

Kalischuk, M., Paret, M. L., Freeman, J. H., Raj, D., Da Silva, S., Eubanks, S., Wiggins, D. J., Lollar, M., Marois, J. J., Mellinger, H. C. and Das, J. (2019). An improved crop scouting technique incorporating unmanned aerial vehicle-assisted multispectral crop imaging into conventional scouting practice for gummy stem blight in watermelon, *Plant Dis.* 103(7), 1642–1650.

Karar, M. E., Abdel-Aty, A.-H., Algarni, F., Hassan, M. F., Abdou, M. A. and Reyad, O. (2022). Smart IoT-based system for detecting RPW larvae in date palms using mixed depthwise convolutional networks, *Alex. Eng. J.* 61 5319, 3509.

Karar, M. E., Alsunaydi, F., Albusaymi, S. and Alotaibi, S. (2021). A new mobile application of agricultural pests recognition using deep learning in cloud computing system, *Alex. Eng. J.* 60(5), 4423–4432.

Karp, D. S., Chaplin-Kramer, R., Meehan, T. D., Martin, E. A., DeClerck, F., Grab, H., Gratton, C., Hunt, L., Larsen, A. E., Martínez-Salinas, A., O'Rourke, M. E., Rusch, A., Poveda, K., Jonsson, M., Rosenheim, J. A., Schellhorn, N. A., Tscharntke, T., Wratten, S. D., Zhang, W., Iverson, A. L., Adler, L. S., Albrecht, M., Alignier, A., Angelella, G. M., Zubair Anjum, M., Avelino, J., Batáry, P., Baveco, J. M., Bianchi, F. J. J. A., Birkhofer, K., Bohnenblust, E. W., Bommarco, R., Brewer, M. J., Caballero-López, B., Carrière, Y., Carvalheiro, L. G., Cayuela, L., Centrella, M., Ćetković, A., Henri, D. C., Chabert, A., Costamagna, A. C., De la Mora, A., de Kraker, J., Desneux, N., Diehl, E., Diekötter, T., Dormann, C. F., Eckberg, J. O., Entling, M. H., Fiedler, D., Franck, P., Frank van Veen, F. J., Frank, T., Gagic, V., Garratt, M. P. D., Getachew, A., Gonthier, D. J., Goodell, P. B., Graziosi, I., Groves, R. L., Gurr, G. M., Hajian-Forooshani, Z., Heimpel, G. E., Herrmann, J. D., Huseth, A. S., Inclán, D. J., Ingrao, A. J., Iv, P., Jacot, K., Johnson, G. A., Jones, L., Kaiser, M., Kaser, J. M., Keasar, T., Kim, T. N., Kishinevsky, M., Landis, D. A., Lavandero, B., Lavigne, C., Le Ralec, A., Lemessa, D., Letourneau, D. K., Liere, H., Lu, Y., Lubin, Y., Luttermoser, T., Maas, B., Mace, K., Madeira, F., Mader, V., Cortesero, A. M., Marini, L., Martinez, E., Martinson, H. M., Menozzi, P., Mitchell, M. G. E., Miyashita, T., Molina, G. A. R., Molina-Montenegro, M. A., O'Neal, M. E., Opatovsky, I., Ortiz-Martinez, S., Nash, M., Östman, Ö, Ouin, A., Pak, D., Paredes, D., Parsa, S., Parry, H., Perez-Alvarez, R., Perović, D. J., Peterson, J. A., Petit, S., Philpott, S. M., Plantegenest, M., Plećaš, M., Pluess, T., Pons, X., Potts, S. G., Pywell, R. F., Ragsdale, D. W., Rand, T. A., Raymond, L., Ricci, B., Sargent, C., Sarthou, J. P., Saulais, J., Schäckermann, J., Schmidt, N. P., Schneider, G., Schüepp, C., Sivakoff, F. S., Smith, H. G., Stack Whitney, K., Stutz, S., Szendrei, Z., Takada, M. B., Taki, H., Tamburini, G., Thomson, L. J., Tricault, Y., Tsafack, N., Tschumi, M., Valantin-Morison, M., Van Trinh, M., van der Werf, W., Vierling, K. T., Werling, B. P., Wickens, J. B., Wickens, V. J., Woodcock, B. A., Wyckhuys, K., Xiao, H., Yasuda, M., Yoshioka, A. and Zou, Y. (2018). Crop pests and predators exhibit inconsistent responses to surrounding landscape composition, *Proc. Natl. Acad. Sci. U. S. A.* 115(33), E7863–E7870.

Kasinathan, T., Singaraju, D. and Uyyala, S. R. (2021). Insect classification and detection in field crops using modern machine learning techniques, *Inform. Process. Agric.* 8(3), 446–457.

Katzanek, J. (2020). Riverside startup taking its bug-detector pitch to Paris, *Press Enterpr.*, MediaNews Group, 3/5/2020.

Kawakita, S. and Ichikawa, K. (2019). Automated classification of bees and hornet using acoustic analysis of their flight sounds, *Apidologie* 50(1), 71–79.

Kennedy, R., Yang, Z., Gorelick, N., Braaten, J., Cavalcante, L., Cohen, W. B. and Healey, S. P. (2018). Implementation of the LandTrendr algorithm on google earth engine, *Remote Sens.* 10(5), 691.

Khan, Z. R., Midega, C. A. O., Pittchar, J. O. and Pickett, J. A. (2014a). Push-pull: A novel IPM strategy for the green revolution in Africa. In: Peshin, R. and Pimental, D. (Eds.), *Integrated Pest Management*, Springer, Dordrecht, The Netherlands, pp. 333–348.

Khan, Z. R., Midega, C. A. O., Pittchar, J. O., Murage, A. W., Birkett, M. A., Bruce, T. J. A. and Pickett, J. A. (2014b). Achieving food security for one million sub-Saharan African poor through push-pull innovation by 2020, *Phil. Trans. Roy. Soc. B Biol. Sci.* 369(1639), 20120284. DOI: 10.1098/rstb.2012.0284.

Kim, H., Lee, G., Song, J. and Kim, S. G. (2022). Real-time visualization of scent accumulation reveals the frequency of floral scent emissions, *Front. Plant Sci.* 13, 835305 DOI: 10.3389/fpls.2022.835305.

Kollasch, A. M., Abdul-Kafi, A. R., Body, M. J. A., Pinto, C. F., Appel, H. M. and Cocroft, R. B. (2020). Leaf vibrations produced by chewing provide a consistent acoustic target for plant recognition of herbivores, *Oecologia* 194(1–2), 1–13.

Koubaa, A., Aldawood, A., Saeed, B., Hadid, A., Ahmed, M., Saad, A., Alkhouja, H., Ammar, A. and Alkanhal, M. (2020). Smart palm: An IoT framework for red palm weevil early detection, *Agronomy* 10(7), 987. DOI: 10.3390/agronomy10070987.

Labatte, J. M., Got, B., Chartier, M. and Ruget, F. (1992). Validation of the European corn borer damage model on maize using X rays, *Ann. Appl. Biol.* 121(3), 483–491.

Laumann, R. A., Maccagnan, D. H. B., Čokl, A., Blassioli-Moraes, M. C. and Borges, M. (2018). Substrate-borne vibrations disrupt the mating behaviors of the neotropical brown stink bug, Euschistus heros: Implications for pest management, *J. Pest Sci.* 91(3), 995–1004. DOI: 10.1007/s10340-018-0961-5.

Laurance, W. F., Camargo, J. L. C., Fearnside, P. M., Lovejoy, T. E., Williamson, G. B., Mesquita, R. C. G., Meyer, C. F. J., Bobrowiec, P. E. D. and Laurance, S. G. W. (2018). An Amazonian rainforest and its fragments as a laboratory of global change, *Biol. Rev. Camb. Philos. Soc.* 93(1), 223–247.

LeCun, Y., Bengio, Y. and Hinton, G. (2015). Deep learning, *Nature* 521(7553), 436–444. 10.1038/nature14539.

Legaspi, J. C., Miller, N., Wolaver, D., Kanga, L., Haseeb, M.. and Zanuncio, J. C. (2016). 'Repellency of mustard (*Brassica juncea*) and arugula (*Eruca sativa*) plants, and plant oils against the sweetpotato whitefly, *Bemisia tabaci* (Hemiptera: Aleyrodidae),' *Subtrop, Agric. Environ.* 67, 28–34.

Li, W., Zheng, T., Yang, Z., Li, M., Sun, C. and Yang, X. (2021). Classification and detection of insects from field images using deep learning for smart pest management: A systematic review, *Ecol. Inform.* 66, 101460.

Li, Y. F., Zhong, S. T., Qin, Y. C., Zhang, S. Q., Gao, Z. L., Dang, Z. H. and Pan, W. L. (2014). Identification of plant chemicals attracting and repelling whiteflies, *Arthropod Plant Interact.* 8(3), 183–190. DOI: 10.1007/s11829-014-9302-7.

Liao, Y.-C., Percy, D. M. and Yang, M.-M. (2022). Biotremology: Vibrational communication of Psylloidea, *Arthropod Struct. Dev.* 66, 101138.

Lima, M. C. F., Leandro, M. E. D. de A., Valero, C., Coronel, L. C. P. and Bazzo, C. O. G. (2020). Automatic detection and monitoring of insect pests–A review, *Agriculture* 10(5), 161, DOI: 10.3390/agriculture10050161.

Liu, J. and Wang, X. (2020). Tomato diseases and pests detection based on improved Yolo V3 convolutional neural network, *Front. Plant Sci.* 11, 898 DOI: 10.3389/fpls.2020.00898.

Low, M. L., Naranjo, M. and Yack, J. E. (2021). Survival sounds in insects: Diversity function, and evolution, *Front. Ecol. Evol.* 9, 64170 DOI: 10.3389/fevo.2021.641740.

Lubanga, U. K., Taylor, G. S. and Steinbauer, M. J. (2021). Developmental biology and seasonal phenology of *Aacanthocnema dobsoni* (Hemiptera: Triozidae) and the influence of climate-mediated changes in body size on vibrational signals, *Austral. Entomol.* 60(1), 234-243.

Lujo, S., Hartman, E., Norton, K., Pregmon, E. A., Rohde, B. B. and Mankin, R. W. (2016). Disrupting mating behavior of *Diaphorina citri* (Liviidae), *J. Econ. Entomol.* 109(6), 2373-2379.

Maeder, M., Guo, X., Neff, F., Schneider Mathis, D. and Gossner, M. M. (2022). Temporal and spatial dynamics in soil acoustics and their relation to soil animal diversity, *PLoS ONE* 17(3), e0263618. DOI: 10.1371/journal.pone.0263618.

Mahanti, N. K., Pandiselvam, R., Kothakota, A., Ishwarya, P., Chakraborty, S. K., Kumar, M. and Cozzolino, D. (2022). Emerging non-destructive imaging techniques for fruit damage detection: Image processing and analysis, *Trends Food Sci. Technol.* 120, 418-438.

Mahmood, A., Hu, Y., Tanny, J.. and Asante, E. A. (2018). Effects of shading and insect-proof screens on crop microclimate and production: A review of recent advances, *Sci. Hortic.* 241, 241-251.

Maluta, N., Castro, T. and Lopes, J. R. S. (2022). Entomopathogenic fungus disrupts the phloem-probing behavior of *Diaphorina citri* and may be an important biological control tool in citrus, *Sci. Rep.* 12(1), 7959.

Mankin, R. (2022). Subterranean arthropod biotremology: Ecological and economic contexts. In: Hill, P. S. M., Mazzoni, V., Stritih-Peljhan, N., Virant-Doberlet, M. and Wessels, A. (Eds.), *Biotremology: Physiology, Ecology, and Evolution, Animal Signals and Communication* (Vol. 8). Springer, Cham, pp. 511-527. DOI: 10.1007/978-3-030-97419-0_21.

Mankin, R., Hagstrum, D., Guo, M., Eliopoulos, P. and Njoroge, A. (2021). Automated applications of acoustics for stored product insect detection, monitoring, and management, *Insects* 12(3), 259. DOI: 10.3390/insects12030259.

Mankin, R. W., Mizrach, A., Hetzroni, A., Levsky, S., Nakache, Y. and Soroker, V. (2008a). Temporal and spectral features of sounds of wood boring beetle larvae.: Identifiable patterns of activity enable improved discrimination from background noise, *Fla. Entomol.* 91(2), 241-248.

Mankin, R. W., Smith, M. T., Tropp, J. M., Atkinson, E. B. and Jong, Y. (2008b). Detection of Anoplophora glabripennis (Coleoptera: Cerambycidae) larvae in different host trees and tissues by automated analyses of sound-impulse frequency and temporal patterns, *J. Econ. Entomol.* 101(3), 836-849.

Mankin, R. W., Hagstrum, D. W., Smith, M. T., Roda, A. L. and Kairo, M. T. K. (2011). Perspective and promise: A century of insect acoustic detection and monitoring, *Am. Entomol.* 57(1), 30-44.

Mankin, R. W., Hodges, R. D., Nagle, H. T., Schal, C., Pereira, R. M. and Koehler, P. G. (2010). Acoustic indicators for targeted detection of stored product and urban insect pests by inexpensive infrared, acoustic, and vibrational detection of movement, *J. Econ. Entomol.* 103(5), 1636-1646. DOI: 10.1603/Ec10126.

Mankin, R. W., Hubbard, J. L. and Flanders, K. L. (2007). Acoustic indicators for mapping infestation probabilities of soil invertebrates, *J. Econ. Entomol.* 100(3), 790-800.

Mankin, R. W., Jetter, E., Rohde, B. and Yasir, M. (2020a). Performance of a low-cost acoustic insect detector system with *Sitophilus oryzae* (Coleoptera: Curculionidae)

Published by Burleigh Dodds Science Publishing Limited, 2023.

in stored grain and *Tribolium castaneum* (Coleoptera: Tenebrionidae) in flour, *J. Econ. Entomol.* 113(6), 3004–3010. DOI: 10.1093/jee/toaa203.

Mankin, R. W., Patel, R., Grugnale, M. and Jetter, E. (2020b). Effects of *Diaphorina citri* population density on daily timing of vibrational communication calls: Potential benefits in finding forage, *Insects* 11(3), 182. DOI: 10.3390/insects11030182.

Mankin, R. W., Mayer, M. S. and Grant, A. J. (1991). Sensitivity of *Trichoplusia ni* (Hübner) pheromone receptor neurons: Relationships between neural thresholds and behavioral responses, *J. Comp. Physiol. A* 168, 739–747.

Mankin, R. W., Moore, A., Samson, P. R. and Chandler, K. J. (2009a). Acoustic characteristics of Dynastid beetle stridulations, *Fla. Entomol.* 92(1), 123–133.

Mankin, R. W., Samson, P. R. and Chandler, K. J. (2009b). Acoustic detection of Melolonthine larvae in Australian sugarcane, *J. Econ. Entomol.* 102(4), 1523–1535.

Mankin, R. W., Stanaland, D., Haseeb, M., Rohde, B., Menocal, O. and Carrillo, D. (2018). Assessment of plant structural characteristics, health, and ecology using bioacoustic tools, *Proc. Meet. Acoust.* 33, 010003.

Martinez, D., Arhidi, L., Demondion, E., Masson, J. B. and Lucas, P. (2014). Using electroantennogram sensors on autonomous robots for olfactory searches, *J. Vis. Exp.* (90), e51704 DOI: 10.3791/51704.

Martini, X., Pelz-Stelinski, K. S. and Stelinski, L. L. (2014). Plant pathogen-induced volatiles attract parasitoids to increase parasitism of an insect vector, *Front. Ecol. Evol.* 2, 8. DOI: 10.3389/fevo.2014.00008.

Mayer, M. S., Mankin, R. W., and Grant, A. J. (1987). Quantitative comparison of behavioral and neurophysiological responses to odorants: inferences about central nervous system processes. *J. Chem. Ecol.* 13, 509–531.

Mazzoni, V., Nieri, R., Eriksson, A., Virant-Doberlet, M., Polajnar, J., Anfora, G. and Lucchi, A. (2019). Mating disruption by vibrational signals: State of the field and perspectives. In: Hill, P., Lakes-Harlan, R., Mazzoni, V., Narins, P., Virant Doberlet, M., Wessel, A. (Eds.), *Biotremology: Studying Vibrational Behavior: Animal Signals and Communication* (Vol. 6), Springer, Cham, pp. 331–354. DOI: 10.1007/978-3-030-22293-2_17.

Meagher, R. L., Watrous, K. M., Fleischer, S. J., Nagoshi, R. N., Brown, J. T., Bowers, K., Miller, N., Hight, S. D., Legaspi, J. C. and Westbrook, J. K. (2019). Documenting potential sunn hemp (Crotalaria *juncea* L.) (Fabaceae) pollinators in Florida, *Environ. Entomol.* 48(2), 343–350. DOI: 10.1093/ee/nvy190.

Mercer, R., Alaee, S., Abdoli, A., Singh, S., Murillo, A. and Keogh, E. (2021). Matrix profile XXIII: Contrast profile: A novel time series primitive that allows real world classification international conference Data Mining. IEEE Publications, pp. 1240–1245.

Miller, J. R., and Gut, L. J. (2015). Mating disruption for the 21st century: matching technology with mechanism. *Environ. Entomol.* 44(3), 427–453.

Montgomery, G. A., Belitz, M. W., Guralnick, R. P. and Tingley, M. W. (2021). Standards and best practices for monitoring and benchmarking insects, *Front. Ecol. Evol.* 8, 579193. DOI: 10.3389/fevo.2020.579193.

Mortimer, B. (2019). A spider's vibration landscape; Adaptations to promote vibrational information transformation in orb webs, *Integr. Comp. Biol.* 59(6), 1636–1645.

Murali-Baskaran, R. K., Sharma, K. C., Kaushal, P., Kumar, J., Parthiban, P., Senthil-Nathan, S. and Mankin, R. W. (2018). Role of kairomone in biological control of crop pests - A review, *Physiol. Mol. Plant Pathol.* 101(1), 3–15.

Nagoshi, R. N., Fleischer, S. and Meagher, R. L. (2017). Demonstration and quantification of restricted mating between fall armyworm host strains in field collections by SNP comparisons, *J. Econ. Entomol.* 110(6), 2568–2575.

Nagoshi, R. N., Silvie, P. and Meagher Jr., R. L. (2007a). Comparison of haplotype frequencies differentiate fall armyworm (Lepidoptera: Noctuidae) corn-strain populations from Florida and Brazil, *J. Econ. Entomol.* 100(3), 954–961.

Nagoshi, R. N., Silvie, P., Meagher Jr., R. L., Lopez, J. and Machado, V. (2007b). Identification and comparison of fall armyworm (Lepidoptera: Noctuidae) host strains in Brazil, Texas, and Florida, *Ann. Entomol. Soc. Am.* 100(3), 394–402.

Nakano, R. and Mason, A. C. (2018). Early erratic flight response of the lucerne moth to the quiet echolocation calls of distant bats, *PLoS ONE* 13(8), e0202679. DOI: 10.1371/journal.pone.0202679.

Nanni, L., Manfè, A., Maguolo, G., Lumini, A., and Brahnam, S. (2022). High performing ensemble of convolutional neural networks for insect pest image detection. *Ecol. Informatics* 67, 101515.

Naranjo, S. E. (2001). Conservation and evaluation of natural enemies in IPM systems for *Bemisia tabaci*, *Crop Prot.* 20(9), 835–852.

Norin, T. (2007). Semiochemicals for insect pest management, *Pure Appl. Chem.* 79(12), 2129–2136.

Oji, F. L. and Mohamed, S. S. (2005). Whitefly management by exploiting biodiversity in a "push-pull" strategy on *Phaseolus vulgaris* L. In: *First International Edible Legume Conference in* Conjunction With *the IVth World Cowpea Congress*, Durban, South Africa, pp. 1–9.

Olsson, S. B. and Hansson, B. S. (2013). Electroantennogram and single sensillum recordings in insect antennae, In: Touhara (Ed.), *Pheromone Signaling: Methods and Protocols*, Springer Science+Business Media, LLC, Clifton, NJ, Vol. 1068, pp. 157–177.

Osco, L. P., Furuya, D. E. G., Furuya, M. T. G., Corrêa, D. V., Gonçalvez, W. N., Junior, J. M., Borges, M., Moraes, M. C. B., Michereff, M. F. F., Aquino, M. F. S., Laumann, R. A., Lisenberg, V., Ramos, A. P. M. and Jorge, L. A. de C. (2022). An impact analysis of pre-processing techniques in spectroscopy data to classify insect-damaged in soybean plants with machine and deep learning methods, *Inf. Phys. Tech.* 123, 104203.

Paris, T. M., Allan, S. A., Udell, B. J. and Stansly, P. A. (2017). Wavelength and polarization affect phototaxis of the Asian citrus psyllid, *Insects* 8(3), 88. DOI: 10.3390/insects8030088.

Partel, V., Nunes, L., Stansly, P. and Ampatzidis, Y. (2019). Automated vision-based system for monitoring Asian citrus psyllid in orchards utilizing artificial intelligence, *Agriculturists El (Comp)* 162, 328–336.

Perring, T. M., Stansly, P. A., Liu, T. X., Smith, H. A. and Andreason, S. A. (2018). Whiteflies: Biology, ecology, and management. In: Wakil, W., Brust, G. G. and Perring, T. M. (Eds.), *Sustainable Management of Arthropod Pests of Tomato*, Academic Press, Elsevier, pp. 73–110.

Phung, Q. V., Ahmad, I., Habibi, D. and Hinckley, S. (2017). Automated insect detection using acoustic features based on sound generated from insect activities, *Acoust. Aust.* 45(2), 445–451.

Pijanowski, B. C., Villanueva-Rivera, L. J., Dumyahn, S. L., Farina, A., Krause, B. L., Napoletano, B. M., Gage, S. H. and Pieretti, N. (2011). Soundscape ecology: The science of sound and landscape, *BioScience* 61(3), 203–216.

Polajnar, J., Eriksson, A., Lucchi, A., Anfora, G., Virant-Doberlet, M. and Mazzoni, V. (2015). Manipulating behaviour with substrate-borne vibrations-potential for insect pest control, *Pest Manag. Sci.* 71, 15–23.

Potamitis, I. (2013). Classifying insects on the fly, *Ecol. Inform.* 21, 40–49.

Potamitis, I., Eliopoulos, P. and Rigakis, I. (2017). Automated remote insect surveillance at a global scale and the internet of things, *Robotics* 6(3), 19.

Potamitis, I., Ganchev, T. and Kontodimas, D. (2009). On automatic bioacoustic detection of pests: The cases of *Rhynchophorus ferrugineus* and *Sitophilus oryzae*, *J. Econ. Entomol.* 102(4), 1681–1690.

Potamitis, I. and Rigakis, I. (2014). The electronic McPhail trap, *PeerJ Prepr.* 2, e510. DOI: 10.7287/peerj.preprints.510v1.

Potamitis, I., Rigakis, I., Tatlas, N. A. and Potirakis, S. (2019). In-vivo vibroacoustic surveillance of trees in the context of the IoT, *Sensors (Basel)* 19(6), 1366. DOI: 10.3390/s19061366.

Potting, R. P. J., Perry, J. N. and Powell, W. (2005). Insect behavioural ecology and other factors affecting the control efficacy of agro-ecosystem diversification strategies, *Ecol. Modell.* 182(2), 199–216.

Preti, M., Favaro, R., Knight, A. L. and Angeli, S. (2021). Remote monitoring of Cydia pomonella adults among an assemblage of nontargets in sex pheromone-kairomone-baited smart traps, *Pest Manag. Sci.* 77(9), 4084–4090.

Raboin, M. (2021). Inaudible noise pollution of the invertebrate world, *Acoust. Today* 17(2), 32–39.

Rathore, D. S., Ram, B., Pal, B. L. and Malviya, S. (2019). Analysis of classification algorithms for insect detection using MATLAB, SSRN journal. In: *Proceedings of the 2nd International Conference on Advanced Computing and Software Engineering (ICACSE)*, Sultanpur, India, pp. 299–302.

Reynolds, D. R., Chapman, J. W., Edwards, A. S., Smith, A. D., Wood, C. R., Barlow, J. F. and Woiwod, I. P. (2005). Radar studies of the vertical distribution of insects migrating over southern Britain: The influence of temperature inversions on nocturnal layer concentrations, *Bull. Entomol. Res.* 95(3), 259–274.

Rigakis, I., Potamitis, I., Tatlas, N. A., Potirakis, S. M. and Ntalampiras, S. (2021a). TreeVibes: Modern tools for global monitoring of trees against borers, *Smart Cities* 4(1), 271–285.

Rigakis, I. I., Varikou, K. N., Nikolakakis, A. E., Skarakis, Z. D., Tatlas, N. A. and Potamitis, I. G. (2021b). The e-funnel trap: Automatic monitoring of lepidoptera; a case study of tomato leaf miner, *Agriculturists Elec (Comp)* 185, 106154.

Riley, J. R. (1989). Remote sensing in entomology, *Annu. Rev. Entomol.* 34(1), 247–271.

Roch, M. A., Gerstoft, P., Kostek, B. and Michalopoulou, Z.-H. (2021). How machine learning contributes to solve acoustical problems, *Acoust. Today* 17(4), 48–56.

Romano, J. M., Brindza, J. P. and Kuchenbecker, K. J. (2013). ROS open-source audio recognizer: ROAR environmental sound detection tools for robot programming, *Auton. Robot.* 34(3), 207–215.

Rudman, S. M., Greenblum, S. I., Rajpurohit, S., Betancourt, N. J., Hanna, J., Tilk, S., Yokoyama, T., Petrov, D. A. and Schmidt, P. (2022). Direct observation of adaptive tracking on ecological time scales in *Drosophila*, *Science* 375(6586), eabj7484.

Rydhmer, K., Bick, E., Still, L., Strand, A., Luciano, R., Helmreich, S., Beck, B. D., Grónne, C., Malmros, L., Poulsen, K., Elbæk, F., Brydegaard, M., Lemmich, J. and Nikolajsen, T.

(2022). Automating insect monitoring using unsupervised nearinfrared sensors, *Sci. Rep.* 12(1), 2603.
Senf, C., Seidl, R. and Hostert, P. (2017). Remote sensing of forest insect disturbances: Current state and future directions, *Int. J. Appl. Earth Obs. Geoinf.* 60, 49–60.
Shaw, T. J., Paula-Moraes, S. V., Hahn, P. G. and Specht, A. (2021). Seasonal flight patterns of *Chrysodeixis includens* (Lepidoptera: Noctuidae) in the Florida panhandle and inventory of Plusiine species cross-attracted to synthetic pheromone, *J. Econ. Entomol.* 114(6), 2315–2325.
Silva, G., Tomlinson, J., Onkokesung, N., Sommer, S., Mrisho, L., Legg, J., Adams, I. P., Gutierrez-Vazquez, Y., Howard, T. P., Laverick, A., Hossain, O., Wei, Q., Gold, K. M. and Boonham, N. (2021). Plant pest surveillance: From satellites to molecules, *Emerg. Top. Life Sci.* 5(2), 275–287.
Singerman, A. and Rogers, M. E. (2020). The economic challenges of dealing with citrus greening: The case of Florida, *J. Integr. Pest Manag.* 11(1), 1–7.
Siriwardena, K. A. P., Fernando, L. C. P., Nanayakkara, N., Perera, K. F. G., Kumara, A. D. N. T. and Nanayakkara, T. (2010). Portable acoustic device for detection of coconut palms infested by *Rhynchophorus ferrugineus* (Coleoptera: Curculionidae), *Crop Prot.* 29(1), 25–29.
Skendžić, S., Zovko, M., Živković, I. P., Lešić, V. and Lemić, D. (2021). The impact of climate change on agricultural insect pests, *Insects* 12(5), 440. DOI: 10.3390/insects12050440.
Snyder, J., Dickens, K. L., Halbert, S. E., Dowling, S., Russell, D., Henderson, R., Rohrig, E., and Ramadugu, C. (2022) The development and evaluation of insect traps for the Asian citrus psyllid, Diaphorina citri (Hemiptera: Psyllidae), vector of citrus huanglongbing. *Insects* 13, 295.
Sparks, T. C., Wessels, F. J., Lorsbach, B. A., Nugent, B. M. and Watson, G. B. (2019). The new age of insecticide discovery-the crop protection inducstry and the impact of natural products, *Pestic. Biochem. Physiol.* 161, 12–22.
Spiesman, B. J., Gratton, C., Hatfield, R. G., Hsu, W. H., Jepsen, S., McCornack, B., Patel, K. and Wang, G. (2021). Assessing the potential for deep learning and computer vision to identify bumble bee species from images, *Sci. Rep.* 11(1), 7580.
Stelinski, L. L., Willet, D., Rivera, M. J. and Ali, J. G. (2019). 'Tuning' communication among four trophic levels of the root biome to facilitate biological control, *Biol. Control* 131, 49–53.
Stepanian, P. M., Entrekin, S. A., Wainwright, C. E., Mirkovic, D., Tank, J. L. and Kelly, J. F. (2020). Declines in an abundant aquatic insect, the burrowing mayfly, across major North American waterways, *Proc. Natl. Acad. Sci. U. S. A.* 117(6), 2987-2992.
Stowasser, A., Mohr, S., Buschbeck, E. and Vilinsky, I. (2015). Electrophysiology meets ecology: Investigating how vision is tuned to the life style of an animal using electroretinography, *J. Undergrad. Neurosci. Educ.* 13(3), A234–A243.
Šturm, R., López Díez, J. J., Polajnar, J., Sueur, J. and Virant-Doberlet, M. (2022). Is it time for ecotremology?, *Front. Ecol. Evol.* 10, 828503. DOI: 10.3389/fevo.2022.828503.
Suckling, D. M., Daly, J. M., Chen, X. and Karg, G. (2007). Field electroantennogram and trap assessments of aerosol pheromone dispensers for disrupting mating in *Epiphyas postvittana*, *Pest Manag. Sci.* 63(2), 202–209.
Sugai, L. S. M., Desjonquéres, C., Silva, T. S. F. and Llusia, D. (2020). A roadmap for survey designs in terrestrial acoustic monitoring, *Remote Sens. Ecol. Conserv.* 6(3), 220–235.

Sun, X., Guo, M., Ma, M. and Mankin, R. W. (2018). Identification and classification of damaged corn kernels with impact acoustics multi-domain patterns, *Agriculturists Elec (Comp)* 150, 152–161.

Sutin, A., Yakubovskiy, A., Salloum, H. R., Flynn, T. J., Sedunov, N. and Nadel, H. (2019). Towards and automated acoustic detection algorithm for wood -boring beetle larvae (Coleoptera: Cerambycidae and Buprestidae), *J. Econ. Entomol.* 112(3), 1327–1336.

Takanashi, T., Uechi, N. and Tatsuta, H. (2019). Vibrations in hemipteran and coleopteran insects: Behaviors and application in pest management, *Appl. Entomol. Zool.* 54(1), 21–29.

Tauc, M. J., Fristrup, K. M., Repasky, K. S. and Shaw, J. A. (2019). Field demonstration of a wing-beat modulation lidar for the 3D mapping of flying insects, *OSA Continuum* 2(2), 332–348.

Thenmozhi, K. and Reddy, U. S. (2019). Crop pest classification based on deep convolutional neural network and transfer learning, *Agriculturists Elec (Comp)* 164, 104906.

Thomas, M. D., Moriarty, T. M., Cook, D. D. and Mazzero, B. A. (2021). Electrical capacitance measurements to assess European corn borer infestation in maize, *Agriculturists Elec (Comp)* 187, 106305.

Thompson, M. N., Medina, R. F., Helms, A. M. and Bernal, J. S. (2022). Improving natural enemy selection in biological control through greater attention to chemical ecology and host associated differentiation of target arthropod pests, *Insects* 13(2), 160. DOI: 10.3390/insects13020160.

Tuthill, J. C. and Wilson, R. I. (2016). Mechanosensation and adaptive motor control in insects, *Curr. Biol.* 26(20), R1022–R1038.

Tyler-Julian, K., Funderburk, J., Srivastava, M., Olson, S., and Adkins, S. (2018) Evaluation of a push-pull system for the management of Frankliniella species (Thysanoptera: Thripidae) in tomato. *Insects* 9, 187.

Ulloa, J. S., Haupert, S., Latorre, J. F., Aubin, T. and Sueur, J. (2021). Scikit-maad: An open-source and modular toolbox for quantitative soundscape analysis in Python, *Methods Ecol. Evol.* 1–7.

Vandervoet, T. (2022). 'Entomologists' role in a food insecure world, *Am. Entomol.* 68(1), 64.

Veen, G. F., Wubs, E. R. J., Bardgett, R. D., Barrios, E., Bradford, M. A., Carvalho, S., De Deyn, G. B., de Vries, F. T., Giller, K. E., Kleijn, D., Landis, D. A., Rossing, W. A. H., Schrama, M., Six, J., Struik, P. C., van Gils, S., Wiskerke, J. S. C., van der Putten, W. H. and Vet, L. E. M. (2019). Applying the aboveground-belowground interaction concept in agriculture: Spatio-temporal scales matter, *Front. Ecol. Evol.* 7. DOI: 10.3389/fevo.2019.00300.

Virant-Doberlet, M., Kuhelj, A., Polajnar, J. and Šturm, R. (2019). Predator-prey interactions and eavesdropping in vibrational communication networks, *Front. Ecol. Evol.* 7, 203. DOI: 10.3389/fevo.2019.00203.

Walgenbach, J. F., Schoof, S. C., Bosch, D., Escudero-Colomar, L.-A., Lingren, B. and Krawczyk, G. (2021). Comparison of sex pheromone and kairomone-enhanced pheromone lures for monitoring oriental frout moth (Lepidoptera: Tortricidae) in mating disruption and non-disruption tree fruit orchards, *Environ. Entomol.* 50(5), 1063–1074.

Westbrook, J., Fleischer, S., Jairam, S., Meagher, R. and Nagoshi, R. (2019). Multigenerational migration of fall armyworm, a pest insect, *Ecosphere* 10(11), e02919. DOI: 10.1002/ecs2.2919.

WFP and FAO. (2022). *Hunger Hotspots. FAO-WFP Early Warnings on Acute Food Insecurity: June to September 2022 Outlook*. World Food Program and Food and Agriculture Organization of the United Nations, Rome, Italy. Available at: https://www.wfp.org/publications/hunger-hotspots-fao-wfp-early-warnings-acute-food-insecurity-june-september-2022.

Wignall, A. E. and Herberstein, M. E. (2022). Male courtship reduces the risk offemale aggression in web-building spider but varies in structure, *Behav. Ecol.* 33(1), 280-287. DOI: 10.1093/beheco/arab140.

Wignall, A. E. and Taylor, P. W. (2011). Assassin bug uses aggressive mimicry to lure spider prey, *Proc. Biol. Sci.* 278(1710), 1427–1433. DOI: 10.1098/rspb.2010.2060.

Willet, D. S., George, J., Willet, N. S., Stelinski, L. L. and Lapointe, S. L. (2016). Machine learning for characterization of insect vector feeding, *PLoS Comput. Biol.* 12(11), e1005158. DOI: 10.1371/journal.pcbi.1005158.

Wu, S.-H., Chang, H.-W., Lin, R.-S. and Tuanmu, M.-N. (2022). SILIC: A cross database framework for automatically extracting robust biodiversity information from soundscape recordings based on object detection and a tiny training dataset, *Ecol. Inform.* 68, 101534.

Wu, X., Zhan, C., Lai, Y.-K., Cheng, M.-M. and Yang, J. (2019). Ip102: A large-scale benchmark dataset for insect pest recognition. In: *Proceedings of the IEEE Conference on Computing, and Vision Pattern Recognition*, Long Beach, CA, June 16–20, 2019, pp. 8787–8796.

Xia, J., Latchininsky, A., Hadi, B. and Elkahky, M. (2022). Sustainable plant pest management through optimization and minimization, *Front. Agr. Sci. Eng.* 9(1), 161–166.

Xie, C., Wang, R., Zhang, J., Chen, P., Dong, W., Li, R., Chen, T. and Chen, H. (2018). Multi-level learning features for automatic classification of field crop pests, *Comput. Electron. Agric.* 152, 233–241. DOI: 10.1016/j.compag.2018.07.014.

Yang, C., Cheng, J., Lin, J., Zheng, Y., Yu, X. and Sun, J. (2022). Sex pheromone receptors of Lepidopteran insects, *Front. Ecol. Evol.* 10, 797287.

Yang, L., Hu, X. P., Allan, S. A., Alborn, H. T. and Bernier, U. R. (2019). Electrophysiological and behavioral responses of the kudzu bug, (*Megacopta cribaria*) (Hemiptera: Plataspidae), to volatile compounds from kudzu and soybean plants, *J. Agric. Food Chem.* 67(15), 4177–4183.

Yinon, U. (1971). The electroretinogram of the compound eye indeveloping holometabolic and hemimetabolic insects, *Comp. Biochem. Physiol.* 39A, 457–463.

Yu, Z., de Vries, A., Ampatzidis, Y. and Sokol, D. D. (2021). Who owns and controls darming data?, *EDIS* 2021(5). DOI: 10.32473/edis-AE564-2021.

Zaffaroni Caorsi, V., Wells, K. E., Moser, D., Berado, A., Miselli, R., Torriani, M., Pugno, N. M., Maistrello, M. and Mazzoni, V. (2021). Design of ideal vibrational signals for stinkbug male attraction, through vibrotaxis experiments, *Pest Man. Sci.* 77(12), 5948–5508.

Zhang, J., Huang, Y., Pu, R., Gonzalez-Moreno, P., Yuan, L., Wu, K. and Huang, W. (2019). Monitoring plant diseases and pests through remote sensing technology: A review, *Comput. Electron. Agric.* 165, 104943. DOI: 10.1016/j.compag.2019.104943.

Zheng, Z. and Zhang, C. (2022). Electronic noses based on metal oxide semiconductor sensors for detecting crop diseases and insect pests, *Agriculturists Elec (Comp)* 197, 106988.

Zhong, Z., Li, G., Sanders, D., Wang, D., Holt, R. D. and Zhang, Z. (2022). A rodent herbivore reduces its predation risk through ecosystem engineering, *Curr. Biol.* 32, 1–6.

Zhou, J., Lai, J., Menda, G., Stafstrom, J. A., Miles, C. I. and Hoy, R. (2022). Outsourced hearing in an orb-weaving spider that uses its web as an auditory sensor, *Proc. Nat. Acad. Sci. U. S. A.* 119(14), e2122789119.

Zhu, P., Wen, L., Du, D., Bian, X., Ling, H., Hu, Q., Wu, H., Nie, Q., Cheng, H., Liu, C., Liu, X., Ma, W., Wang, L., Schumann, A., Wang, D., Ortego, D., Luna, E., Michail, E., Bochinski, E., Ni, F., Bunyak, F., Zhang, G., Seetharaman, G., Li, G., Yu, H., Kompatsiaris, I., Zhao, J., Gao, J., Martínez, J. M., San Miguel, J. C., Palaniappan, K., Avgerinakis, K., Sommer, L., Lauer, M., Liu, M., Al-Shakarji, N. M., Acatay, O., Giannakeris, P., Zhao, Q., Ma, Q., Huang, Q., Vrochidis, S., Sikora, T., Senst, T., Song, W., Tian, W., Zhang, W., Zhao, Y., Bai, Y., Wu, Y., Wang, Y., Li, Y., Pi, Z. and Ma, Z. (2019). VisDrone-VDT2018: The vision meets drone video detection and tracking challenge results. In: *Proceedings of the European Conference on Computer Vision (ECCV) Workshops*, Munich, pp. 496–518.

Published by Burleigh Dodds Science Publishing Limited, 2023.

Chapter 5

Monitoring airborne movement of crop insect pests and beneficials

V. Alistair Drake, University of Canberra and University of New South Wales, Australia

1 Introduction

Whenever the weather is fine and temperatures are sufficiently high, insects will be found in flight above almost any vegetated terrestrial habitat. The composition of this 'aerofauna', and the abundance of the various species, will of course vary from hour to hour, place to place, season to season, and year to year. Some of these insects will be highly visible, such as butterflies by day or moths in the beam of a car's headlights at night; others, such as hoverflies and aphids, may be noticed only by enthusiasts and experts or when present in exceptional numbers. Insects rising more than a few metres above the surface are soon lost from sight, and those that ascend at dusk and spend hours at heights of a few hundred metres will be observed visually only by a patient, telescope-equipped, observer as they cross the face of a full moon (Pruess and Pruess, 1971).

http://dx.doi.org/10.19103/AS.2022.0113.04

Movements covering hundreds of kilometres, often windborne and referred to here as *migrations*, can indeed be 'massive, influential, and overlooked' (Satterfield et al., 2020). Flying typically between 100 m and 1 km above the ground, wind-assisted migrants can achieve displacements of a few hundred kilometres overnight, potentially taking a population from one climatic zone to another within days if they fly multiple times (Drake and Gatehouse, 1995). Shorter-distance movements in less consistent directions, termed here *dispersal*, move populations among wild hosts, between wild hosts and crops, and from one crop to another as the season progresses (Stinner et al., 1983; Kennedy and Storer, 2000). Insects within a suitable habitat will make short *foraging* flights in search of food and to enable mating and egg-laying (Hassell and Southwood, 1978). All three types of movement play critical roles in the lives of the insects undertaking them; the first two are key processes, at different scales, in the spatial dynamics of populations (Chapman et al., 2012; Parry, 2013; Gao et al., 2020) and of central importance for the design of strategies to manage populations (Cerritos et al., 2012; Schellhorn et al., 2014). Integrated pest management (IPM), at least in its most developed ('level III') form (Kogan, 1998), and the associated concept of area-wide management (Hendrichs et al., 2007), implicitly recognise the importance of movements between different hosts. To understand population processes, and to forecast population sizes, information on the timing, spatial scale, and intensity of these movements will be required. Moreover, day-to-day decision-making for crop protection and pollination requires information about population sizes and phenologies (Rossi et al., 2012); monitoring of insects in flight is one means of providing this. Methods for obtaining this information, through detection, quantification, and characterisation of insect flight above vegetation, form the central topic of this chapter, which also considers the utility of the various techniques, and of the information they provide, for agriculturalists.

Migration and dispersal flights are important to crop production because many of the insects involved are pests (species that damage a crop directly; Pedgley, 1993), vectors of plant-disease agents such as viruses, mycoplasmas, and bacteria (Reynolds et al., 2006), or 'beneficials' (i.e. predators or parasites of pest species, or pollinators; Schellhorn et al., 2014; Wotton et al., 2019). From an agricultural perspective, migratory or dispersal flights are of concern primarily when they terminate in a crop: insects that are departing or 'passing through' will do no harm. Colonisation of a crop can arise from nearby or distant sources, involving flights over short (Topper, 1987) or long (Irwin and Thresh, 1988; Chapman et al., 2015) distances. Wind and weather modulate the incidence and character of these flights and their distance and direction (Drake and Farrow, 1988; Pedgley, 1999). Insects are winged ('alate') only in their adult stage, so when immigration occurs damage will often arise primarily from larvae of the subsequent generation (Cerritos et al., 2012).

Migration and dispersal occur in a wide range of insect species and a wide variety of environments and circumstances (Johnson, 1969; Drake and Gatehouse, 1995; Reynolds et al., 2018). Both are the result of *migratory behaviour*, in which an animal is unresponsive to immediate appetitive cues (food items, shelter, reproduction) and embarks on sustained and directed locomotory activity (Kennedy, 1985; Dingle, 2014). Movement arises either from the insect's flight through the air, or by transport on the wind, or a combination of these; but the insect initiates take-off, and its wing-flapping to keep itself in the air is a critical factor in the resulting displacement. Migratory flights are therefore in no sense accidental. Wind-tunnel tests have shown that settled insects can resist being blown away by a strong breeze, though in extreme weather involuntary displacements will likely occur (Antolinez et al., 2021).

Insect flight speeds through the air are typically low ($<10\ \text{m s}^{1}$; Dudley, 2000) and movement upwind will be impossible when the wind speed exceeds the airspeed; at higher wind speeds, movement in a broad band of directions around upwind will be unachievable (Drake et al., 2001). However, wind speeds decrease markedly near the ground (Stull, 1988), and there will usually be a zone, termed the *flight boundary layer* (FBL), within which an insect can move in any direction (Taylor, 1974; Antolinez et al., 2021). Much insect migration occurs well above the FBL and is predominantly wind-borne; long distances can be covered but reaching a suitable destination depends on the flight being undertaken when the wind is in a favourable direction. However, some insects remain within the FBL while exhibiting migratory behaviour, and during the day such directed movements may be evident and studied relatively easily (Walker, 1985). Foraging movements occur almost entirely within the FBL. They result from *appetitive behaviour* and generally follow erratic trajectories extending over a limited area as the insect searches for food items, oviposition sites, etc. and from time to time settles on vegetation or the ground in response to these (Evans et al., 2020).

Methods of detecting insects in flight above crops and other vegetation, estimating their numbers, and characterising their activities (i.e. recording height, speed, direction, straightness of movement, etc.) fall into four broad categories: trapping, visual observation, remote sensing, and radiotelemetry (Drake and Bruderer, 2018). *Remote sensing* here encompasses instrument-based techniques that detect and track insects that are in flight some distance away – up to about 2 km for individuals and 100 km for ensembles. The various methods are described in the following sections, with remote sensing divided into radio-based and optical techniques. There is an emphasis on technology-based solutions as these show most promise for advancing knowledge and are the focus of much current research. In many of the examples that follow, instrument development remains at the prototype stage and field deployments have had the form of trials rather than biological investigations; however, there are also technologies, notably radar in various forms, with well-established entomological applications and a record of producing biologically useful information. A case study of a programme that employed a variety of methods follows, and the chapter concludes with a discussion of the potential of the various methods for agricultural research and decision-making.

The chapter's coverage extends beyond 'monitoring' in its strict sense of detecting and quantifying populations and includes methods for studying flight activities and observing or inferring trajectories. Both research and 'operational' (i.e. routine decision-support) applications are considered, though the former predominate. However, only techniques relating to airborne activity, i.e. to the *process* of foraging, dispersal, or migration by flight, are considered; to obtain information on the *outcomes* of those movements, i.e. displacements, population mixing, and colonisation, a more extensive set of methods is required (Drake, 1991; Osborne et al., 2002). Researchers working in related contexts, e.g. conservation (Lahoz-Monfort and Magrath, 2021) and general entomology (Rhodes et al., 2022), have overlapping aims and are adopting similar technical approaches, some of which will be adaptable to agricultural applications.

2 Trapping

In comparison with visual observations and remote-sensing methods, trapping has the enormous advantage of providing specimens for examination, either directly or via an image recorded at close range, so that firm identifications can be made. If specimens are retained, sex and physiological state (e.g. of oocyte development, McDonald and Cole, 1991) can be determined and genetic analyses (e.g. of microsatellite markers, Scott et al., 2005) are possible. A range of methods for trapping flying insects is available (Muirhead-Thomson, 1991; Epsky et al., 2008; Miller et al., 2015; Henderson and Southwood, 2016; McCravy, 2018; Montgomery et al., 2021); the choice of a trap type depends on the data required and the behaviour and abundance of the species of interest. Farmers and crop-protection organisations are often concerned only with a few species and with whether population densities have reached thresholds for management interventions. Researchers studying insect flight activity and its consequences for crop ecosystems will have broader requirements. When visual or remote-sensing observations are being made, trapping plays an important complementary role of supporting the discrimination of target and non-target species (see Case study in Section 7).

Trapping methods can be divided into two broad categories: attraction and interception. The former relies on insect behaviour and locomotion to bring specimens to a lure, while the latter passively catches insects undertaking their normal flights. Attracting insects to traps disturbs normal behaviour but provides a useful measure of timing of appetitive flight and of abundance within the nearby area. Attraction draws in insects from some distance, enabling efficient detection of populations even at low densities. Sex-pheromone lures are species-specific and widely used for crop-pest monitoring as their catch requires minimal sorting and identification (Epsky et al., 2008). Light traps attract a wide range of species, allowing assessment of the broader entomofauna (though only at night). Light traps incorporating lamps emitting strongly in the ultra-violet (UV, e.g. 395 nm; Zemel and Houghton, 2017; Pan et al., 2021) are most effective and sometimes produce large catches. Both sex-pheromone and light traps were used in an 11-year study of populations of two *Helicoverpa* moth species in a cotton-growing region of Australia (Baker et al., 2011). A food lure proved more effective than either pheromones or UV light for monitoring oriental armyworm *Mythimna separata* in Japan (Hirai, 1988). Insects flying well above the surface can be trapped with a light mounted high on a tower, with a shield to obscure the light from insects flying near the ground (Gregg et al., 1993). Intense vertically directed 'searchlight' beams attract migrating insects from heights of 100 m or more (Feng et al., 2003); when deployed on isolated islands, this technique has revealed over-sea migration routes of important pest-insect species such as the Asian corn borer *Ostrinia furnacalis* (Shen et al., 2020) and the fall armyworm *Spodoptera frugiperda* (Zhou et al., 2021).

Interception of flying insects can be undertaken with static or moving apertures and either near the surface using ground-based equipment or at altitude with nets or suction traps carried by kites, balloons, or aircraft. Large-aperture semi-permanent funnel nets, constructed in pairs with the two openings facing opposite directions (approximately north and south) were used to monitor seasonal butterfly migrations (Walker, 2001). A demountable net with 8 m^2 aperture proved effective for monitoring low-flying hoverflies migrating through an alpine pass (Aubert et al., 1976); hoverflies marked with minute droplets of fluorescent lacquer were recorded at a second pass 50 km away (Aubert et al., 1969; Aubert and Goeldlin de Tiefenau, 1981). For smaller insects such as aphids, suction traps are effective. For example, since 1964, a network of traps with apertures at a height of 12 m has monitored the longer-distance movements of crop pests in Britain (Bell et al., 2015); this network now extends across Europe (e.g. Luquet et al., 2019). The suction trap aperture faces upwards to eliminate directional bias. In a recent study, catches from these traps revealed the aerial incidence, diversity, and phenology of Psylloidea species (psyllids), including one that is a vector of a pathogen of potatoes, over Britain (Greenslade et al., 2020). For small, weaker-flying insects, catches from suction traps can be coupled with meteorological data on wind direction to infer direction of movement. Likely source regions of aphids arriving in Sweden and Finland have been identified in this way (Wiktelius, 1981; Leskinen et al., 2011).

A trap combining attraction (yellow colour) and interception (sticky surface turned upwind by a wind-vane) showed that Asian citrus psyllids *Diaphorina citri*, a vector of the presumed causal agent of huanglongbing disease of citrus trees, moved predominantly downwind – an unsurprising result as the maximum airspeed reported for this species is 0.33 ms^{1} (Antolinez et al., 2021).

To monitor insects flying near crops and in unpredictable directions, nets mounted on vehicles can be used (e.g. Greenstone et al., 1991; Svenningsen et al., 2021). Trapping runs at moderate speeds over distances of a few kilometres, ideally on farm tracks immediately adjacent to the crops of interest, provide samples of the aerofauna unbiased by trap-induced behavioural responses. For larger, less-numerous insects, a net mounted above the vehicle and extending over its full width (Fig. 1; Drake, 1991) can obtain adequate sample sizes. The sampling and sorting effort required is generally too great for long-term monitoring, but limited campaigns timed for periods of peak flight can be very effective (see Case study in Section 7).

For smaller insects flying at high densities, an unmanned aerial vehicle (UAV, or 'drone') can be deployed to carry a sampling device and make low passes over a crop. For example, thrips flying at 50–60 m above an onion field were sampled with sticky traps suspended below a UAV (Smith et al., 2015). A drone fitted with two small nets and flown at 5–100 m over rice fields captured delphacid planthoppers and thrips, major pests of this crop; the nets were turned to a non-collecting angle during the ascent and descent stages of each flight (Kim et al., 2018). Both studies employed commercial off-the-shelf multirotor UAVs, which makes the technique relatively affordable and accessible

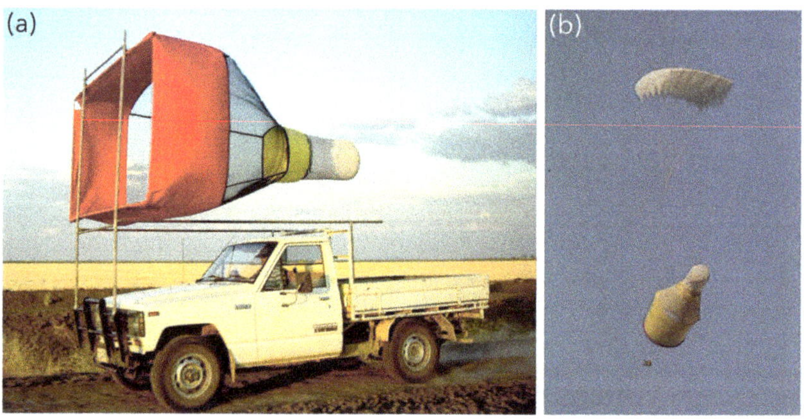

Figure 1 (a) Giant vehicle-mounted net for interception trapping of moths flying at heights of 2–4 m. (b) Insect net with aperture 1 m^2 carried to a height of about 100 m by a parafoil kite. The small box below the net operates a radio-controlled noose system to close the net before descent. Both developed by Commonwealth Scientific and Industrial Research Organisation (CSIRO), Australia. Source: Photos by V. A. Drake.

to entomologists. Piloting was manual, via radio control, but autonomous operation has been demonstrated in similar contexts (e.g. Cliff et al., 2018) and is probably feasible. Kakutani et al. (2021) have tested a live-capture insect sampler developed for deployment on a UAV. It uses electrostatic attraction to hold the insects and works for stronger-flying species like flies that would easily escape from an open net.

Insects flying above the FBL will usually be moving downwind and can be captured in a windsock-type net held stationary below a kite or helium-filled balloon (Fig. 1; Farrow and Dowse, 1984; Chapman et al., 2004a). To prevent insects escaping as the net is brought down, or insects flying nearer the surface entering and contaminating the sample, the net's rear section can be closed with a noose via a radio-control system. Alternatively, for small insects, sticky panels can be substituted for the net (Florio et al., 2020). A study of windborne movements, at a height of 150 m, of a range of agricultural pest species flying over northeast India employed a commercially available helium-filled 'kytoon', which provides more reliable sampling than a kite as the net remains aloft even in wind lulls (Reynolds et al., 1999). Net apertures are limited by drag to about 1 m^2 and catch rates for the less numerous larger species will be modest in all but the most severe outbreak situations. For example, in two campaigns in inland Australia, during periods of high insect activity, catches of 28 and 36 large- and medium-sized insects were obtained in, respectively, 48 h and 20 h of night-time kite-net sampling (Drake and Farrow, 1983, 1985). This may seem a poor return for the considerable effort required, but confirmation of the identity of the insects undertaking these movements was critical to the interpretation of accompanying radar, meteorological, and ecological data. For small species, sample sizes are often large enough to compute quantitative measures of migration intensity such as densities and fluxes (e.g. Reynolds et al., 1999) and perform statistical tests of significance (Florio et al., 2020).

An aircraft allows sampling to be extended to greater heights than is practicable with tethered systems, for example, to 1.5 km in the pioneering investigations of Glick (1939) in the USA and Mexico. If a fixed-wing aircraft is used, a type that can fly at low speeds is preferable, to avoid the sample becoming damaged in the net (Greenstone et al., 1991). Alternatively, a helicopter can be used (e.g. Beerwinkle et al., 1989), though this is likely to be more costly, or the net can be replaced by a rigid sampling system incorporating an expansion chamber to slow the airflow (Spillman, 1980). As with tethered systems, the aperture of the net or intake is limited by the drag it produces and sampling rates for the less numerous larger insects are usually low. A helicopter-mounted system incorporating both expansion and isokinetic flow (a requirement for accurate estimation of the volume of air sampled) and flown over farmland in Illinois, USA, captured a variety of small insects, including three aphid species that are agricultural pests, at heights up to 1.3 km (Isard et al., 1990). Densities were mainly in the range 0–5 × 10^{-3} m^{-3}.

Servicing traps and sorting and identifying the samples can be prohibitively time-consuming. In response to the servicing problem, automated electro-mechanical systems for partitioning the catch at regular intervals were developed, enabling good time resolution (e.g. hourly) without requiring frequent trap retrievals or visits (Johnson, 1950). With modern video and information technology, more sophisticated solutions are possible. A common aim is to record trap catches automatically and send timely reports to crop managers and researchers via an internet connection; systems with these capabilities are becoming commercially available (Sciarretta and Calabrese, 2019; Cardim Ferreira Lima et al., 2020; Preti et al., 2021). For example, the 'Trapview' product from EFOS (https://trapview.com, accessed 15 January 2022) was used in a study of diamondback moth *Plutella xylostella* infestations in *Brassica* crops in the UK (Wainwright et al., 2020a). This study relied on a species-specific lure and manual checking of downloaded images, but automated identification of mixed-species samples, using the artificial intelligence (AI) technologies of machine vision and deep learning (Høye et al., 2021), is also being explored. Analysis of images of flying nocturnal insects attracted to a light achieved a 69% average precision for identification of eight target noctuid moth species and revealed night-to-night variations in their numbers over a 1-month period (Bjerge et al., 2021).

An alternative approach to the identification task is to employ the recently developed molecular-genetics method of DNA metabarcoding to identify the species in a trap sample (Yu et al., 2012). Over 2000 species were identified in this way from an extensive 1-month programme of car-net sampling on public roads throughout Denmark (Svenningsen et al., 2021). In a trial in Portugal that used metabarcoding to analyse samples caught in simple, low-cost, light traps, over 100 lepidopteran species or species groups were identified by metabarcoding, a similar total to that for morphological examination (Mata et al., 2021). An advantage of the molecular method that was exploited in both these studies is that damage to specimens during capture and handling does not compromise identification, so more efficient bulk methods can be used. Automated monitoring stations combining both image analysis and DNA methods have been proposed, though the envisaged application is general biodiversity monitoring rather than flights over crops (Wägele et al., 2022).

3 Visual observation

Observing insect flight visually is feasible for larger, day-flying insects (Montgomery et al., 2021). Flights of dragonflies and butterflies moving in an approximately consistent direction within a few metres of the surface and in open terrain are highly visible (Williams, 1930; Russell et al., 1998; Dingle et al., 1999). Hoverflies are smaller and less colourful, but their movements can be evident if numbers are sufficiently large (Menz et al., 2019). Many butterflies and dragonflies can be identified in flight, with complementary hand netting to confirm identities and obtain voucher specimens (Srygley and Dudley, 2008). A study at an inland site in Florida, USA, used visual observations, distance markers, and a stopwatch to determine track directions and speeds of three butterfly species flying at a height of about 0.5 m over a ploughed field (Balciunas and Knopf, 1977). The three species were all moving in the same general direction, which hardly changed even when there was a shift in the wind direction. Movements near the surface can be funnelled by terrain, such as a high mountain pass or a coastline, rendering them more visible and practicable to study (Lack and Lack, 1951; Russell et al., 1998).

At dusk, the initial ascent phase of night-time long-distance flights can be detected by viewing the 'twilight arch' above the western horizon through binoculars. This can be effective for small insects such as planthoppers (Riley et al., 1987) as well as large species such as locusts (Clark, 1971). After dark, insects in flight within about 3 m of the ground can be seen in the beam of a torch, a method used to identify the schedule of nocturnal flight activity of cotton bollworm *Helicoverpa armigera* moths over cotton, groundnut, and sorghum crops in the Gezira, Sudan (Topper, 1987). Night-vision goggles, which use optoelectronic technology to enhance vision in low light levels (down to starlight), make studies of this type more practicable. For example, take-off flights of corn earworm *Helicoverpa zea* moths were followed to heights of around 100 m with goggles and supplementary near-infrared (nIR, 600-900 nm) illumination, revealing an initial spiral phase followed by a slower and straighter ascent (Lingren et al., 1995). nIR was used as almost all insects are insensitive to it (van der Kooi et al., 2021). A bright hand-held spotlight with a narrow beam will allow observation of larger species to a height of about 10 m (VAD, unpublished). If the beam is moved continuously, the flights of the insects that it briefly illuminates are usually undisturbed and any general direction of movement soon becomes apparent. Visual identification at night usually depends on knowledge of the species present and observer experience and is not always achievable; ancillary trapping or hand-netting, to obtain a sample for careful examination, will often be necessary.

4 Radar

Detection of insects by radar was first documented in 1949 (Crawford, 1949), and development and use of radars specifically for insect observation commenced in 1968 (Schaefer, 1976). Research employing special purpose 'entomological' radars has continued since then, with ongoing technical development and application to a wide range of entomological questions (Reynolds and Riley, 1997; Chapman et al., 2004b, 2011; Drake and Reynolds, 2012). Insects were occasionally detected on weather radars from about 1970 (e.g. Gray et al., 1975) and are now frequently recorded by the highly sensitive units being incorporated into the operational weather-forecasting systems of most advanced-economy countries (e.g. Martin and Shapiro, 2007; Rennie, 2013, and see below). Radars used for insect observation are of four types: (a) scanning entomological radars (SERs), (b) vertical-looking radars (VLRs), (c) harmonic entomological radars, and (d) weather surveillance radars (WSRs). Harmonic radars require an electronic tag to be attached to the insect and are considered in Section 6. Radar designs suited to insect observation, and methods for analysing the data they produce, are well documented (Chapman et al., 2002a; Riley and Smith, 2002; Drake and Reynolds, 2012; Stepanian et al., 2016) and continue to evolve (e.g. Drake et al., 2020; Gauthreaux and Diehl, 2020; Long et al., 2020).

Radars can operate over a range of transmission frequencies. The signal's wavelength, which is inversely proportional to the frequency, is a critical factor determining the strength of an echo signal from a small target like an insect. If the ratio of target size to wavelength decreases much below one, the reflecting power of the target (its 'radar cross-section') becomes low (Riley, 1985; Drake and Reynolds, 2012). If the radar is purpose-built, the transmission band can be chosen (within regulatory constraints; see Appendix) to maximise the echo signal from individual insects of the species of focal interest by choosing a wavelength comparable to or smaller than the size of the insect. X-band (9.4 GHz, 3 cm wavelength) is suitable for grasshoppers and larger moths and has the advantage that equipment produced for the marine-navigation market can be adapted and procured relatively easily. With the beam energy concentrated into a narrow 'pencil' by a dish antenna, a detection range for individual large insects of around 2 km is achievable. Ka-band (34 GHz, 9 mm wavelength) has been used for studies of planthoppers (Riley et al., 1991), which are only a few millimetres in length, and some use has also been made of W-band (95 GHz, 3 mm wavelength; Geerts and Miao, 2005; Wood et al., 2009a) though so far only by employing equipment constructed for other fields of research. Weather radars use wavelengths of 5 cm or 10 cm, which are optimal for their intended application but not ideal for insect observation; nevertheless, these powerful and sensitive units can detect insects in many circumstances.

4.1 Scanning entomological radars

The purpose-built entomological radar developed by Schaefer (1976) transmitted at X-band and employed a 1.5° pencil beam which was scanned in azimuth at a series of elevation angles, providing coverage of the surrounding airspace out to 1–2 km. Following successful demonstration of this system in a region of the African sahel where daytime butterfly and nocturnal locust flights were observed, radars with this 'configuration' were predominant in entomological research for the next two decades (Fig. 2; Drake and Reynolds, 2012). These units have usually been deployed for periods of 10–20 days in intensive field campaigns timed to coincide with the period of flight activity of the study population. For example, nocturnal migratory movements of Australian plague locusts *Chortoicetes terminifera* over a wheat/livestock farming region in eastern Australia were documented with this technology (Drake and Farrow, 1983), and a variant transmitting at Ka-band revealed similar night-time movements of planthoppers, a much smaller radar target, over intensive cropping regions of east-central China (Riley et al., 1991). Modernised units with digital data acquisition (Cheng et al., 2002) and automated operation are still in use (e.g. Sun et al., 2022).

Insect echoes appear as discrete 'paints' on an SER's plan position indicator (PPI) display, and these can be followed for a few scans of the antenna (Schaefer, 1976; Cheng et al., 2002). Counts of paints within a defined area of the PPI provide estimates of insect density and analysis of a succession of images allows speeds and directions to be determined. By combining these measured values, the intensity of the migration can be determined. In northeast China, for example, over an 8-night period, a total of 5000 *M. separata* moths were estimated to have crossed each metre of a line perpendicular to their general direction of movement, flying mainly at heights of 100–500 m (Chen et al., 1989). This immigration was sufficient to account for the damaging infestations of maize and other crops that were then occurring regularly in that region. SER observations have also been incorporated into a long-term

Figure 2 Scanning entomological radar deployed for observation of moth flight over irrigated cotton fields near Narrabri, New South Wales, Australia. This was a 'second generation' unit with a low-height antenna. Antenna on mobile scanning pedestal (housing the transceiver) at right, observation cabin with displays and controls at left; electrical generators and other ancillary equipment were located further to the left. Radar developed by CSIRO, Australia. Source: Photo by V. A. Drake, reproduced with permission of CABI Publishing.

study of insects migrating across the Bohai Gulf, which separates the extensive croplands of east-central and of northeast China. Sited on a small offshore island and supported by searchlight-trapping to provide identifications, this radar documented migrations of moths of the pest species *H. armigera* and *M. separata* (Feng et al., 2008, 2009) and of beneficial dragonflies, predominantly *Pantala flavescens* (Feng et al., 2006).

When scanning at a low elevation angle, an SER can reveal insect concentrations associated with specific habitat types. Exodus flights at dusk from a groundnut crop in the Gezira, Sudan, were detected in this way and attributed, via hand-net catching and daytime surveys, to moths of two pest species, *H. armigera* and the Egyptian cotton leafworm *Spodoptera littoralis* (Schaefer, 1976). In a study in rural Kenya, plumes of African armyworm moths *Spodoptera exempta*, an outbreak pest of cereals and pastures, were observed taking off from groups of trees where they had previously gathered (Riley et al., 1983). Mating flights of European honey bee *Apis mellifera* males have been observed with an SER operating in flat terrain with little vegetation (Loper et al., 1992). A research program developed around the low-height observing capability of SERs is described as a case study (Section 7).

4.2 Vertical-looking radars

A radar with a fixed vertical beam has an advantage of design simplicity that makes it much easier to automate than an SER. VLRs are therefore capable of unattended operation and can be deployed into long-term observing programs. An early model with a very simple design recorded insect migration over agricultural lands in east-central Texas during 1990 and 1991, revealing both diurnal and seasonal patterns of migratory activity at heights between 450 m and 2300 m (Beerwinkle et al., 1993, 1995). Only larger species were detectable at these heights, and it was inferred from trapping programs and earlier research results that the migrants were adults (moths) of various noctuid lepidopteran species that are important crop pests in this region. As the beam of a VLR is stationary, observations take the form of transits of the beam by the insect. These last a few seconds, allowing acquisition of an extended recording of the echo from which information about the target's identity can be derived (Drake et al., 2002), adding to the advantages of this radar type over SERs. Since around 2000, VLRs have become the predominant configuration among purpose-built entomological radars, and years of almost continuous observations have been accumulated at sites in Britain, Australia, and China. Modern units usually employ beams with rapidly rotating linear polarisation and a small 'wobble' or 'nutation' – a combination termed the 'ZLC configuration' (Drake and Reynolds, 2012). These beam modulations 'interrogate' the insect during its transit and produce a rather complicated echo signal from which the size of the insect, its direction of movement, its alignment direction, and a measure of its shape can be extracted (Chapman et al., 2003; Harman and Drake, 2004). These characters allow the detected insects to be differentiated, but only into a small number of broad classes (Hao et al., 2020).

Current VLRs cannot observe below about 150 m and their observations are therefore almost entirely of windborne migratory movements. The Australian units are located within the inland region where *C. terminifera* populations develop, and they perform, inter alia, a locust-detecting role. Significant migrations of this species are reported to the Australian Plague Locust Commission (APLC), the organisation with principal responsibility for locust forecasting and management. Both the intensity and the direction of movement can be estimated, and likely source and destination regions inferred. For example, between February and April of 2008, there were nine occasions when adult locusts flew in large numbers over a VLR at Bourke, New South Wales (Drake and Wang, 2013); these nocturnal movements were predominantly from the northeast, with source regions estimated to be 200–400 km away. The migrants could be confidently identified as *C. terminifera* because echoes from this species have distinctive shape and wing-beat characters that distinguish them from most other large insects; in addition, the migratory behaviour of this species is well established and contemporary data on locust populations is available from APLC's routine survey and report-gathering procedures (Adriaansen et al., 2016). Units in Britain have documented immigrations from the nearby European landmass of both the pest *P. xylostella* (Chapman et al., 2002b) and beneficial hoverflies (mainly *Episyrphus balteatus*; Wotton et al., 2019). A second beneficial species, the green lacewing *Chrysoperla carnea*, was estimated to move slightly shorter distances, covering 40–150 km in a night (Chapman et al., 2006). The hoverflies were observed moving predominantly north in late spring and south in early autumn, at heights up to 750 m. It was estimated that, on average, approximately 0.5 billion hoverflies, corresponding to a biomass of about 10 t, enter southern England in spring, with these subsequently producing an additional 2 billion adults through local breeding. In these studies, identification of the targets detected by the VLR relied on estimates, from the echo signal, of size and shape, these being combined with information from trapping programs, citizen-science recording schemes, and established knowledge of the likely incidence of the various species.

4.3 Weather surveillance radars

Weather surveillance radars (WSRs) observe over a wide area, out to ranges of 100 km or more. At these ranges their beams are wide, and they invariably detect 'multiple-target echo', i.e. the combined signals returned by an ensemble of targets rather than a single individual (Chilson et al., 2018). Their 5 cm or 10 cm wavelengths favour detection of birds and bats rather than insects. Modern WSRs employ Doppler processing and dual polarisation, which provide additional information that can help to distinguish biological target types (Stepanian et al., 2016; Gauthreaux and Diehl, 2020). When present in large numbers, and not obscured by more dominant bird or bat echoes, insects can produce obvious features on a WSR display that reveal their location and direction and speed of movement. Examples include ascent at dusk of beet armyworm *Spodoptera exigua* and cabbage looper *Trichoplusia ni* from an area of cotton and other crops in southern Texas, USA, and their subsequent downwind movement (Westbrook and Eyster, 2017). Ascent on individual nights was concentrated into areas only 5–10 km across, though the migrants soon diffused into a broader 'cloud' which was estimated, from analysis of meteorological data on winds, to have moved up to 450 km north over the following 10 h of darkness. Flights of moths of the spruce budworm *Choristoneura fumiferana*, a pest of coniferous forests in eastern Canada, have been tracked with a WSR (Boulanger et al., 2017), and concentrations of mayflies above rivers and lakes (Stepanian et al., 2020) and grasshoppers above city lights (Tielens et al., 2021) have been documented. Even aphids, despite their small size, can be detected if present in sufficient numbers (Irwin and Thresh, 1988). WSR observations have been integrated into a pest-insect immigration alarm system operating in Finland, where infestations of aphids in cereals and *P. xylostella* in *Brassica* crops are often initiated by flights on warm southerly winds across the 70–130 km width of the Gulf of Finland (Leskinen et al., 2011).

4.4 Strengths, limitations, and prospects

As the previous examples make clear, radar is very effective at detecting insects at heights above about 100 m, providing coverage through to the upper limit of insect activity at around 2 km. Radar observations typically encompass the intensity of the movements, their direction and speed, how these vary with height, the times of day or night when they occur, and their duration. It therefore has great utility for observing, quantifying, and monitoring the long-distance windborne migratory movements that occur well above the surface. This capability is lost in rain and (WSRs only) during intense bird migrations, as echo from these targets overwhelms that from insects. SERs and VLRs detect individual insects and can provide quantitative measures of migration intensities, speeds, etc. (Schaefer, 1976; Drake and Reynolds, 2012). Quantitative interpretation of the multiple-target echoes recorded by WSRs is less straightforward, but differences in migration intensity, both in space and in time, will be evident (e.g. Westbrook and Eyster, 2017). The great advantage of WSRs is their wide coverage, in contrast to both SERs and VLRs which essentially provide a point sample. For observations below 150 m, current VLRs are unsuitable for technical reasons related to the very short range of the observations, and the beam of a WSR is well above the ground at most observing ranges. SERs are more capable in this respect, and in the ideal conditions of flat terrain and low or absent vegetation they can detect individual larger insects to within a few metres of the surface.

The critical limitation of radar is that it provides little information about the identity of the targets. ZLC-configuration VLRs are least unsatisfactory in this respect and the size, shape, and wing-beat-frequency values they provide have been invaluable for the biological interpretation of their observations. However, these broad characters are not in themselves sufficient for identification to species or narrow species groups. For that, ancillary biological inputs are needed: knowledge of species likely to be present and their phenologies and typical abundances, and preferably also results from a concurrent trapping programme or recording scheme. A limitation specific to WSR observations is that not all national meteorological organisations provide access to the raw signal data needed to distinguish insects from other sources of echo (Shamoun-Baranes et al., 2021, 2022).

Radar observations predominantly reveal the transmigration phase of long-distance insect movement. Emigration is evident as an increase in numbers aloft, around dusk in the case of nocturnal migrants and either at dawn or mid-morning (when convective air movements become established) for day-flying species (Reynolds et al., 2008; Wood et al., 2009b). If the radar is a scanning type, take-off plumes from local concentrations may also be detected. Unfortunately, immigration, the phase that is of most immediate concern to agriculturalists, is harder to resolve. Generally, it manifests only as a gradual decrease in numbers aloft, and this could be due not to landing but to fewer migrants arriving from more distant source regions where take-off was less intense. Occasionally it is possible to associate a decrease in radar targets with an increase in insects at the surface. For example, SER observations in the Tilemsi Valley of Mali revealed an intense movement of grasshoppers, probably mainly *Diabolocatantops axillaris*, at heights of 240–350 m (Reynolds and Riley, 1988). A decrease in intensity of this layer concentration coincided with an increased catch of *D. axillaris* in a light trap at the radar site. Detection of immigration has been limited by the inability of most entomological radars to detect the vertical component of a target's velocity. However, recent advances in VLR signal processing (Drake and Wang, 2019) and adaptation of newer, more sophisticated, radar technologies to insect observation (Wainwright et al., 2020b; Wang et al., 2021) allow ascent and descent to be observed directly, potentially making this important phase of migration routinely observable.

5 Optoelectronics

Light, like radio waves, is a form of electromagnetic radiation, but its wavelengths are much shorter, leading to different reflection properties when it strikes an object or surface. Optical observation has both advantages and constraints, which arise from light's physical nature and interactions with matter (e.g. Lipson et al., 2010). Note that the term 'light' is to be understood here as including nIR. Light wavelengths are shorter than the dimensions of even small insects, hence the steep decrease in reflecting power with size that occurs with radar is not experienced. This makes light much more suitable than radio waves for detecting mosquitoes and other similar-sized aerofauna. Laser technology enables the generation of beams with very narrow divergence, making horizontally directed observation practicable as the beam can be aimed to avoid vegetation and other obstructions. Imaging using cameras, a technique not available at radio wavelengths, provides a second means of distinguishing airborne insects from nearby vegetation. Hence optical methods are particularly suited to observation of insects near to the ground, especially if the species of interest are small.

A further advantage of light is that it produces specular reflections from wings, so that wing-beat modulations are much stronger than with radar, are rich in harmonics, and can be distinguished from a steadier body component of the echo signal (Brydegaard, 2015; Genoud et al., 2018; Kirkeby et al., 2021). This additional information content increases the capacity of modulation data to discriminate between species or species groups. Differences in the polarisation of the transmitted and reflected light are also potentially measurable and may have value for target identification (e.g. Genoud et al., 2019; Jansson et al., 2019). If two light frequencies are used, an indication of the extent of pigmentation ('melanisation') may be obtainable (Gebru et al., 2018). Light is of course present naturally, both by day and by night - though at very varying intensity levels. This 'ambient' light can constitute either a resource (a source of illumination) or an impediment (an unwanted background signal), depending on its intensity and whether the observing system incorporates its own illuminator.

Optoelectronics is the technique of converting incoming light into an outgoing electronic signal. A variety of forms of optoelectronics-based equipment have been developed for detecting insects in flight and observing their behaviour. In these devices, the electronic signal is recorded to provide a permanent record that can subsequently be analysed to retrieve information of biological interest; in current units, recording is invariably to a digital file. Night-vision goggles or telescopes, which enhance the vision of a human observer but do not record data, were discussed in Section 3.

5.1 Short-range monitoring

A device that monitors insects in flight at distances of up to 1.5 m is, at the time of writing, undergoing field trials with a view to commercial production as a pest-management tool (Rydhmer et al., 2022; https://www.faunaphotonics.com/technology/, accessed 27 January 2022). The device employs nIR illumination and a crossed-beam configuration: the illuminator is angled upwards and the receiving optics are mounted above it and directed horizontally. The volume within which insects were detected varies from 0.005 m^3 to 0.1 m^3, depending mainly on the size of the insect. The nIR is produced by light-emitting diodes (LEDs) that emit at two different frequencies (808 nm and 980 nm); the emissions are modulated at 119 kHz and 79 kHz, respectively, and detection of signals with these precise frequencies distinguishes echoes from ambient illumination. Different echo intensities at the two frequencies may indicate the degree of melanisation and a high sampling rate captures wing-beat frequencies to 1 kHz. These weatherproof devices can be mounted on posts in or next to crops, operate unattended, and report activity throughout the day and night via a mobile telephone network. Data will be accessible to growers, pest managers, and researchers both instantly and via an archival database on the 'cloud'.

5.2 Long-range monitoring

Insects can be detected optically at longer distances in the field of view of a high-magnification telescope. Astronomical telescopes produced for the amateur market are suitable and readily available at reasonable cost. At night, illumination will be required, and for longer distances the minimally diverging beam of a laser will be needed. Passive observation, using ambient light, has advantages of simplicity and lower cost (Brydegaard et al., 2016). An instrument capable of detecting low-flying insects during daytime employed an astronomical telescope in which the eyepiece was replaced by a 'quadrant' (four-segment) optoelectronic detector (Jansson and Brydegaard, 2018). As an insect transits the beam, the reflected light falls first on one segment of the detector and then on the segment opposite. The difference in the timing of the two signals provides an imprecise measure of the insect's range. To reduce the background illumination, the telescope is aimed at a black termination cavity. In a trial lasting 3 h at a site in southern Sweden, with the telescope directed horizontally over low vegetation, almost 1300 insects were recorded over a 140 m transect. A range value was retrieved for 39% of these and wing-beat frequencies in the range of 100–200 Hz were observed. A limitation of this system is that the need for a physical termination restricts coverage to a single fixed linear transect, usually within a few metres of the ground.

A system employing nIR laser illumination and a large-aperture (25 cm) telescope observed insects out to 36 m but did not attempt to determine their range or to alter the observing direction (Genoud et al., 2021). An nIR filter eliminated most of the ambient-light background. Insect echoes were separated from other signals by the presence of a wing-beat frequency and then assigned to one of four classes based on this frequency and the ratio of the wing and body components of the echo signal. Insect densities were estimated from the proportion of time that insect echoes were being detected, a method that avoids biases arising from insects transiting the beam at different speeds. The system operated continuously for 80 days with very little down time, observing over tall grass bordered by trees. It revealed differing diurnal activity patterns for the four classes and a general decrease in insect numbers over the autumn observation period, with short-term peaks and troughs attributable to changing weather and perhaps population cycles.

Systems that employ laser illumination and determine the target's range are termed 'lidars'. In most types, pulses of illumination are emitted and the range of a target is determined from the time-of-flight of the pulse and its returning echo, in the same manner as a radar. A system using this method, and again employing a modified astronomical telescope to capture the reflected light, was able to detect insects in daylight as well as darkness (Tauc et al., 2019). By using very short pulses, a range resolution of 0.75 m was achieved. Pulse operation obviates the need for a dark termination, and this allowed the observation direction to be varied in both azimuth and elevation (using a pan-tilt mechanism), providing three-dimensional spatial coverage. The beam was expanded to 5 cm to increase the probability of an insect transit occurring and to lengthen the transit's duration so that wing beats could be detected. The presence of wing beats in a signal was used to distinguish insects from vegetation and the ground. In a trial in a national park in Wyoming, USA, insects were detected to about 100 m in range and 30 m in height, and an increase in numbers around sunset was evident.

An alternative laser-based design uses continuous transmission and obtains range information via the Scheimpflug principle, in which the image plane of a camera is slanted so that objects at different ranges remain in focus (Steger, 2017; Brydegaard et al., 2017). A system of this type for detecting insects employs a receiving telescope that is slightly offset from the laser beam and directed at a very slight angle to it so that a length of the beam is in the field of view; images from sources at points along this length are focussed on different regions of a slant-mounted detector array (Brydegaard et al., 2014). Range resolution is poorer than with a time-of-flight lidar, and increasingly so at longer ranges, but continuous transmission greatly increases signal strength. A system operating in Kenya over 5 days detected 300 000 insects (Brydegaard et al., 2020b). The laser beam was expanded to a width of about 10 cm (its form varying with range) and directed horizontally at a height of about 4 m above almost flat farmland, terminating on a dark board at 600 m. Modulation frequencies in the range 100–800 Hz, due to wing-beating (fundamental and overtones) were observed in many of the recorded echo signals. A cluster analysis of the modulation spectra identified about 20 classes, of which those with higher wing-beat frequencies (300–800 Hz) were deemed to be mosquitoes. All classes of insects exhibited peaks in activity around dawn and/or dusk and an additional, smaller peak was observed when a solar eclipse was near its maximum (Fig. 3). Scheimpflug lidars can operate without a termination screen, and thus the beam can be directed upwards, allowing monitoring of the airspace well above the ground. A unit was employed in this way to detect insects in the vicinity of an 80 m high wind turbine in Sweden (Jansson et al., 2020).

5.3 Imagers

Video technology provides an image of the subject area allowing a segment of the track of a moving insect to be recorded. A pair of video cameras and an nIR illuminator were used to track moths flying over a pigeon-pea crop at night (Riley et al., 1990). Tracks (in three dimensions) were estimated stereoscopically from the locations of the moths and of fixed markers (on poles erected in the field of view) on two image frames; with a camera separation of 35 m, coverage

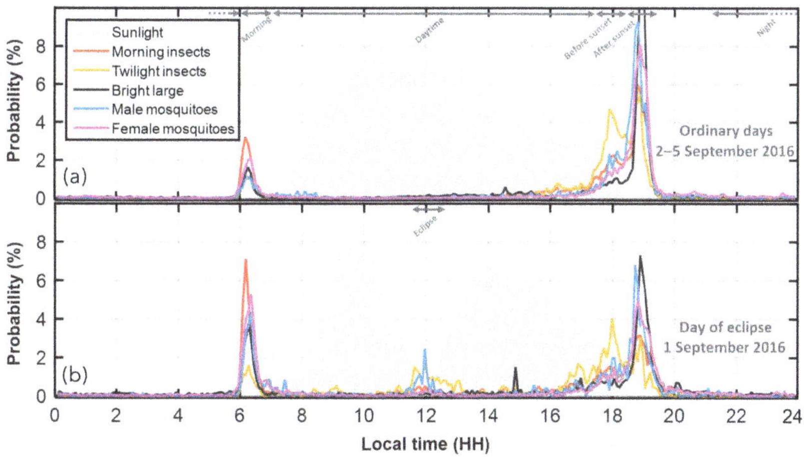

Figure 3 Variation with time of the number of insects of 5 different types flying above fields in Kenya over a 5-day period, observed with a Scheimpflug lidar. Top: average for 4 consecutive 'ordinary' days. Bottom: day with a solar eclipse around midday. Source: Reproduced from Brydegaard et al. (2020b), with permission of American Association for the Advancement of Science.

extended over a volume of 90 m^3. A broadly similar system using modern digital video technology and tracking algorithms has been developed but so far used only for birds and bats (Betke et al., 2018). Instead of video, Wallace et al. (2021) used low-cost wildlife cameras and nIR illumination to record single-shot time exposures. At night, exposure times were long enough to record not only the track of an insect as it moves across the field of view, but also modulations in the size of the image due to wing-beating. Image processing techniques were adapted to count the number of brightness peaks along the track, and division by the exposure time then gave an estimate of the wing-beat frequency. In a trial over a boulder field in the Snowy Mountains, Australia, Bogong moths *Agrotis infusa* flying at dusk were found to have wing-beat frequencies predominantly in the range of 40–60 Hz.

5.4 Strengths, limitations and prospects

From the above examples it is evident that light-based methods are effective near the ground, including when there is tall vegetation nearby, and that they can detect species as small as mosquitoes. In comparison with radar, they provide stronger wing-beat signals with greater information content and, at least potentially, an estimate of melanisation, a character completely inaccessible at radio frequencies. Many optical systems have small sensitive volumes or employ narrow beams, so insects may pass through the sensitive volume too quickly for wing-beating to be fully characterised; moreover, species present at low densities will be recorded only infrequently. However, good sample sizes have been obtained in trials, indicating that volumes can be adequate for observation of small insects near the surface as these are often numerous.

Automated digital video and image-interpretation technologies are now established and widely used, and systems have been developed for entomological applications such as monitoring pollinator visits to flowers (Pegoraro et al., 2020) and locating settled insects marked with a fluorescent dye (Stumph et al., 2019). Technologies of these types are developing rapidly and warrant further investigation for possible adaptation to detection of insects flying above the vegetation layer. Small and light-weight (12.5 mg) photoluminescent tags, which draw their energy from sunlight, have been shown to enable camera-tracking of insects in a cage; mounting the detector system on a pursuing autonomous UAV is proposed for tracking insects that move over a wide area (Walter et al., 2021).

Video and laser technologies have been combined in a prototype system that aims not only to detect and identify insects but also to control them (Mullen et al., 2016). A single camera is aimed at a retro-reflective sheet 8 m away, the sheet being illuminated with nIR. Mosquito-sized insects are detected from their silhouette. Image processing is used to track lateral motion of the insect and direct a green-light laser beam onto the insect to enable recording of wing-beat modulation via a lens and photodiode. If an undesirable species is identified, a pulse from a second, more powerful laser is directed at it to kill or at least disable it.

Some of the systems described here are clearly prototypes and field experiments have mostly taken the form of trials rather than biological studies. However, recent Scheimpflug-lidar observation campaigns continuing for around a week, in one case through both day and night, and addressing phenomena of biological interest or practical value (Brydegaard et al., 2020b; Malmqvist et al., 2019; Jansson et al., 2020), suggest that this technology is transitioning from development to application. A concern is that optical systems involving telescopes and lasers need careful adjustment on installation, and occasional readjustment and cleaning, and this may limit their application for long-term unattended monitoring. The non-ranging laser-based system of Genoud et al. (2021) was kept in operation for almost 3 months, but check-up visits were required every 3 or 4 days. Concerns about eye safety, for livestock and wildlife as well as humans, will also constrain instrument deployment and introduce regulatory overheads. Short-range observing systems, using either video or a non-imaging detector and incorporating LED rather than laser illumination, can avoid many of these difficulties. They may therefore prove to be the preferred technology for researchers and pest managers despite their limited spatial coverage. If the cost can be reduced sufficiently, deploying a network of multiple devices of this type to cover different locations within a study area might be a practicable approach.

6 Radiotelemetry and harmonic radar

The methods described so far provide brief observations of wild insects as they go about their normal activities. For each insect the data obtained comprise a single record of its presence at a particular location plus, with some of the methods, a single estimate of its track: i.e. three position coordinates and sometimes also three velocity components. Most methods also provide an indication of the insect's identity. To address many research questions, information about activity over an extended period is required: i.e. a sequence of position coordinates is needed. Biologists studying vertebrate species can now obtain such data almost routinely, using a variety of telemetry methods or recoverable dataloggers (Kissling et al., 2014; Drake and Bruderer, 2018). These methods require the animal to be tagged with a device containing active electronic circuits, powered by an on-board battery. In earlier systems, the animal was located from detections of a simple signal from the device at one or more receiving stations. Modern units record positional information, and often also physiological or environmental data, which is then either transmitted periodically or downloaded after the animal has been retrapped and the device recovered. Any tag constitutes a burden for the animal carrying it, and the small size of insects limits application of these methods in entomology. However, in a few cases, tracking of insects in flight has been achieved and produced valuable data. Transmitter tags have been more widely used to relocate settled insects to determine their pattern of movements within a habitat, often over their adult lifetime, or to study insect flight in laboratory wind tunnels (e.g. Hedin and Ranius, 2002; Kutsch, 2002; Ando et al., 2002). These applications employ similar technology to those described here but fall outside the scope of the present review.

A few large, strong flying, insect species can carry the lightest radio transmitters available, which have a mass of about 0.2 g and battery lives of about three weeks (Naef-Daenzer et al., 2005). Dragonflies and monarch butterflies *Danaus plexippus* undertaking long-distance migrations have been tracked for hundreds of kilometres using a network of receiving towers (which provide only sparse location information; Knight et al., 2019) or handheld receiving units operated from a following vehicle or light aircraft (Wikelski et al., 2006). In the latter study, 1.2 g green darner *Anax junius* dragonflies carrying a 0.3 g tag (25% of their body mass), were able to complete flights of up to 150 km and continue active flight for up to 12 days. They moved predominantly to the south, consistent with expected migratory behaviour for the location (eastern USA) and autumn season. The extended tracking that this technology enables provides information on migration routes on a scale of order 100 km that is not available by any other means. A transmitter tag of mass only 80 mg has been produced by replacing the battery with a piezo-electric generator powered by the insect's wing-flapping (Shearwood et al., 2021). Preliminary tests of the tag and a purpose-built receiver demonstrated detection and determination of direction at 10–20 m, which may be sufficient for pursuit of a tagged insect by an autonomous UAV. In a field trial of a prototype UAV-based insect-tracking system, stationary 0.2 g battery-powered radio transmitters were located to within 20–50 m from starting distances of about 300 m (Kim et al., 2022).

An alternative tag design, the harmonic transponder, can be made extremely light (down to 1 mg) as it employs a very simple circuit and does not require a battery (Brazee et al., 2005; Drake and Reynolds, 2012; Ala and Colpitts, 2021). These devices receive a radio signal at a microwave frequency via a simple antenna and rectify it, to produce signals at integer multiples of the incoming frequency ('harmonics'). These are then emitted via the same antenna. Detection of one of the outgoing frequencies (usually the second harmonic, at double the frequency of the illuminating signal) allows the tag to be located even in a highly cluttered environment, because vegetation, soil, other insects, etc. do not contain rectifying materials and therefore reflect only at the illuminating frequency. The method has two significant drawbacks: the illuminator, receiver, and tag are all specialised equipment and not readily available; and the tag's simple circuit provides no identification information. A related technology, radio-frequency identification (RFID), does provide individual identifications and tags light enough to be carried by large insects are available (Pope et al., 2015); however, detection distances are of order 10 cm, and they therefore have no utility for observing insects in flight.

Harmonic insect detectors fall into two categories: portable direction-finders and true radars. Direction-finders do not determine range but are effective for relocating tagged insects periodically, usually once or twice a day, when they are settled; individuals are then identified visually from a number on the tag or similar conventional mark. They have a detection range of 10–50 m (Drake and Reynolds, 2012; Storz and Lavrenko, 2020), which makes them practicable for insects that disperse over distances of order 100 m. For example, attraction to cash and trap crops by the brown marmorated stink bug *Halyomorpha halys*, a pest of fruit and vegetable production, was examined using a handheld harmonic direction finder (Blaauw et al., 2017). A possible application of this technology to flying insects that involves mounting units on pursuing autonomous UAVs is being explored (Lavrenko et al., 2021).

Harmonic radars are larger, trailer-mounted devices that operate much like a conventional SER except there is a separate receiving antenna which directs the echo into a receiver tuned to the harmonic frequency (Riley and Smith, 2002). These units transmit short high-power pulses and determine range by the time-of-flight method. Tags can be detected out to about 900 m and are positioned to within a few metres both in range and laterally. Insects are commonly followed for several minutes and tracked over several hundred metres. Harmonic radars are therefore excellent tools for observing insect flights that extend over tens or hundreds of metres but stay inside an area of a few hectares. At least for current units, the insect must remain within a few metres of the ground, as while the antennas scan azimuthally through 360°, the beam is directed horizontally and has only a small vertical extent. Although the frequency doubling very effectively eliminates ground clutter, the beam is still subject to shadowing and therefore the units must be operated in flat and open terrain. Another limitation is that, as tags are not individually identifiable on the radar's PPI, tracks may become confused if multiple tagged insects are present within radar range.

Harmonic radar is a specialised technology requiring significant investment of funds and engineering expertise, and so far only four units have been constructed worldwide. These have been used mainly in studies of navigation and homing by *A. mellifera* (e.g. Menzel et al., 2019) and of foraging by this species (Riley et al., 2003; Reynolds et al., 2009; Lihoreau et al., 2013). The effects of low concentrations of neonicotinoid insecticides on *A. mellifera* flight behaviours were assessed with this technique (Tison et al., 2016).

Studies of foraging by butterflies (Cant et al., 2005; Ovaskainen et al., 2008) and bumblebees (Fig. 4; Osborne et al., 2013), and of the responses of turnip moth *Agrotis segetum* to pheromones (Reynolds et al., 2007), have also used this methodology. Qualitative visual observations of tagged and untagged butterflies of two species indicated that the 12 mg tag (4–8% of body mass) had no effect on flight activity (Cant et al., 2005).

Harmonic radar is also being used to track yellow-legged hornets *Vespa velutina,* an invasive predator of *A. mellifera* (Maggiora et al., 2019). The aim is to locate hornet nests, not easily found by conventional searching, so that these can be destroyed. Hornets attacking a beehive are caught, tagged, and released; tag detections are recorded and presented on a PPI, revealing flightpaths to and from the nest. It is not necessary to identify individuals, so multiple tagged hornets can be released. This procedure needs to be carried out in all types of terrain and in the presence of trees and buildings, so the radar's two antennas are mounted on a 6 m tall extendable mast and are designed to produce a beam with a large vertical extent. The maximum detection range is, at best, 500 m, and the radar usually must be repositioned one or more times before the nest is located. In a trial, nests were found in both rural and built-up areas; however, the procedure was time-consuming, and the present unit is too heavy to be operated in areas inaccessible to vehicles.

7 Case study: moths over a cotton crop

Specific applications of various individual detection and monitoring methods have been briefly described above. In this section, an attempt to combine

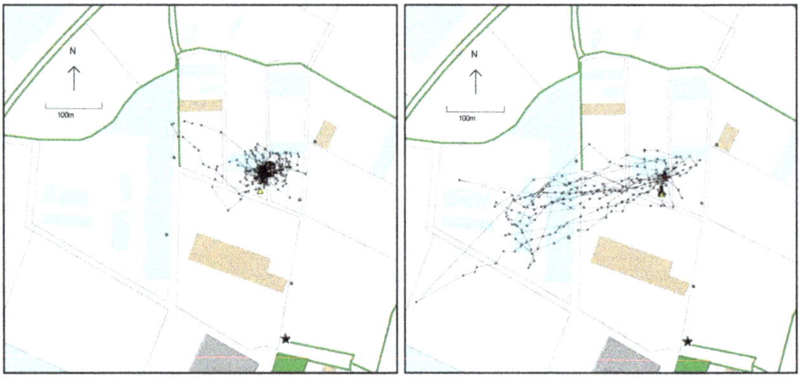

Figure 4 Tracks of foraging buff-tailed bumblebees, *Bombus terrestris*, in an agricultural landscape (Hertfordshire, UK), revealed by harmonic entomological radar observations of tagged individual workers. Left: first trip from colony (14 individuals). Right: experienced bees (more than 6 previous trips; 8 individuals). The star indicates the location of the radar. Source: Extracted from a larger figure in Osborne et al. (2013) and reproduced under the terms of the Creative Commons Attribution licence.

several of these methods to study the aerial movements of two pest Lepidoptera species is recounted. The study took place in Australia between 1989 and 1991, when larvae of *H. armigera* and the native budworm *Helicoverpa punctigera* were severely impacting cotton growing. The programme was based around a scanning radar (Fig. 2) supplemented by a range of meteorological and ancillary entomological observations (Fig. 5; Drake et al., 1993); it built on experience from previous radar-based campaigns studying these and other species. The need for accompanying meteorological data had been recognised from the earliest days of radar entomology (Schaefer, 1976) and interception trapping, as a means of obtaining definite identifications for the pests detected by the radar, was gradually introduced during the 1980s (e.g. Drake and Farrow, 1985; Riley et al., 1987).

Identification of the insects detected by the various observing methods was particularly important, as other moth species, similar in size to *Helicoverpa* spp., were likely to be present and needed to be excluded from the dataset. Preliminary investigations had indicated that the *Helicoverpa* moths sometimes flew at heights of only a few metres, and interception traps operating at 2–4 m and 8–12 m were therefore developed for the 1989–1991 fieldwork (Fig. 1; Drake, 1991; Drake et al., 1993). The radar was modified to reduce the height

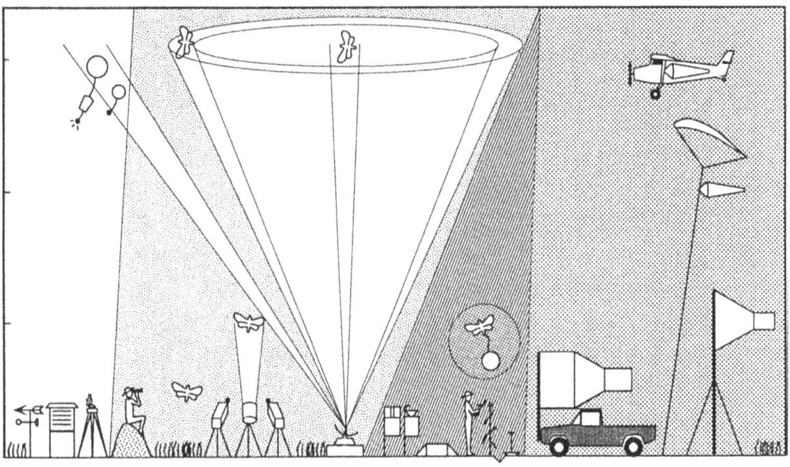

Figure 5 Methods used to investigate movement of *Helicoverpa* spp. moths over crops, at both low and high altitudes. From left: meteorological observations – surface wind, temperature, and humidity, and radar- and theodolite-tracked wind-finding balloons and temperature sondes; direct observations – observer with binoculars or night-vision goggles, stereoscopic video system, entomological radar in scanning and vertical-beam modes; population monitoring – light and pheromone traps, emergence traps, crop surveys, and transect searches; mark-and-capture study; interception trapping – vehicle-mounted net (2–4 m), tower-mounted net (8–12 m), kite- or kytoon-borne net (100–200 m), aircraft-borne net (400 m). Source: Reproduced from Drake et al. (1993), Drake and Reynolds (2012), with permission of CSIRO Publishing, CABI Publishing.

of the antenna (Fig. 2), as this decreases clutter echo from vegetation when observing insects flying immediately above a crop (Drake and Reynolds, 2012). The unit was sited overlooking a fallow field almost clear of vegetation and an adjacent cotton crop. Ancillary observations included dual-video recordings, handheld nIR spotlighting with night-vision goggles over the fallow field and the fallow-cotton boundary, and observations with night-vision goggles and a vertical nIR beam (Drake and Fitt, 1990; Drake et al., 1993). An accompanying project used a mark-recapture method to estimate the distance to which newly emerged moths dispersed (Fitt et al., 1995).

Observation campaigns were timed for periods when *Helicoverpa* adults were predicted to emerge from pupae and observations and sampling were scheduled mainly for the hours from dusk to midnight when flight activity was strongest. Host plants favoured by the two species were grown in small plots, without pest control, to provide a study population. A compendium of sample results obtained by these methods is presented in Fig. 6. Movements across crop boundaries were detected mainly by interception trapping and visual observation methods, while the radar was effective at monitoring migration above 100 m; dusk ascent was recorded by both methods (Drake and Fitt, 1990). These results contributed to the choice of parameter values for the movement component of a life-system model for the two moth species, and of a model of moth movements over long distances, that were developed in the following years (Fitt et al., 1995; Rochester et al., 1996). Cotton growing in eastern Australia has subsequently adapted to the challenge of these and other insect pest species through application of a range of technologies incorporated into an IPM approach in which the role of movement is implicit (for example in its use of trap, attractant, and refuge crops) though no longer emphasised (Wilson et al., 2018).

Although this campaign was designed around the radar, the ancillary observing and sampling methods, and the mark-recapture experiment, played increasingly significant roles (Drake and Reynolds, 2012). The radar observations provided confirmation that the movement patterns and behaviours observed at a few locations by visual means were typical of what was happening over a wide area, and continually reminded the researchers of the larger scale movements aloft, which otherwise would easily have been overlooked. A simplistic assessment might be that the remote-sensing components produced good quality quantitative estimates of insect movement and the trapping components produced firm identifications of the target species (the *Helicoverpa* moths), but integration of these two types of information was not fully achievable. Low capture rates of insects in the kite- and aircraft-borne nets, and the difficulty of confidently identifying low-flying moths seen in light beams or on video or radar, were specific weaknesses. Because of these limitations, application of these methods to other, lower-value cropping systems was not

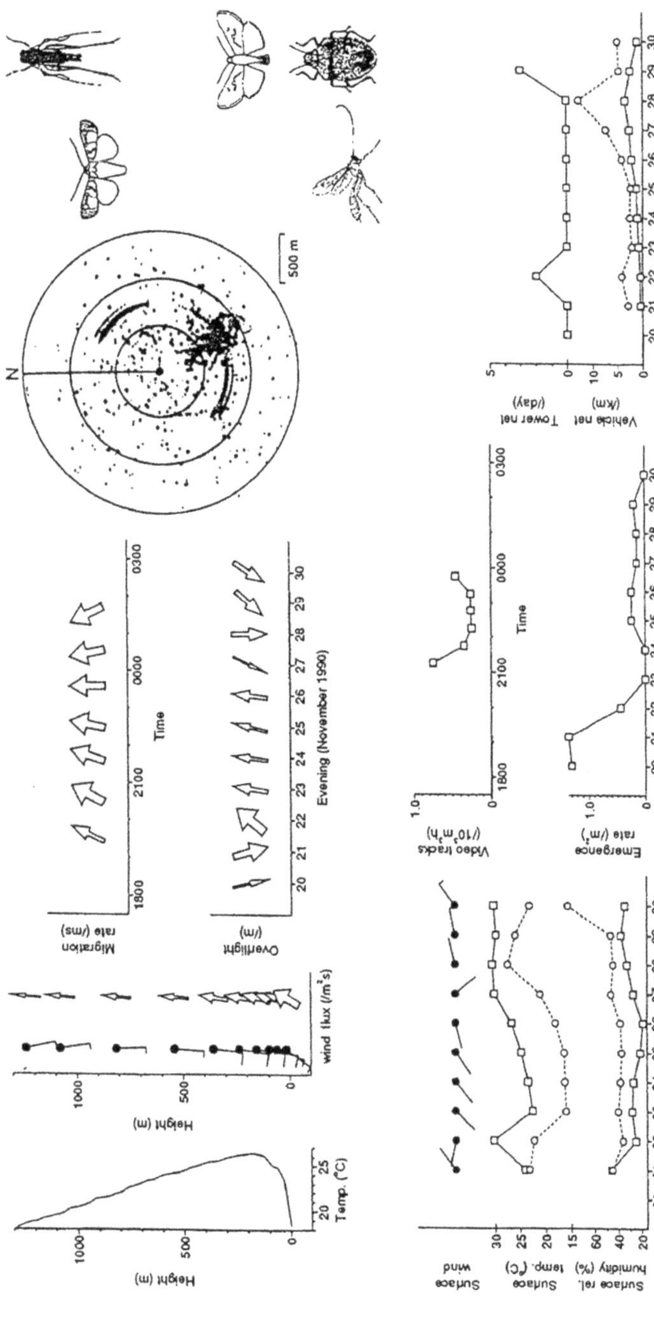

Figure 6 Example results from comprehensive study of airborne movement of two moth species. Top, left to right: temperature profile, wind profile, insect flux (rate per unit area, and direction) profile, variation of migration intensity and direction through a night, radar display showing dusk ascent flight of insects from a source crop (to SSE), insects trapped in aircraft-borne (upper pair) and kite-borne (lower three) nets. Bottom: variation of surface wind, temperature, and humidity at dusk over 11 nights, variation of low-level flight activity estimated from video recordings during one evening (upper graph) and of emergence of *Helicoverpa* moths from pupae in a source crop (lower), variation over 11 nights of captures of *Helicoverpa* moths at about 10 m in a tower trap and of *Helicoverpa* moths (solid line) and all larger insects (dotted line) at about 3 m in a vehicle trap. Source: Reproduced from Drake et al. (1993), with permission of CSIRO Publishing.

considered worthwhile. A broadly similar scenario played out in a comparable research programme on *H. armigera* undertaken in India (Riley et al., 1992), and use of conventional entomological radars for observations of this type has not developed further.

It is interesting to speculate how the new methods now available, especially automated optoelectronic sensing, would add to field studies of this nature, or allow them to be conducted differently and perhaps at less expense. Many of the questions the researchers were asking appear suitable for investigation with harmonic radar, but, with the exception of the *A. segetum* study described in Section 6, this technique has not been applied to pest species, perhaps due to cost and logistical constraints.

8 Conclusion

Growers, pest managers, and agroecosystem entomologists are aware that insects, in their adult forms, fly over, into, and out of crops and orchards, and that species they know to be either pests or beneficials participate in these movements. Unsurprisingly, therefore, there have been persistent efforts to investigate movement: specifically, to identify the species, to estimate numbers and distances, and to determine which crops and other habitats are sources or sinks. Flight above a crop renders an insect more detectable as it is no longer hidden among vegetation, so observing insects in flight appears a potentially effective means for monitoring populations. It might therefore be expected that by now methods for doing this would be well established and in wide use, but as this chapter demonstrates, this is not the case. Most methods remain at the trial and demonstration stage, and those that have advanced to application - most obviously radar and harmonic radar - are used only by a handful of research groups worldwide, and often in curiosity-driven investigations rather than to support agricultural production and sustainability.

A possible explanation for the limited monitoring of airborne movement above and around crops is that often the real interest of stakeholders is not the movement phase of the insect but its outcome, i.e. the arrival of a pest or beneficial in a crop. More specifically, the primary concern is with the number of insects arriving and the timing in relation to the insects' reproductive status and the growth stage of the crop. Systems for detecting arrivals, predominantly pheromone traps and 'scouting' for eggs and early-instar larvae, are well established and widely used; they are supported by a thriving industry that provides both materials (pheromones, traps) and services (scouting, advice) and is, in turn, supported by government (advisories, research and extension, and general maintenance of specialist expertise).

Monitoring for presence provides firm information on which to base a decision about a management intervention. Monitoring airborne flight might provide slightly earlier warning, but linking the information provided to a prediction of yield loss or gain requires an additional step: establishing whether the insects are resident or just passing through. A further factor influencing operational monitoring of airborne activity is cost: radars and lidars are far more expensive than a network of pheromone traps. The advanced electronic and information technologies they employ, and analysis and interpretation of their data, fall outside the professional expertise of most biological scientists; their use has therefore generally been restricted to major interdisciplinary programmes able to attract engineering as well as biological and agricultural researchers - such as the case study of Section 7. Recently, initiatives have been taken to transition some of these technologies to commercial production and application by non-specialist users, a welcome development that has the potential to transform crop-protection entomology. Data from new-generation weather radars has similar potential for monitoring and forecasting longer-distance movements. At the farm and field level, operating costs for interception trapping near the surface might not be much greater than for scouting as both methods are quite labour intensive but draw only on basic entomological and field skills. They both provide identification to species, a critical attribute for decision-making, but scouting also detects initiation of crop damage by larvae of the next generation. For some crops, unavoidable trampling by scouts or plant-quarantine concerns might shift the balance of costs and benefits towards trapping or even remote-sensing methods.

Remote-sensing observations, especially those made with radar, have been very valuable for alerting crop-protection entomologists and pest managers to the intensity and scale of pest movements, and the frequency with which they occur, leaving no doubt that migration and dispersal play key roles in the development of economically significant infestations (Drake and Reynolds, 2012; Hu et al., 2016). Long-term observation programmes with more open-ended objectives have recognised additional species as migratory (e.g. Feng et al., 2006; Jeffries et al., 2013; Wotton et al., 2019). They have also enabled researchers to move beyond case studies to investigation of seasonal patterns and weather effects and to estimate both short-term and year-to-year variability (Drake et al., 2001; Hao et al., 2017): quantities required to parameterise forecasting and risk-assessment procedures.

Up to now, use of radar monitoring to provide alerts when major immigration events are occurring has been limited to programmes in Finland (Leskinen et al., 2011) and Australia (Drake and Wang, 2013). Such movements can cross international borders and the internal jurisdictional boundaries that often define the geographical scope of pest-management authorities and programmes. Management of migratory pests is institutionally and socially challenging because the best way to prevent losses in destination areas may be to intervene in source areas, where the intervention may incur a cost (e.g. through pesticide contamination) and provide little local benefit (Adriaansen et al., 2016). Intergovernmental coordination has been achieved for locust control, though with external support in both documented instances: internationally, from the United Nations for desert locusts *Schistocerca gregaria* (Cressman, 2016); and from the national government for Australian states affected by *C. terminifera* (Adriaansen et al., 2016). At the lesser level of coordinated monitoring and exchange of information, there are effective programmes for aphids in both the USA (Lagos-Kutz et al., 2020) and Europe (Luquet et al., 2019). Data from WSR networks, with their wide coverage, appear particularly valuable in these geographically extended contexts - provided the data are available (Shamoun-Baranes et al., 2021, 2022). Combining radar observations with wind and temperature information from numerical weather-forecasting systems enables estimation of migration trajectories (e.g. Chapman et al., 2010; Westbrook and Eyster, 2017; Sun et al., 2022), and hence identification of likely source and destination regions. To be useful for pest-management decision-making, however, real-time data would be needed rather than the open-source archival data (e.g. NCEP, https://psl.noaa.gov/data/gridded/data.ncep.reanalysis.html, accessed 3 February 2022; Kalnay et al., 1996) used by researchers for retrospective investigations.

It is evident from the material presented in this chapter that advances in electronics and information technologies continually provide new ways of tackling old problems, and that researchers are quick to identify and explore such opportunities. Two themes emerge from the accounts presented here: first, that the small size of insects necessitates specific and novel engineering solutions; and secondly, that identification to species is difficult or even unachievable. Nevertheless, new methods are appearing, and they show potential for enabling new types of investigations, or at least for providing much larger and more informative datasets than those available to earlier generations of researchers and managers. Perhaps the best example is harmonic radar, which has provided answers to long-standing fundamental questions about *A. mellifera* navigation. Another is the ability of vertical-beam entomological radars to monitor insect migration continuously, to heights of 2.5 km and year after year. Short-range nIR detectors, with AI data interpretation and direct reporting via the internet, show promise and are on course for commercial production. Of the other monitoring methods introduced in this chapter, a few may make the transition from demonstration and trial to application, in either research or operations - or both. Others will probably be abandoned, in some cases to reappear in a decade or so when further technological progress makes their implementation practicable, or when recognition of a need for a particular type of information provides an incentive and prompts adequate funding.

Investigation of insect migratory flight well above the surface with conventional radars and use of harmonic radars to study insect foraging flights just above the vegetation layer remain active fields of research, albeit with objectives that extend well beyond agricultural concerns. Technological development is continuing for both radar types (e.g. Drake et al., 2020; Milanesio et al., 2020; Storz and Lavrenko, 2020; Wang et al., 2021), and for optoelectronic detection of smaller insects flying near the ground (e.g. Brydegaard et al., 2020a; Genoud et al., 2021; Rydhmer et al., 2022). In addition to the continued operation of suction-trapping networks already noted, there has recently been renewed interest in other forms of interception trapping of insects, both near the ground and above 100 m (e.g. Florio et al., 2020; Svenningsen et al., 2021). While the research and technology-development programmes addressing this topic are mostly small, it is encouraging to see new groups arising and exploring novel approaches. As agriculture seeks to adapt to a changing climate and evolving social and economic norms, the importance of pest and beneficial insects, and of their movements into, over, and out of crops and pastures, is unlikely to diminish. A continuing need to detect and monitor these movements, and to do so more accurately and precisely than current methods allow, and with less expenditure of effort and funds, can confidently be predicted.

9 Where to look for further information

9.1 Publications

The following publications present more comprehensive and detailed accounts of aspects of the chapter topic:

- Chapman, J. W., Drake, V. A. and Reynolds, D. R. (2011). Recent insights from radar studies of insect flight. *Annual Review of Entomology* 56, 337-356. doi:10.1146/annurev-ento-120709-144820.
- Reynolds, D. R., Chapman, J. W. and Drake, V. A. (2018). Riders on the wind: the aeroecology of insect migrants. In: Chilson, P. B., Frick, W. F., Kelly, J. F. and Liechti, F. (Eds) *Aeroecology*, pp. 145–178 (Ch. 7). Springer, Cham. doi:10.1007/978-3-319-68576-2_7.
- Drake, V. A. and Bruderer, B. (2018). Aeroecological observation methods. In: Chilson, P. B., Frick, W. F., Kelly, J. F. and Liechti, F. (Eds) *Aeroecology*, pp. 201–237 (Ch. 9). Springer, Cham. doi:10.1007/978-3-319-68576-2_9.
- Drake V. A. and Gatehouse A. G. (1995). *Insect Migration: Tracking Resources Through Space and Time*. Cambridge University Press, Cambridge.
- Drake, V. A. and Reynolds, D. R. (2012). *Radar Entomology: Observing Insect Flight and Migration*. CABI Publishing, Wallingford.
- Epsky, N. D., Morrill, W. L. and Mankin, R. W. (2008). Traps for capturing insects. In: Capinera, J. L. (Ed.) *Encyclopedia of Entomology*, pp. 3887-3901. Springer, Dordrecht. doi:10.1007/978-1-4020-6359-6_2523.

- Henderson, P. A. and Southwood, T. R. E. (2016). *Ecological Methods*, 4th edn. Wiley-Blackwell, Oxford (Chapters 4, 7).
- Kissling, W. D., Pattemore, D. E. and Hagen, M. (2014). Challenges and prospects in the telemetry of insects. *Biological Reviews* 89(3), 511-530. doi:10.1111/brv.12065.
- Riley, J. R. and Drake, V. A. (Eds) (2002). Special issue: technologies for insect movement and migration research. *Computers and Electronics in Agriculture* 35(2, 3).

9.2 Research groups

Research on this broad topic is undertaken by small groups led by individual scientists and specialising in one or two particular aspects. These can best be identified from their publications above and in the reference list. The following organisations have slightly wider programmes:

- Rothamsted Research, UK. www.rothamsted.ac.uk;
- BioDAR project biodarproject.org; and
- Faunaphotonics www.faunaphotonics.com.

10 Appendix: safety and regulatory issues

Several of the methods described in this chapter are both potentially hazardous and subject to regulation – though the details of the latter will differ between jurisdictions (Drake and Reynolds, 2012). Flying of kites and tethered balloons presents a collision danger to aviators and will likely require approval by the local aviation regulatory authority and be subject to constraints (location, height) and requirements (advance notice, warning lights). A site well clear of power lines is also essential. Mounting a net or intake on an aircraft may require engineering assessment and testing prior to regulatory approval. Safety and regulatory issues must also be considered before flying a UAV or operating a road vehicle modified to carry a net or trap.

Radar transmissions will require approval from the local spectrum management authority and installations must be designed to ensure that personnel and the public are not subjected to radiation levels exceeding the locally prescribed exposure limits. Laser beams present a particular hazard and at the intensities required for insect observation are likely to require regulatory approval, including from aviation authorities in the case of upward-pointing beams.

11 Acknowledgements

The book editors and Dr D. R. Reynolds alerted me to several relevant publications and made valuable comments on a draft.

12 References

Adriaansen, C., Woodman, J. D., Deveson, E. and Drake, V. A. (2016). The Australian plague locust - risk and response. In: Shroder, J. F. and Sivanpillai, R. (Eds) *Biological and Environmental Hazards, Risks, and Disasters*, pp. 67-86. Elsevier, Amsterdam, Netherlands. doi:10.1016/B978-0-12-394847-2.00005-X.

Ala, R. and Colpitts, B. G. (2021). Development of extra low mass harmonic radar transponders. In: *IEEE 19th International Symposium on Antenna Technology and Applied Electromagnetics (ANTEM), Virtual*, 8-11 August 2021, 2 pp. IEEE Publications, Piscataway, NJ. doi:10.1109/ANTEM51107.2021.9518747.

Ando, N., Shimoyama, I. and Kanzaki, R. (2002). A dual-channel FM transmitter for acquisition of flight muscle activities from the freely flying hawkmoth, *Agrius convolvuli*. *Journal of Neuroscience Methods* 115(2), 181-187. doi:10.1016/S0165-0270(02)00013-4.

Antolinez, C. A., Martini, X., Stelinski, L. L. and Rivers, M. J. (2021). Wind speed and direction drive assisted dispersal of Asian citrus psyllid. *Environmental Entomology* 51(2), 305-312. doi:10.1093/ee/nvab140.

Aubert, J., Aubert, J.-J. and Goeldlin, P. (1976). Douze ans de captures systématiques de Syrphides (Diptères) au col de Bretolet (Alpes valaisannes). *Bulletin de la Société Entomologique Suisse* 49(1-2), 115-142. doi:10.5169/seals-401808. [In French].

Aubert, J. and Goeldlin de Tiefenau, P. (1981). Observations sur les migrations de Syrphides (Dipt.) dans les Alpes de Suisse occidentale. *Bulletin de la Société Entomologique Suisse* 54(4), 377-388. doi:10.5169/seals-402013. [In French].

Aubert, J., Goeldlin, P. and Lyon, J. P. (1969). Essais de marquage et reprises d'insectes migrateurs en automne 1968. *Bulletin de la Société Entomologique Suisse* 42(1-2), 140-166. doi:10.5169/seals-401588. [In French].

Baker, G. H., Tann, C. R. and Fitt, G. P. (2011). A tale of two trapping methods: *Helicoverpa* spp. (Lepidoptera, Noctuidae) in pheromone and light traps in Australian cotton production systems. *Bulletin of Entomological Research* 101(1), 9-23. doi:10.1017/S0007485310000106.

Balciunas, J. and Knopf, K. (1977). Orientation, flight speeds, and tracks of three species of migrating butterflies. *Florida Entomologist* 60(1), 37-39. doi:10.2307/349448.

Beerwinkle, K. R., Lopez, J. D., Bouse, L. F. and Brusse, J. C. (1989). Large helicopter-towed net for sampling airborne arthropods. *Transactions of the American Society of Agricultural Engineers* 32(6), 1847-1852. doi:10.13031/2013.31233.

Beerwinkle, K. R., Lopez, J. D. Jr., Schleider, P. G. and Lingren, P. D. (1995). Annual patterns of aerial insect densities at altitudes from 500 to 2400 meters in east-central Texas indicated by continuously-operating vertically-oriented radar. *Southwestern Entomologist. Supplement* 18, 63-79.

Beerwinkle, K. R., Witz, J. A. and Schleider, P. G. (1993). An automated, vertical looking, X-band radar system for continuously monitoring aerial insect activity. *Transactions of the American Society of Agricultural Engineers* 36(3), 965-970. doi:10.13031/2013.28423.

Bell, J. R., Alderson, L., Izera, D., Kruger, T., Parker, S., Pickup, J., Shortall, C. R., Taylor, M. S., Verrier, P. and Harrington, R. (2015). Long-term phenological trends, species accumulation rates, aphid traits and climate: five decades of change in migrating aphids. *Journal of Animal Ecology* 84(1), 21-34. doi:10.1111/1365-2656.12282.

Betke, M., Hedrick, T. and Theriault, D. (2018). (Chapter 10). Multi-camera videography methods for aeroecology. In: Chilson, P. B., Frick, W. F., Kelly, J. F. and

Liechti, F. (Eds) *Aeroecology*, pp. 239-257. Springer, Cham, Switzerland. doi:0.1007/978-3-319-68576-2_10.

Bjerge, K., Nielsen, J. B., Sepstrup, M. V., Helsing-Nielsen, F. and Høye, T. T. (2021). An automated light trap to monitor moths (Lepidoptera) using computer vision-based tracking and deep learning. *Sensors* 21(2), article 343 (18 pp.). doi:10.3390/s21020343.

Blaauw, B. R., Morrison, W. R., Mathews, C., Leskey, T. C. and Nielsen, A. L. (2017). Measuring host plant selection and retention of *Halyomorpha halys* by a trap crop. *Entomologia Experimentalis et Applicata* 163(2), 197-208. doi:10.1111/eea.12571.

Boulanger, Y., Fabry, F., Kilambi, A., Pureswaran, D. S., Sturtevant, B. R. and Saint-Amant, R. (2017). The use of weather surveillance radar and high-resolution three-dimensional weather data to monitor a spruce budworm mass exodus flight. *Agricultural and Forest Meteorology* 234-235, 127-135. doi:10.1016/j.agrformet.2016.12.018.

Brazee, R. D., Miller, E. S., Reding, M. E., Klein, M. G., Nudd, B. and Zhu, H. (2005). A transponder for harmonic radar tracking of the black vine weevil in behavioral research. *Transactions of the American Society of Agricultural Engineers B* 48(2), 831-838. doi:10.13031/2013.18306.

Brydegaard, M. (2015). Towards quantitative optical cross sections in entomological laser radar - potential of temporal and spherical parameterizations for identifying atmospheric fauna. *PLoS ONE* 10(8), e0135231 (15 pp.). doi:10.1371/journal.pone.0135231.

Brydegaard, M., Gebru, A. and Svanberg, S. (2014). Super resolution laser radar with blinking atmospheric particles – application to interacting flying insects. *Progress in Electromagnetics Research* 147, 141-151. doi:10.2528/PIER14101001.

Brydegaard, M., Kouakoiu, B., Jansson, S., Rydell, J. and Zoueu, J. (2020a). High dynamic range in entomological Scheimpflug lidars. *IEEE Journal of Selected Topics in Quantum Electronics* 27(4), article 6900711 (11 pp.). doi:10.1109/JSTQE.2021.3062088.

Brydegaard, M., Jansson, S., Malmqvist, E., Mlacha, Y. P., Gebru, A., Okumu, F., Killeen, G. F. and Kirkeby, C. (2020b). Lidar activity reveals activity anomaly of malarial mosquitoes during Pan-African eclipse. *Science Advances* 6(20), article eeay5487 (9 pp.). doi:10.1126/sciadv.aay5487.

Brydegaard, M., Malmqvist, E., Jansson, S., Larsson, J., Török, S. and Zhao, G. (2017). The Scheimpflug lidar method. In: *Lidar Remote Sensing for Environmental Monitoring 2017*, San Diego, CA, 6-10 August 2017. SPIE Proceedings Volume 10406, article 104060I. doi:10.1117/12.2272939.

Brydegaard, M., Merdasa, A., Gebru, A., Jayaweera, H. and Svanberg, S. (2016). Realistic instrumentation platform for active and passive optical remote sensing. *Applied Spectroscopy* 70(2), 372-385. doi:10.1177/0003702815620564.

Cant, E. T., Smith, A. D., Reynolds, D. R. and Osborne, J. L. (2005). Tracking butterfly flight paths across the landscape with harmonic radar. *Proceedings of the Royal Society of London B, Biological Sciences* 272(1565), 785-790. doi:10.1098/rspb.2004.3002.

Cardim Ferreira Lima, M., Damascena de Almeida Leandro, M. E., Valero, C., Pereira Coronel, L. C. and Gonçalves Bazzo, C. O. (2020). Automatic detection and monitoring of insect pests–a review. *Agriculture* 10(5), article 161. doi:10.3390/agriculture10050161.

Cerritos, R., Wegier, A. and Alavez, V. (2012). Towards the development of novel long-term pest control strategies based on insect ecological and evolutionary dynamics. In: Soloneski, S. (Ed.) *Integrated Pest Management and Pest Control - Current and Future Tactics*, pp. 35-62. doi:10.5772/29900.

Chapman, J. W., Bell, J. R., Burgin, L. E., Reynolds, D. R., Pettersson, L. B., Hill, J. K., Bonsall, M. B. and Thomas, J. A. (2012). Seasonal migration to high latitudes results in major reproductive benefits in an insect. *Proceedings of the National Academy of Sciences of the United States of America* 109(37), 14924–14929. doi:10.1073/pnas.1207255109.

Chapman, J. W., Drake, V. A. and Reynolds, D. R. (2011). Recent insights from radar studies of insect flight. *Annual Review of Entomology* 56, 337-356. doi:10.1146/annurev-ento-120709-144820.

Chapman, J. W., Nesbit, R. L., Burgin, L. E., Reynolds, D. R., Smith, A. D., Middleton, D. R. and Hill, J. K. (2010). Flight orientation behaviors promote optimal migration trajectories in high-flying insects. *Science* 327(5966), 682–685. doi:10.1126/science.1182990.

Chapman, J. W., Reynolds, D. R., Brooks, S. J., Smith, A. D. and Woiwod, I. P. (2006). Seasonal variation in the migration strategies of the green lacewing *Chrysoperla carnea* species complex. *Ecological Entomology* 31(4), 378–388. doi:10.1111/j.1365-2311.2006.00797. x.

Chapman, J. W., Reynolds, D. R. and Smith, A. D. (2003). Vertical-looking radar: a new tool for monitoring high-altitude insect migration. *BioScience* 53(5), 503–511. doi:10.1641/0006-3568(2003)053[0503:VRANTF]2.0.CO;2.

Chapman, J. W., Reynolds, D. R., Smith, A. D., Smith, E. T. and Woiwod, I. P. (2004a). An aerial netting study of insects migrating at high altitude over England. *Bulletin of Entomological Research* 94(2), 123–136. doi:10.1079/BER2004287.

Chapman, J. W., Reynolds, D. R. and Smith, A. D. (2004b). Migratory and foraging movements in beneficial insects: a review of radar monitoring and tracking methods. *International Journal of Pest Management* 50(3), 225–232. doi:10.1080/09670870410001731961.

Chapman, J. W., Reynolds, D. R. and Wilson, K. (2015). Long-range seasonal migration in insects: mechanisms, evolutionary drivers and ecological consequences. *Ecology Letters* 18(3), 287-302. doi:10.1111/ele.12407.

Chapman, J. W., Smith, A. D., Woiwod, I. P., Reynolds, D. R. and Riley, J. R. (2002a). Development of vertical-looking radar technology for monitoring insect migration. *Computers and Electronics in Agriculture* 35(2–3), 95–110. doi:10.1016/S0168-1699(02)00013-3.

Chapman, J. W., Reynolds, D. R., Smith, A. D., Riley, J. R., Pedgley, D. E. and Woiwod, I. P. (2002b). High-altitude migration of the diamondback moth *Plutella xylostella* to the U.K.: a study using radar, aerial netting and ground trapping. *Ecological Entomology* 27(6), 641–650. doi:10.1046/j.1365-2311.2002.00472.x.

Chen, R. L., Bao, X. Z., Drake, V. A., Farrow, R. A., Wang, S. Y., Sun, Y. J. and Zhai, B. P. (1989). Radar observations of the spring migration into northeastern China of the oriental armyworm moth, *Mythimna separata*, and other insects. *Ecological Entomology* 14(2), 149–162. doi:10.1111/j.1365-2311.1989.tb00765.x.

Cheng, D. F., Wu, K. M., Tian, Z., Wen, L. P. and Shen, Z. R. (2002). Acquisition and analysis of migration data from the digitised display of a scanning entomological radar. *Computers and Electronics in Agriculture* 35(2–3), 63–75. doi:10.1016/S0168-1699(02)00012-1.

Chilson, P. B., Stepanian, P. M. and Kelly, J. F. (2018). Radar aeroecology. In: Chilson, P. B., Frick, W. F., Kelly, J. F. and Liechti, F. (Eds) *Aeroecology*, pp. 277–309. Springer, Cham, Switzerland. doi:10.1007/978-3-319-68576-2_12.

Clark, D. P. (1971). Flights after sunset by the Australian plague locust, *Chortoicetes terminifera* (Walk.), and their significance in dispersal and migration. *Australian Journal of Zoology* 19(2), 159–176. doi:10.1071/ZO9710159.

Cliff, O. M., Saunders, D. L. and Fitch, R. (2018). Robotic ecology: tracking small dynamic animals with an autonomous aerial vehicle. *Science Robotics* 3(23), article eaat8409. doi:10.1126/scirobotics.aat8409.

Crawford, A. B. (1949). Radar reflections in the lower atmosphere. *Proceedings of the IRE* 37(4), 404–405. doi:10.1109/JRPROC.1949.231967.

Cressman, K. (2016). Desert locust. In: Shroder, J. F. and Sivanpillai, R. (Eds) *Biological and Environmental Hazards, Risks, and Disasters*, pp. 87–105. Elsevier, Amsterdam, Netherlands. doi:10.1016/B978-0-12-394847-2.00006-1.

Dingle, H. (2014). *Migration: the Biology of Life on the Move* (2nd edn.). Oxford University Press, Oxford, UK.

Dingle, H., Zalucki, M. P. and Rochester, W. A. (1999). Season-specific directional movement in migratory Australian butterflies. *Australian Journal of Entomology* 38(4), 323–329. doi:10.1046/j.1440-6055.1999.00117.x.

Drake, V. A. (1991). Methods for studying adult movement in *Heliothis*. In: Zalucki, M. P. (Ed.) *Heliothis: Research Methods and Prospects*, pp. 109–121. Springer-Verlag, New York.

Drake, V. A. and Bruderer, B. (2018). Chapter 9: Aeroecological observation methods. In: Chilson, P. B., Frick, W. F., Kelly, J. F. and Liechti, F. (Eds) *Aeroecology*, pp. 201–237. Springer, Cham, Switzerland. doi:10.1007/978-3-319-68576-2_9.

Drake, V. A. and Farrow, R. A. (1983). The nocturnal migration of the Australian plague locust, *Chortoicetes terminifera* (Walker) (Orthoptera, Acrididae): quantitative radar observations of a series of northward flights. *Bulletin of Entomological Research* 73(4), 567–585. doi:10.1017/S0007485300009172.

Drake, V. A. and Farrow, R. A. (1985). A radar and aerial-trapping study of an early spring migration of moths (Lepidoptera) in inland New South Wales. *Australian Journal of Ecology* 10(3), 223–235. doi:10.1111/j.1442-9993.1985.tb00885.x.

Drake, V. A. and Farrow, R. A. (1988). The influence of atmospheric structure and motions on insect migration. *Annual Review of Entomology* 33, 183–210. doi:10.1146/annurev.en.33.010188.001151.

Drake, V. A. and Fitt, G. P. (1990). Studies of *Heliothis* mobility at Narrabri, summer 1989/90. In: *Fifth Australian Cotton Conference, Broadbeach, Queensland*, 8–9 August 1990, pp. 295–304. Australian Cotton Growers Research Association, Narrabri, Australia.

Drake, V. A. and Gatehouse, A. G. (1995). *Insect Migration: Tracking Resources Through Space and Time*. Cambridge University Press, Cambridge, UK.

Drake, V. A. and Reynolds, D. R. (2012). *Radar Entomology: Observing Insect Flight and Migration*. CABI Publishing, Wallingford, UK.

Drake, V. A. and Wang, H. (2013). Recognition and characterization of migratory movements of Australian plague locusts, *Chortoicetes terminifera*, with an insect monitoring radar. *Journal of Applied Remote Sensing* 7(1), article 075095 (17 pp.). doi:10.1117/1.JRS.7.075095.

Drake, V. A. and Wang, H. (2019). Ascent and descent rates of high-flying insect migrants determined with a non-coherent vertical-beam entomological radar. *International Journal of Remote Sensing* 40(3), 883–904. doi:10.1080/01431161.2018.1519283.

Drake, V. A., Fitt, G. P. and Rochester, W. A. (1993). Methodology for a comprehensive field study of adult movement in *Helicoverpa* spp. In: Corey, S. A., Dall, D. J. and Milne, W. M. (Eds) *Pest Control and Sustainable Agriculture*, pp. 456–459. CSIRO Publications, Melbourne, Australia.

Drake, V. A., Gregg, P. C., Harman, I. T., Wang, H. K., Deveson, E. D., Hunter, D. M. and Rochester, W. A. (2001). Characterizing insect migration systems in inland Australia with novel and traditional methodologies. In: Woiwod, I., Reynolds, D. R. and Thomas, C. D. (Eds) *Insect Movement: Mechanisms and Consequences*, pp. 207–233. CABI Publishing, Wallingford, UK. doi:10.1079/9780851994567.0207.

Drake, V. A., Harman, I. T. and Wang, H. K. (2002). Insect monitoring radar: stationary-beam operating mode. *Computers and Electronics in Agriculture* 35(2–3), 111–137. doi:10.1016/S0168-1699(02)00014-5.

Drake, V. A., Hatty, S., Symons, C. and Wang, H. (2020). Insect Monitoring Radar: maximizing performance and utility. *Remote Sensing* 12(4), article 596 (21 pp.). doi:10.3390/rs12040596.

Dudley, R. (2000). *The Biomechanics of Insect Flight: Form, Function, Evolution*. Princeton University Press, Princeton, NJ.

Epsky, N. D., Morrill, W. L. and Mankin, R. W. (2008). Traps for capturing insects. In: Capinera, J. L. (Ed.) *Encyclopedia of Entomology*, pp. 3887-3901. Springer, Dordrecht, Netherlands. doi:10.1007/978-1-4020-6359-6_2523.

Evans, L. C., Sibly, R. M., Thorbek, P., Sims, I., Oliver, T. H. and Walters, R. J. (2020). The importance of including habitat-specific behaviour in models of butterfly movement. *Oecologia* 193(2), 249–259. doi:10.1007/s00442-020-04638-4.

Farrow, R. A. and Dowse, J. E. (1984). Method of using kites to carry tow nets in the upper air for sampling migrating insects and its application to radar entomology. *Bulletin of Entomological Research* 74(1), 87–95. doi:10.1017/S0007485300009950.

Feng, H. Q., Wu, K. M., Cheng, D. F. and Guo, Y. Y. (2003). Radar observations of the autumn migration of the beet armyworm *Spodoptera exigua* (Lepidoptera: Noctuidae) and other moths in northern China. *Bulletin of Entomological Research* 93(2), 115–124. doi:10.1079/BER2002221.

Feng, H.-Q., Wu, K.-M., Ni, Y.-X., Cheng, D.-F. and Guo, Y. Y. (2006). Nocturnal migration of dragonflies over the Bohai Sea in northern China. *Ecological Entomology* 31(5), 511–520. doi:10.1111/j.1365-2311.2006.00813.x.

Feng, H. Q., Wu, X. F., Wu, B. and Wu, K. M. (2009). Seasonal migration of *Helicoverpa armigera* (Lepidoptera: Noctuidae) over the Bohai Sea. *Journal of Economic Entomology* 102(1), 95–104. doi:10.1603/029.102.0114.

Feng, H. Q., Zhao, X. C., Wu, X. F., Wu, B., Wu, K. M., Cheng, D. F. and Guo, Y. Y. (2008). Autumn migration of *Mythimna separata* (Lepidoptera: Noctuidae) over the Bohai Sea in northern China. *Environmental Entomology* 37(3), 774–781. doi:10.1093/ee/37.3.774.

Fitt, G. P., Dillon, M. L. and Hamilton, J. G. (1995). Spatial dynamics of *Helicoverpa* populations in Australia: simulation modelling and empirical studies of adult movement. *Computers and Electronics in Agriculture* 13(2), 177–192. doi:10.1016/0168-1699(95)00024-X.

Florio, J., Verú, L. M., Dao, A., Yaro, A. S., Diallo, M., Sanogo, Z. L., Samaké, D., Huestis, D. L., Yossi, O., Talamas, E., Chamorro, M. L., Frank, J. H., Biondi, M., Morkel, C., Bartlett, C., Linton, Y-M., Strobach, E., Chapman, J. W., Reynolds, D. R., Faiman, R., Krajacich, B. J., Smith, C. S. and Lehmann, T. (2020). Diversity, dynamics, direction, and magnitude of high-altitude migrating insects in the Sahel. *Scientific Reports* 10, article 20523 (14 pp.). doi:10.1038/s41598-020-77196-7.

Gao, B., Hedlund, J., Reynolds, D. R., Zhai, B., Hu, G. and Chapman, J. W. (2020). The 'migratory connectivity' concept, and its applicability to insect migrants. *Movement Ecology* 8(1), article 48 (13 pp.). doi:10.1186/s40462-020-00235-5.

Gauthreaux, S. A. and Diehl, R. (2020). Discrimination of biological scatterers in polarimetric weather radar data: opportunities and challenges. *Remote Sensing* 12(3), article 545 (34 pp.). doi:10.3390/rs12030545.

Gebru, A., Jansson, S., Ignell, R., Kirkeby, C., Prangsma, J. C. and Brydegaard, M. (2018). Multiband modulation spectroscopy for the determination of sex and species of mosquitoes in flight. *Journal of Biophotonics* 11(8), article e201800014 (13 pp.). doi:10.1002/jbio.201800014.

Geerts, B. and Miao, Q. (2005). Airborne radar observations of the flight behaviour of small insects in the atmospheric convective boundary layer. *Environmental Entomology* 34(2), 361–377. doi:10.1603/0046-225X-34.2.361.

Genoud, A. P., Basistyy, R., Williams, G. M. and Thomas, B. P. (2018). Optical remote sensing for monitoring flying mosquitoes, gender identification and discussion on species identification. *Applied Physics. B, Lasers and Optics* 124(3), article 46 (11 pp.). doi:10.1007/s00340-018-6917-x.

Genoud, A. P., Gao, Y., Williams, G. M. and Thomas, B. P. (2019). Identification of gravid mosquitoes from changes in spectral and polarimetric backscatter cross sections. *Journal of Biophotonics* 12(10), article e201900123 (9 pp.). doi:10.1002/jbio.201900123.

Genoud, A. P., Williams, G. M. and Thomas, B. P. (2021). Continuous monitoring of aerial density and circadian rhythms of flying insects in a semi-urban environment. *PLoS ONE* 16(11), article e0260167. doi:10.1371/journal.pone.0260167.

Glick, P. A. (1939). *The Distribution of Insects, Spiders, and Mites in the Air. United States Department of Agriculture Technical Bulletin 673.* United States Department of Agriculture, Washington.

Gray, G. R., Serafin, R. J., Atlas, D., Rinehart, R. E. and Boyajian, J. J. (1975). Real-time color Doppler radar display. *Bulletin of the American Meteorological Society* 56(6), 580–588. doi:10.1175/1520-0477(1975)056<0580:RTCDRD>2.0.CO;2.

Greenslade, A. F. C., Carnegie, M., Ouvrard, D., Sjölund, M. J., Highet, F., Sigvald, R., Kenyon, D. and Bell, J. (2020). Aerial psyllid (Hemiptera: Psylloidea) detection and monitoring using suction traps in Britain: population observations, new species found and a revised British checklist. *Entomologist's Gazette* 71(3), 151–163. doi:10.31184/G00138894.713.1749.

Greenstone, M. H., Eaton, R. R. and Morgan, C. E. (1991). Sampling aerially dispersing arthropods: A high-volume, inexpensive, automobile- and aircraft-borne system. *Journal of Economic Entomology* 84(6), 1717–1724. doi:10.1093/jee/84.6.1717.

Gregg, P. C., Fitt, G. P., Coombs, M. and Henderson, G. S. (1993). Migrating moths (Lepidoptera) collected in tower-mounted light traps in northern New South Wales, Australia: species composition and seasonal abundance. *Bulletin of Entomological Research* 83(4), 563–578. doi:10.1017/S0007485300039997.

Hao, Z. H., Drake, V. A., Sidhu, L. and Taylor, J. R. (2017). Locust displacing winds in eastern Australia reassessed with observations from an insect monitoring radar. *International Journal of Biometeorology* 61(12), 2073–2084. doi:10.1007/s00484-017-1404-3.

Hao, Z., Drake, V. A., Taylor, J. R. and Warrant, E. (2020). Insect target classes discerned from entomological radar data. *Remote Sensing* 12(4), article 673 (18 pp.). doi:10.3390/rs12040673.

Harman, I. T. and Drake, V. A. (2004). Insect monitoring radar: analytical time-domain algorithm for retrieving trajectory and target parameters. *Computers and Electronics in Agriculture* 43(1), 23–41. doi:10.1016/j.compag.2003.08.005.

Hassell, M. P. and Southwood, T. R. E. (1978). Foraging strategies in insects. *Annual Review of Ecology and Systematics* 9, 75–98. doi:10.1146/annurev.es.09.110178.000451.

Hedin, J. and Ranius, T. (2002). Using radio telemetry to study dispersal of the beetle *Osmoderma eremita*, an inhabitant of tree hollows. *Computers and Electronics in Agriculture* 35(2-3), 171–180. doi:10.1016/S0168-1699(02)00017-0.

Henderson, P. A. and Southwood, T. R. E. (2016). *Ecological Methods* (4th edn.). Wiley-Blackwell, Oxford, UK.

Hendrichs, J., Kenmore, P., Robinson, A. S. and Vreysen, M. J. B. (2007). Area-wide integrated pest management (AW-IPM): principles, practice and prospects. In: Vreysen, M. J. B., Robinson, A. S. and Hendrichs, J. (Eds) *Area-Wide Control of Insect Pests. From Research to Field Implementation*, pp. 3–33. Springer, Dordrecht, Netherlands.

Hirai, K. (1988). Sudden outbreaks of the armyworm, *Pseudaletia separata* WALKER and its monitoring systems in Japan. *Japan Agricultural Research Quarterly* 22(5), 166–174.

Høye, T. T., Ärje, J., Bjerge, K., Hansen, O. L. P., Iosifidis, A., Leese, F., Mann, H. M. R., Meissner, K., Melvad, C. and Raitoharju, J. (2021). Deep learning and computer vision will transform entomology. *Proceedings of the National Academy of Sciences of the United States of America* 118(2), article e2002545117 (10 pp.). doi:10.1073/pnas.2002545117.

Hu, G., Lim, K. S., Horvitz, N., Clark, S. J., Reynolds, D. R., Sapir, N. and Chapman, J. W. (2016). Mass seasonal bioflows of high-flying insect migrants. *Science* 354(6319), 1584–1587. doi:10.1126/science.aah4379.

Irwin, M. E. and Thresh, J. M. (1988). Long range aerial dispersal of cereal aphids as virus vectors in North America. *Philosophical Transactions of the Royal Society of London B, Biological Sciences* 321(1207), 421–446. doi:10.1098/rstb.1988.0101.

Isard, S. A., Irwin, M. E. and Hollinger, S. E. (1990). Vertical distribution of aphids (Homoptera: Aphididae) in the planetary boundary layer. *Environmental Entomology* 19(5), 1473–1484. doi:10.1093/ee/19.5.1473.

Jansson, S. and Brydegaard, M. (2018). Passive kHz lidar for the quantification of insect activity and dispersal. *Animal Biotelemetry* 6, article 6 (10 pp.). doi:10.1186/s40317-018-0151-5.

Jansson, S., Atkinson, P., Ignell, R. and Brydegaard, M. (2019). First polarimetric investigation of malaria mosquitoes as lidar targets. *IEEE Journal of Selected Topics in Quantum Electronics* 25(1) article 7201808 (8 pp.). doi:10.1109/JSTQE.2018.2859225.

Jansson, S., Malmqvist, E., Brydegaard, M., Åkesson, S. and Rydell, J. (2020). A Scheimpflug lidar used to observe insect swarming at a wind turbine. *Ecological Indicators* 117, article 106578 (7 pp.). doi:10.1016/j.ecolind.2020.106578.

Jeffries, D. L., Chapman, J. W., Roy, H. E., Humphries, S., Harrington, R., Brown, P. M. and Handley, L. J. (2013). Characteristics and drivers of high-altitude ladybird flight: insights from vertical-looking entomological radar. *PLoS ONE* 8(12) (12 pp.). doi:10.1371/journal.pone.0082278.

Johnson, C. G. (1950). A suction trap for small airborne insects which automatically segregates the catch into successive hourly samples. *Annals of Applied Biology* 37(1), 80–91. doi:10.1111/j.1744-7348.1950.tb00952.x.

Johnson, C. G. (1969). *Migration and Dispersal of Insects by Flight*. Methuen, London, UK.

Kakutani, K., Matsuda, Y., Nonomura, T., Takikawa, Y., Osamura, K. and Toyoda, H. (2021). Remote-controlled monitoring of flying pests with an electrostatic insect capturing apparatus carried by an unmanned aerial vehicle. *Agriculture* 11(2), article 176 (15 pp.). doi:10.3390/agriculture11020176.

Kalnay, E., Kanamitsu, M., Kistler, R., Collins, W., Deaven, D., Gandin, L., Iredell, M., Saha, S., White, G., Woollen, J., Zhu, Y., Leetmaa, A., Reynolds, R., Chelliah, M., Ebisuzaki, W., Higgins, W., Janowiak, J., Mo, K. C., Ropelewski, C., Wang, J., Jenne, R. and Joseph, D. (1996). The NCEP/NCAR 40-year reanalysis project. *Bulletin of the American Meteorological Society* 77(3), 437–471. doi:10.1175/1520-0477(1996)077<0437:TNYRP>2.0.CO;2.

Kennedy, G. G. and Storer, N. P. (2000). Life systems of polyphagous arthropod pests in temporally unstable cropping systems. *Annual Review of Entomology* 45, 467–493. doi:10.1146/annurev.ento.45.1.467.

Kennedy, J. S. (1985). Migration: behavioral and ecological. In: *Contributions in Marine Science, Suppl.* Rankin, M. A. (Ed.) *Migration: Mechanisms and Adaptive Significance*. Marine Science Institute, The University of Texas at Austin, Port Aransas, TX, vol. 27, 5–26.

Kim, B., Ju, C. and Son, H. I. (2022). Field evaluation of UAV-based tracking method for localization of small insects. *Entomological Research* 52(3), 135–147. doi:10.1111/1748-5967.12573.

Kim, H. G., Park, J.-S. and Lee, D.-H. (2018). Potential of unmanned aerial sampling for monitoring insect populations in rice fields. *Florida Entomologist* 101(2), 330–334. doi:10.1653/024.101.0229.

Kirkeby, C., Rydhmer, K., Cook, S. M., Strand, A., Torrance, M. T., Swain, J. L., Prangsma, J., Johnen, A., Jensen, M., Brydegaard, M. and Græsbøll, K. (2021). Advances in automatic identification of flying insects using optical sensors and machine learning. *Scientific Reports* 11(1), article 1555 (8 pp.). doi:10.1038/s41598-021-81005-0.

Kissling, W. D., Pattemore, D. E. and Hagen, M. (2014). Challenges and prospects in the telemetry of insects. *Biological Reviews of the Cambridge Philosophical Society* 89(3), 511–530. doi:10.1111/brv.12065.

Knight, S. M., Pitman, G. M., Flockhart, D. T. T. and Norris, D. R. (2019). Radio-tracking reveals how wind and temperature influence the pace of daytime insect migration. *Biology Letters* 15(7), article 20190327. doi:10.1098/rsbl.2019.0327.

Kogan, M. (1998). Integrated pest management: historical perspectives and contemporary developments. *Annual Review of Entomology* 43, 243–270. doi:10.1146/annurev.ento.43.1.243.

Kutsch, W. (2002). Transmission of muscle potentials during free flight of locusts. *Computers and Electronics in Agriculture* 35(2–3), 181–199. doi:10.1016/S0168-1699(02)00018-2.

Lack, D. and Lack, E. (1951). Migration of insects and birds through a Pyrenean pass. *Journal of Animal Ecology* 20(1), 63–67.

Lagos-Kutz, D., Voegtlin, D. J., Onstad, D., Hogg, D., Ragsdale, D., Tilmon, K., Hodgson, E., Difonzo, C., Groves, R., Krupke, C., Laforest, J., Seiter, N. J., Duerr, E., Bradford, B. and Hartman, G. L. (2020). The soybean aphid suction trap network: sampling the aerobiological "soup". *American Entomologist* 66(1), 48–55. doi:10.1093/ae/tmaa009.

Lahoz-Monfort, J. J. and Magrath, M. J. L. (2021). A comprehensive overview of technologies for species and habitat monitoring and conservation. *BioScience* 71(10), 1038–1062. doi:10.1093/biosci/biab073.

Lavrenko, A., Barry, Z., Norman, R., Frazer, C., Ma, Y., Woodward, G. and Pawson, S. (2021). Autonomous swarm of UAVs for tracking of flying insects with harmonic radar. In: *The 2021 IEEE 93rd Vehicular Technology Conference (VTC2021-Spring), Virtual*, 25–28 April 2021. IEEE Publications, Piscataway, NJ. (5 pp.) doi:10.1109/vtc2021-spring51267.2021.9449074.

Leskinen, M., Markkula, I., Koistinen, J., Pylkkö, P., Ooperi, S., Siljamo, P., Ojanen, H., Raiskio, S. and Tiilikkala, K. (2011). Pest insect immigration warning by an atmospheric dispersion model, weather radars and traps. *Journal of Applied Entomology* 135(1-2), 55–67. doi:10.1111/j.1439-0418.2009.01480.x.

Lihoreau, M., Raine, N. E., Reynolds, A. M., Stelzer, R. J., Lim, K. S., Smith, A. D., Osborne, J. L. and Chittka, L. (2013). Unravelling the mechanisms of trapline foraging in bees. *Communicative and Integrative Biology* 6(1), article e22701 (4 pp.). doi:10.4161/cib.22701.

Lingren, P. D., Raulston, J. R., Popham, T. W., Wolf, W. W., Lingren, P. S. and Esquivel, J. F. (1995). Flight behaviour of corn earworm (Lepidoptera: Noctuidae) moths under low wind speed conditions. *Environmental Entomology* 24(4), 851–860. doi:10.1093/ee/24.4.851.

Lipson, A., Lipson, S. G. and Lipson, H. (2010). *Optical Physics* (4th edn.). Cambridge University Press, Cambridge, UK. doi:10.1017/CBO9780511763120.

Long, T., Hu, C., Wang, R., Zhang, T., Kong, S., Li, W., Cai, J., Tian, W. and Zeng, T. (2020). Entomological radar overview: system and signal processing. *IEEE Aerospace and Electronic Systems Magazine* 35(1), 20–32. doi:10.1109/MAES.2019.2955575.

Loper, G. M., Wolf, W. W. and Taylor, O. R. (1992). Honey bee drone flyways and congregation areas – radar observations. *Journal of the Kansas Entomological Society* 65(3), 223–230.

Luquet, M., Hullé, M., Simon, J.-C., Parisey, N., Buchard, C. and Jaloux, B. (2019). Relative importance of long-term changes in climate and land-use on the phenology and abundance of legume crop specialist and generalist aphids. *Insect Science* 26(5), 881–896. doi:10.1111/1744-7917.12585.

Maggiora, R., Saccani, M., Milanesio, D. and Porporato, M. (2019). An innovative harmonic radar to track flying insects: the case of *Vespa velutina*. *Scientific Reports* 9, article 11694 (10 pp.). doi:10.1038/s41598-019-48511-8.

Malmqvist, E., Jansson, S., Zhu, S., Li, W., Svanberg, K., Svanberg, S., Rydell, J., Song, Z., Bood, J., Brydegaard, M. and Åkesson, S. (2019). The bat-bird-bug battle: daily flight activity of insects and their predators over a rice field revealed by high-resolution Scheimpflug lidar. *Royal Society Open Science* 5(4), 172303. doi:10.1098/rsos.172303.

Martin, W. J. and Shapiro, A. (2007). Discrimination of bird and insect radar echoes in clear air using high-resolution radars. *Journal of Atmospheric and Oceanic Technology* 24(7), 1215–1230. doi:10.1175/JTECH2038.1.

Mata, V. A., Ferreira, S., Campos, R. M., da Silva, L. P., Verissimo, J., Corley, M. F. V. and Beja, P. (2021). Efficient assessment of nocturnal flying insect communities by combining automatic light traps and DNA metabarcoding. *Environmental DNA* 3(2), 398–408. doi:10.1002/edn3.125.

McCravy, K. W. (2018). A review of sampling and monitoring methods for beneficial arthropods in agroecosystems. *Insects* 9(4), article 170 (27 pp.). doi:10.3390/insects9040170.

McDonald, G. and Cole, P. G. (1991). Factors influencing oocyte development in *Mythimna convecta* (Lepidoptera: Noctuidae) and their possible impact on migration in eastern Australia. *Bulletin of Entomological Research* 81(2), 175-184. doi:10.1017/S0007485300051257.

Menz, M. H. M., Brown, B. V. and Wotton, K. R. (2019). Quantification of migrant hoverfly movements (Diptera: Syrphidae) on the West Coast of North America. *Royal Society Open Science* 6(4), article 190153 (7 pp.). doi:10.1098/rsos.190153.

Menzel, R., Tison, L., Fischer-Nakai,J., Cheeseman, J. F., Balbuena, M. S., Chen, X., Landgraf, T., Petrasch, J., Polster, J. and Greggers, U. (2019). Guidance of navigating honeybees by learned elongated ground structures. *Frontiers in Behavioral Neuroscience* 12, article 322. doi:10.3389/fnbeh.2018.00322.

Milanesio, D., Bottigliero, S., Saccani, M., Maggiora, R., Viscardi, A. and Gallesi, M. M. (2020). An harmonic radar prototype for insect tracking in harsh environments. In: *IEEE International Radar Conference, Virtual*, 28-30 April 2020 (vol. 2020), pp. 648-653. IEEE Publications, Piscataway, NJ. doi:10.1109/RADAR42522.2020.9114540.

Miller, J. R., Adams, C. G., Weston, P. A. and Schenker, J. H. (2015). *Trapping of Small Organisms Moving Randomly. Principles and Applications to Pest Monitoring and Management*. Springer, Cham, Switzerland.

Montgomery, G. A., Belitz, M. W., Guralnick, R. P. and Tingley, M. W. (2021). Standards and best practices for monitoring and benchmarking insects. *Frontiers in Ecology and Evolution* 8, article 579193. doi:10.3389/fevo.2020.579193.

Muirhead-Thomson, R. C. (1991). *Trap Responses of Flying Insects: the Influence of Trap Design on Capture Efficiency*. Academic Press, London, UK.

Mullen, E. R., Rutschman, P., Pegram, N., Patt, J. M., Adamczyk, J. J. Jr. and Johanson, 3. (2016). Laser system for identification, tracking, and control of flying insects. *Optics Express* 24(11), article 11828. doi:10.1364/OE.24.011828.

Naef-Daenzer, B., Früh, D., Stalder, M., Wetli, P. and Weise, E. (2005). Miniaturization (0.2·g) and evaluation of attachment techniques of telemetry transmitters. *Journal of Experimental Biology* 208(21), 4063-4068. doi:10.1242/jeb.01870.

Osborne, J. L., Loxdale, H. D. and Woiwod, I. P. (2002). Monitoring insect dispersal: methods and approaches. In: Bullock, J. M., Kenward, R. E. and Hails, R. S. (Eds). *Dispersal Ecology*. 42nd Symposium of the Royal Ecological Society, pp. 24-49. Cambridge University Press, Cambridge.

Osborne, J. L., Smith, A., Clark, S. J., Reynolds, D. R., Barron, M. C., Lim, K. S. and Reynolds, A. M. (2013). The ontogeny of bumblebee flight trajectories: From naïve explorers to experienced foragers. *PLoS ONE* 8(11) article e78681 (11 pp.). doi:10.1371/journal.pone.0078681.

Ovaskainen, O., Smith, A. D., Osborne, J. L., Reynolds, D. R., Carreck, N. L., Martin, A. P., Niitepõld, K. and Hanski, I. (2008). Tracking butterfly movements with harmonic radar reveals an effect of population age on movement distance. *Proceedings of the National Academy of Sciences of the United States of America* 105(49), 19090-19095. doi:10.1073/pnas.0802066105.

Pan, H., Liang, G. and Lu, Y. (2021). Response of different insect groups to various wavelengths of light under field conditions. *Insects* 12(5), article 427 (11 pp.). doi:10.3390/insects12050427.

Parry, H. R. (2013). Cereal aphid movement: general principles and simulation modelling. *Movement Ecology* 1, article 14 (15 pp.). doi:10.1186/2051-3933-1-14.

Pedgley, D. E. (1993). Managing migratory insects pests – a review. *International Journal of Pest Management* 39(1), 3–12. doi:10.1080/09670879309371751.

Pedgley, D. E. (1999). Weather influences on pest movement. In: Ruberson, J. R. (Ed.) *Handbook of Pest Management*, pp. 57–78. Marcel Dekker, New York.

Pegoraro, L., Hidalgo, O., Leitch, I. J., Pellicer, J. and Barlow, S. E. (2020). Automated video monitoring of insect pollinators in the field. *Emerging Topics in Life Sciences* 4(1), 87–97. doi:10.1042/ETLS20190074.

Pope, T., Gundalai, E., Elliott, L., Blackshaw, R., Hough, G., Wood, A., Bennison, J., Prince, G. and Chandler, D. (2015). Recording the movement of adult vine weevil within strawberry crops using radio frequency identification tags. *Journal of Berry Research* 5(4), 197–206. doi:10.3233/JBR-150102.

Preti, M., Verheggen, F. and Angeli, S. (2021). Insect pest monitoring with camera-equipped traps: strengths and limitations. *Journal of Pest Science* 94(2), 203–217. doi:10.1007/s10340-020-01309-4.

Pruess, K. P. and Pruess, N. C. (1971). Telescopic observation of the moon as a means for observing migration of the army cutworm, *Chorizagrotis auxiliaris* (Lepidoptera: Noctuidae). *Ecology* 52(6), 999–1007. doi:10.2307/1933805.

Rennie, S. J. (2013). Common orientation and layering of migrating insects in southeastern Australia observed with a Doppler weather radar. *Meteorological Applications* 21(2), 218–229. doi:10.1002/met.1378.

Reynolds, A. M., Reynolds, D. R., Smith, A. D., Svensson, G. P. and Löfstedt, C. (2007). Appetitive flight patterns of male *Agrotis segetum* moths over landscape scales. *Journal of Theoretical Biology* 245(1), 141–149. doi:10.1016/j.jtbi.2006.10.007.

Reynolds, A. M., Swain, J. L., Smith, A. D., Martin, A. P. and Osborne, J. L. (2009). Honeybees use a Lévy flight search strategy and odour-mediated anemotaxis to relocate food sources. *Behavioral Ecology and Sociobiology* 64(1), 115–123. doi:10.1007/s00265-009-0826-2.

Reynolds, D. R. and Riley, J. R. (1988). A migration of grasshoppers, particularly *Diabolocatantops axillaris* (Thunberg) (Orthoptera: Acrididae), in the West African Sahel. *Bulletin of Entomological Research* 78(2), 251–271. doi:10.1017/S000748530001302X.

Reynolds, D. R. and Riley, J. R. (1997). Flight behaviour and migration of insect pests: radar studies in developing countries. *NRI Bulletin No. 71*. Natural Resources Institute, Chatham, UK.

Reynolds, D. R., Chapman, J. W. and Drake, V. A. (2018). Riders on the wind. The aeroecology of insect migrants. In: Chilson, P. B., Frick, W. F., Kelly, J. F. and Liechti, F. (Eds) *Aeroecology*, pp. 145–178. Springer, Cham, Switzerland. doi:10.1007/978-3-319-68576-2_7.

Reynolds, D. R., Chapman, J. W. and Harrington, R. (2006). The migration of insect vectors of plant and animal viruses. *Advances in Virus Research* 67, 453–517. doi:10.1016/S0065-3527(06)67012-7.

Reynolds, D. R., Mukhopadhyay, S., Riley, J. R., Das, B. K., Nath, P. S. and Mandal, S. K. (1999). Seasonal variation in the windborne movement of insect pests over northeast India. *International Journal of Pest Management* 45(3), 195–205. doi:10.1080/096708799227798.

Reynolds, D. R., Smith, A. D. and Chapman, J. W. (2008). A radar study of emigratory flight and layer formation by insects at dawn over southern Britain. *Bulletin of Entomological Research* 98(1), 35–52. doi:10.1017/S0007485307005470.

Rhodes, M. W., Bennie, J. J., Spalding, A., ffrench-Constant, R. H. and Maclean, I. M. D. (2022). Recent advances in the remote sensing of insects. *Biological Reviews of the Cambridge Philosophical Society* 97(1), 343–360. doi:10.1111/brv.12802.

Riley, J. R. (1985). Radar cross section of insects. *Proceedings of the IEEE* 73(2), 228–232. doi:10.1109/PROC.1985.13135.

Riley, J. R. and Smith, A. D. (2002). Design considerations for an harmonic radar to investigate the flight of insects at low altitude. *Computers and Electronics in Agriculture* 35(2-3), 151–169. doi:10.1016/S0168-1699(02)00016-9.

Riley, J. R., Armes, N. J., Reynolds, D. R. and Smith, A. D. (1992). Nocturnal observations on the emergence and flight behaviour of *Helicoverpa armigera* (Lepidoptera: Noctuidae) in the post-rainy season in central India. *Bulletin of Entomological Research* 82(2), 243–256. doi:10.1017/S0007485300051798.

Riley, J. R., Cheng, X.-N., Zhang, X.-X., Reynolds, D. R., Xu, G.-M., Smith, A. D., Cheng, J.-Y., Bao, A.-D. and Zhai, B.-P. (1991). The long-distance migration of *Nilaparvata lugens* (Stål) (Delphacidae) in China: radar observations of mass return flight in autumn. *Ecological Entomology* 16(4), 471–489. doi:10.1111/j.1365-2311.1991.tb00240.x.

Riley, J. R., Reynolds, D. R. and Farmery, M. J. (1983). Observations of the flight behaviour of the armyworm moth, *Spodoptera exempta*, at an emergence site using radar and infra-red optical techniques. *Ecological Entomology* 8(4), 395–418. doi:10.1111/j.1365-2311.1983.tb00519.x.

Riley, J. R., Reynolds, D. R. and Farrow, R. A. (1987). The migration of *Nilaparvata lugens* (Stål) (Delphacidae) and other Hemiptera associated with rice during the dry season in the Philippines: a study using radar, visual observations, aerial netting and ground trapping. *Bulletin of Entomological Research* 77(1), 145–169. doi:10.1017/S0007485300011627.

Riley, J. R., Smith, A. D. and Bettany, B. W. (1990). The use of video equipment to record in three dimensions the flight trajectories of *Heliothis armigera* and other moths at night. *Physiological Entomology* 15(1), 73-80. doi:10.1111/j.1365-3032.1990.tb00494.x.

Riley, J. R., Greggers, U., Smith, A. D., Stach, S., Reynolds, D. R., Stollhoff, N., Brandt, R., Schaupp, F. and Menzel, R. (2003). The automatic pilot of honeybees. *Proceedings of the Royal Society of London B, Biological Sciences* 270(1532), 2421–2424. doi:10.1098/rspb.2003.2542.

Rochester, W. A., Dillon, M. L., Fitt, G. P. and Zalucki, M. P. (1996). A simulation model of the long distance migration of *Helicoverpa* spp. moths. *Ecological Modelling* 86(2–3), 151–156. doi:10.1016/0304-3800(95)00043-7.

Rossi, V., Caffi, T. and Salinari, F. (2012). Helping farmers face the increasing complexity of decision-making for crop protection. *Phytopathologia Mediterranea* 51(3), 457–479. doi:10.14601/Phytopathol_Mediterr-11038.

Russell, R. W., May, M. L., Soltesz, K. L. and Fitzpatrick, J. W. (1998). Massive swarm migrations of dragonflies (Odonata) in eastern North America. *American Midland Naturalist* 140(2), 325–342. doi:10.1674/0003-0031(1998)140[0325:MSMODO]2.0.CO;2.

Rydhmer, K., Bick, E., Still, L., Strand, A., Luciano, R., Helmreich, S., Beck, B. D., Grønne, C., Malmros, L., Poulsen, K., Elbæk, F., Brydegaard, M., Lemmich, J. and Nikolajsen,

T. (2022). Automating insect monitoring using unsupervised near-infrared sensors. *Scientific Reports* 12, article 2603 (11 pp.). doi:10.1038/s41598-022-06439-6.
Satterfield, D. A., Sillett, T. S., Chapman, J. W., Altizer, S. and Marra, P. P. (2020). Seasonal insect migrations: massive, influential, and overlooked. *Frontiers in Ecology and the Environment* 18(6), 335–344. doi:10.1002/fee.2217.
Schaefer, G. W. (1976). Radar observations of insect flight. In: Rainey, R. C. (Ed.) *Insect Flight (Symposia of the Royal Entomological Society No. 7)*, pp. 157–197. Blackwell Publishing, Oxford, UK.
Schellhorn, N. A., Bianchi, F. J. J. A. and Hsu, C. L. (2014). Movement of entomophagous arthropods in agricultural landscapes: links to pest suppression. *Annual Review of Entomology* 59, 559–581. doi:10.1146/annurev-ento-011613-161952.
Sciarretta, A. and Calabrese, P. (2019). Development of automated devices for the monitoring of insect pests. *Current Agriculture Research Journal* 7(1), 19–25. doi:10.12944/CARJ.7.1.03.
Scott, K. D., Lawrence, N., Lange, C. L., Scott, L. J., Wilkinson, K. S., Merritt, M. A., Miles, M., Murray, D. and Graham, G. C. (2005). Assessing moth migration and population structuring in *Helicoverpa armigera* (Lepidoptera: Noctuidae) at the regional scale: example from the Darling Downs, Australia. *Journal of Economic Entomology* 98(6), 2210–2219. doi:10.1093/jee/98.6.2210.
Shamoun-Baranes, J., Bauer, S., Chapman, J. W., Desmet, P., Dokter, A. M., Farnsworth, A., Haest, B., Koistinen, J., Kranstauber, B., Liechti, F., Mason, T. H. E., Nilsson, C., Nussbaumer, R., Schmid, B., Weisshaupt, N. and Leijnse, H. (2021). Weather radars' role in biodiversity monitoring. *Science* 372(6539), 248. doi:10.1126/science.abi4680.
Shamoun-Baranes, J., Bauer, S., Chapman, J. W., Desmet, P., Dokter, A. M., Farnsworth, A., van Gasteren, H., Haest, B., Koistinen, J., Kranstauber, B., Liechti, F., Mason, T. H. E., Nilsson, C., Nussbaumer, R., Schmid, B., Weisshaupt, N. and Leijnse, H. (2022). Meteorological data policies needed to support biodiversity monitoring with weather radar. *Bulletin of the American Meteorological Society* 103(4), E1234–E1242. doi:10.1175/BAMS-D-21-0196.1.
Shearwood, J., Aldabashi, N., Eltokhy, A., Franklin, E. L., Raine, N. E., Zhang, C., Palmer, E., Cross, P. and Palego, C. (2021). C-band telemetry of insect pollinators using a miniature transmitter and a self-piloted drone. *IEEE Transactions on Microwave Theory and Techniques* 69(1), 938–946. doi:10.1109/TMTT.2020.3034323.
Shen, X., Fu, X., Huang, Y., Guo, J., Wu., Q., He, L., Yang, X. and Wu., K. (2020). Seasonal migration patterns of *Ostrinia furnacalis* (Lepidoptera: Crambidae) across the Bohai Strait in Northern China. *Journal of Economic Entomology* 113(1), 194–202. doi:10.1093/jee/toz288.
Smith, E. A., Fuchs, M., Shields, E. J. and Nault, B. A. (2015). Long-distance dispersal potential for onion thrips (Thysanoptera: Thripidae) and iris yellow spot virus (Bunyaviridae: Tospovirus) in an onion ecosystem. *Environmental Entomology* 44(4), 921–930. doi:10.1093/ee/nvv072.
Spillman, J. J. (1980). The design of an aircraft-mounted net for catching airborne insects. In: *Trends in Airborne Equipment for Agriculture and Other Areas. Proceedings of a Seminar Organized by the United Nations Economic Commission for Europe, Warsaw*, 18–22 September 1978, pp. 169–180. Pergamon Press, Oxford, UK.
Srygley, R. B. and Dudley, R. (2008). Optimal strategies for insects migrating in the flight boundary layer: mechanisms and consequences. *Integrative and Comparative Biology* 48(1), 119–133. doi:10.1093/icb/icn011.

Steger, C. (2017). A comprehensive and versatile camera model for cameras with tilt lenses. *International Journal of Computer Vision* 123(2), 121–159. doi:10.1007/s11263-016-0964-8.

Stepanian, P. M., Entrekin, S. A., Wainwright, C. E., Mirkovic, M., Tank, J. L. and Kelly, J. F. (2020). Declines in an abundant aquatic insect, the burrowing mayfly, across major North American waterways. *Proceedings of the National Academy of Sciences of the United States of America* 117(6), 2987–2992. doi:10.1073/pnas.1913598117.

Stepanian, P. M., Horton, K. G., Melnikov, V. M., Zrnić, D. S. and Gauthreaux, S. A. Jr. (2016). Dual-polarization radar products for biological applications. *Ecosphere* 7(11), article e01539 (27 pp.). doi:10.1002/ecs2.1539.

Stinner, R. E., Barfield, C. S., Stimac, J. L. and Dohse, L. (1983). Dispersal and movement of insect pests. *Annual Review of Entomology* 28, 319–335. doi:10.1146/annurev.en.28.010183.001535.

Storz, G. and Lavrenko, A. (2020). Compact low-cost FMCW harmonic radar for short range insect tracking. In: *International Radar Conference, Virtual, 28–30 April 2020*, (vol. 2020), pp. 642–647. IEEE Publications, Piscataway, NJ. doi:10.1109/RADAR42522.2020.9114612.

Stull, R. B. (1988). *An Introduction to Boundary Layer Meteorology*. Kluwer Academic Publishers, Dordrecht, Netherlands.

Stumph, B., Virto, M. H., Medeiros, H., Tabb, A., Wolford, S., Rice, K. and Leskey, T. (2019). Detecting invasive insects with unmanned aerial vehicles. In: International Conference on Robotics and Automation (ICRA), Montreal, Canada, May 20–24, 2019 (vol. 2019), pp. 648–654. IEEE Publications, Piscataway, NJ. doi:10.1109/ICRA.2019.8794116.

Sun, W., Hu, G., Su, Q., Wang, Y., Yang, W., Zhou, J. and Gao, Y. (2022). Population source of third-generation oriental armyworm in Jilin, China, determined by entomology radar, trajectory analysis and mitochondrial COI sequences. *Environmental Entomology* 51(3), 621–632. doi:10.1093/ee/nvac020.

Svenningsen, C. S., Frøslev, T. G., Bladt, J., Pedersen, L. B., Larsen, J. C., Ejrnæs, R., Fløjgaard, C., Hansen, A. J., Heilmann-Clausen, J., Dunn, R. R. and Tøttrup, A. P. (2021). Detecting flying insects using car nets and DNA metabarcoding. *Biology Letters* 17(3), article 20200833 (6 pp.). doi:10.1098/rsbl.2020.0833.

Tauc, M. J., Fristrup, K. M., Repasky, K. S. and Shaw, J. A. (2019). Field demonstration of a wing-beat modulation lidar for the 3D mapping of flying insects. *OSA Continuum* 2(2), 332–348. doi:10.1364/OSAC.2.000332.

Taylor, L. R. (1974). Insect migration, flight periodicity and the boundary layer. *Journal of Animal Ecology* 43(1), 225–238. doi:10.2307/3169.

Tielens, E. K., Cimprich, P. M., Clark, B. A., DiPilla, A. M., Kelly, J. F., Mirkovic, D., Strand, A. I., Zhai, M. and Stepanian, P. M. (2021). Nocturnal city lighting elicits a macroscale response from an insect outbreak population. *Biology Letters* 17(3), article 20200808 (5 pp.). doi:10.1098/rsbl.2020.0808.

Tison, L., Hahn, M.-L., Holtz, S., Rößner, A., Greggers, U., Bischoff, G. and Menzel, R. (2016). Honey bees' behavior is impaired by chronic exposure to the neonicotinoid thiacloprid in the field. *Environmental Science and Technology* 50(13), 7218–7227. doi:10.1021/acs.est.6b02658.

Topper, C. P. (1987). Nocturnal behaviour of adults of *Heliothis armigera* (Hübner) (Lepidoptera: Noctuidae) in the Sudan Gezira and pest control implications. *Bulletin of Entomological Research* 77(3), 541–554. doi:10.1017/S0007485300012025.

van der Kooi, C. J., Stavenga, D. G., Arikawa, K., Belušič, G. and Kelber, A. (2021). Evolution of insect color vision: from spectral sensitivity to visual ecology. *Annual Review of Entomology* 66, 435-461. doi:10.1146/annurev-ento-061720-071644.

Wägele, J. W., Bodesheim, P., Bourlat, S. J., Denzler, J., Diepenbroek, M., Fonseca, V., Frommolt, K.-H., Geiger, M. F., Gemeinholzer, B., Glöckner, F. O., Haucke, T., Kirse, A., Kölpin, A., Kostadinov, I., Kühl, H. S., Kurth, F., Lasseck, M., Liedke, S., Losch, F., Müller, S., Petrovskaya, N., Piotrowski, K., Radig, B., Scherber, C., Schoppmann, L., Schulz, J., Steinhage, V., Tschan, G. F., Vautz, W., Velotto, D., Weigend, M. and Wildermann, S. (2022). Towards a multisensor station for automated biodiversity monitoring. *Basic and Applied Ecology* 59, 105-138. doi:10.1016/j.baae.2022.01.003.

Wainwright, C., Jenkins, S., Wilson, D., Elliott, M., Jukes, A. and Collier, R. (2020a). Phenology of the diamondback moth (*Plutella xylostella*) in the UK and provision of decision support for Brassica growers. *Insects* 11(2), article 118 (16 pp.). doi:10.3390/insects11020118.

Wainwright, C. E., Reynolds, D. R. and Reynolds, A. M. (2020b). Linking small-scale flight manoeuvers and density profiles to the vertical movement of insects in the nocturnal stable boundary layer. *Scientific Reports* 10, article 1019 (11 pp.). doi:10.1038/s41598-020-57779-0.

Walker, T. J. (1985). Butterfly migration in the boundary layer. In Rankin, M.A. *Migration: Mechanisms and Adaptive Significance. Contributions in Marine Science, Suppl.* Marine Science Institute, The University of Texas at Austin, Port Aransas, TX, vol. 27, 704-723.

Walker, T. J. (2001). Butterfly migrations in Florida: seasonal patterns and long-term changes. *Environmental Entomology* 30(6), 1052-1060. doi:10.1603/0046-225X-30.6.1052.

Wallace, J. R. A., Reber, T., Beaton, B., Dreyer, D. and Warrant, E. J. (2021). Inexpensive monitoring of flying insect activity and abundance using wildlife cameras. Preprint. *bioRxiv*. doi:10.1101/2021.08.24.457487.

Walter, T., Degen, J., Pfeiffer, K., Stöckl, A., Montenegro, S. and Degen, T. (2021). A new innovative real-time tracking method for flying insects applicable under natural conditions. *BMC Zoology* 6, article 35 (11 pp.). doi:10.1186/s40850-021-00097-3.

Wang, R., Zhang, T., Hu, C., Cai, J. and Li, W. (2021). Digital detection and tracking of tiny migratory insects using vertical-looking radar and ascent and descent rate observation. *IEEE Transactions on Geoscience and Remote Sensing* 60, article 5101615 (15 pp.). doi:10.1109/TGRS.2021.3071934.

Westbrook, J. K. and Eyster, R. S. (2017). Doppler weather radar detects emigratory flights of noctuids during a major pest outbreak. *Remote Sensing Applications: Society and Environment* 8, 64-70. doi:10.1016/j.rsase.2017.07.009.

Wikelski, M., Moskowitz, D., Adelman, J. S., Cochran, J., Wilcove, D. S. and May, M. L. (2006). Simple rules guide dragonfly migration. *Biology Letters* 2(3), 325-329. doi:10.1098/rsbl.2006.0487.

Wiktelius, S. (1981). Wind dispersal of insects. *Grana* 20(3), 205-207. doi:10.1080/00173138109427667.

Williams, C. B. (1930). *The Migration of Butterflies*. Oliver and Boyd, Edinburgh, UK.

Wilson, L. J., Whitehouse, M. E. A. and Herron, G. A. (2018). The management of insect pests in Australian cotton: an evolving story. *Annual Review of Entomology* 63, 215-237. doi:10.1146/annurev-ento-020117-043432.

Wood, C. R., O'Connor, E. J., Hurley, R. A., Reynolds, D. R. and Illingworth, A. J. (2009a). Cloud-radar observations of insects in the UK convective boundary layer. *Meteorological Applications* 16(4), 491–500. doi:10.1002/met.146.

Wood, C. R., Reynolds, D. R., Wells, P. M., Barlow, J. F., Woiwod, I. P. and Chapman, J. W. (2009b). Flight periodicity and the vertical distribution of high-altitude moth migration over southern Britain. *Bulletin of Entomological Research* 99(5), 525–535. doi:10.1017/S0007485308006548.

Wotton, K. R., Gao, B., Menz, M. H. M., Morris, R. K. A., Ball, S. G., Lim, K. S., Reynolds, D. R., Hu, G. and Chapman, J. W. (2019). Mass seasonal migrations of hoverflies provide extensive pollination and crop protection services. *Current Biology* 29(13), 2167–2173. doi:10.1016/j.cub.2019.05.036.

Yu, D. W., Ji, Y., Emerson, B. C., Wang, X., Ye, C., Yang, C. and Ding, Z. (2012). Biodiversity soup: metabarcoding of arthropods for rapid biodiversity assessment and biomonitoring. *Methods in Ecology and Evolution* 3(4), 613–623. doi:10.1111/j.2041-210X.2012.00198.x.

Zemel, R. S. and Houghton, D. (2017). The ability of specific-wavelength LED lights to attract night-flying insects. *The Great Lakes Entomologist* 50(3–4), 79–85.

Zhou, X.-Y., Wu, Q.-L., Jia, H.-R. and Wu, K.-M. (2021). Searchlight trapping reveals seasonal cross-ocean migration of fall armyworm over the South China Sea. *Journal of Integrative Agriculture* 20(3), 673–684. doi:10.1016/S2095-3119(20)63588-2.

Part 2

Identification, modelling and risk assessment

Chapter 6

Advances in image-based identification and analysis of crop insect pests

Daniel Guyer, Michigan State University, USA; and Charles Whitfield, NIAB, UK

1 Introduction

As discussed in other chapters in this book, there are numerous methods, including visual- or image-based techniques, which attempt to identify the presence of insect pests within crop production sites for the purpose of timely control to prevent damage and yield loss. These include *direct* monitoring, such as visualization of pests either in flight or on crops, and *indirect* monitoring, notably the occurrence of specific damage to the crop leaves, seed or fruit. Spatial image information, and/or combined with spectral information, can be utilized in either direct or indirect monitoring. This chapter will focus on the direct identification of pests and beneficial insects.

The primary goal of direct image-based insect identification is to support precision agriculture and, more specifically, integrated pest management (IPM) by aiding or automating regular manual insect identification and quantification tasks generally conducted by trained and experienced human taxonomists. Generally referred to as 'scouting', this is labor intensive and limited spatially and temporally by cost and availability of trained experts.

http://dx.doi.org/10.19103/AS.2022.0113.08

There are many insect morphological features which provide potential information for detection and classification and can be used for automated image-based detection. These include *global* features as follows:

- overall size (area);
- aspect ratio, e.g. if an insect is long and narrow or more rounded or square;
- complexity, estimated by the perimeter-to-area ratio of the insect (achievable after good segmentation and then edge detection routines - discussed later);
- compactness, which indicates if the insect's overall area (size) fits within a small or large circle or rectangle and is an indicator if the insect has long appendages;
- boundary/perimeter tracing pattern or 'signature'; and
- overall color.

More *localized* or specific features include:

- lengths of legs or relationship and angles of legs to each other or abdomen;
- size of the head in relation to the abdomen;
- identification of head/facial characteristics such as mandibles, labrum, antennae, eyes, etc. and the ratio of the size of these features to each other;
- shape of individual wings and/or angle of wing tips;
- specific color relationships, color features or color patterns/markings of wings or body;
- physical texture, including pubescence, which can be determined by color and/or intensity variation across the surface of the object/insect; and
- length of any straight edges.

Ideally, all features for measurement should be rotation and size invariant, meaning the feature can be measured accurately regardless of the field of view and rotation of the image.

By analyzing such features, machine vision image-based electronic insect identification has significant potential to automate insect identification. Within such systems, a digital image is produced of appropriate resolution which can then be input into a wide range of programs to identify the objects or insects. However, while image processing, object recognition and machine vision technologies are highly advanced and widely available, especially in structured or semi-structured environments, the biological complexity and variability that exist in agricultural environments add great complexity in applying such technology in automated insect identification.

2 Challenges and solutions in automated image-based insect identification

As noted earlier, there are many insect morphological features used to identify a particular species, and while machine vision is a highly developed technology, there remain challenges for automated insect identification in agriculture. For example, if systems can identify individual human faces from millions that are relatively similar, it might be assumed that identifying insects should be quite straightforward. However, there are some specific challenges in insect identification, including the following:

- The potential presentation and orientation of an insect for imaging are hugely diverse. Insects may be in free flight or mobile within an arena (e.g. trap) or may be stuck to a sticky surface, resulting in the specimen being presented in an unnatural posture. A similar issue is the angle at which the object is viewed, which will create a variation in its presentation. Overcoming the variation in the presentation of the insect is a significant challenge.
- Possible damage and disfiguration to an insect captured within the system. This often occurs due to the insect's attempts to escape its confines and is particularly common when sticky surfaces are used to retain the insect. Insects will often damage their legs and wings as they attempt to free themselves from the sticky substrate. In 'bucket' or McPhail-type traps, insects can suffer damage from repeatedly bumping into the container walls. Lepidoptera can lose scales and parts of their wings, making visual identification difficult.
- Image-based insect detection is particularly challenging in complex, natural backgrounds, especially if the insect is in free flight within a natural setting. Insect image-based identification is also made difficult by the fact that many insects have evolved to 'blend in' with their surroundings to avoid detection by predators.
- Variations in lighting and shadows can occur if natural light is utilized when images are taken. Atmospheric pollutants such as dust can also compromise image quality.
- Within species variability, including variations between instars (developmental stages) and distinct adult morphologies. For example, *Drosophila suzukii* females have distinct summer and winter morphologies exhibiting different coloration and size; similarly, solitary and gregarious forms of locusts are visually extremely different despite being the same species.

- The size of the target can range from <1 mm (e.g. adult eriophyid mites can be <0.2 mm in length) to much larger insects of >50 mm in length (e.g. locusts).
- Visual similarity between species is common. Even trained taxonomists can find accurate identification between similar species challenging. Accurate identification is crucial for pest monitoring as visually-similar species may or may not be important crop pests. For example, within almost identical-looking Drosophila species, there are some that are important economic pests and some that are not.

Despite the many challenges to image-based insect detection and identification, there are also some unique features of automated systems, which are not available with manual human identification, which include:

- ability to use specific wavelengths via lighting or optical filtering to observe insect characteristics that might be otherwise unobservable to human vision. The use of spectral wavelengths outside of the visible range to humans, such as the ultra-violet, near-infrared or far-infrared ranges can illuminate characteristics. The use of spectral indices and fluorescence can greatly improve the differentiation of the target insect against the background (Xiao et al., 2022);
- use of machine learning (artificial intelligence - AI) as part of the computational classification, which can identify characteristics not commonly used in manual analysis, and analyze a far greater number of variables than humans could manage (Hoye et al., 2020). Combined with additional information from a wider use of the electromagnetic spectrum (unavailable to human observers), AI can be a powerful tool in the detection and classification of target species;
- the ability to conduct a wide range of quantitative measurements easily and rapidly without human error or bias; and
- automated detection can also provide additional information, such as precise timing of insect activity, which may be particularly useful for research.

Identification can be facilitated by limiting variability in some way to simplify the range of potential options to investigate. Options include the following:

- utilize traps that include species-specific attractants to limit the number of insect species captured to only those likely to be the target species, thus greatly limiting choices and also limiting physical overlap/intersection of multiple insects in the field of view. The most commonly used types of species-specific lure are sex pheromones. In addition to being species-specific, sex pheromones are usually extremely attractive over long distances (kilometers in the case of some Lepidoptera). However, they are usually only attractive to one of the sexes. Other semiochemical attractants may be suitable, such as aggregation pheromones or kairomones;

- utilize 'live traps' that lure individual target insects of interest through a tube. This has potential since insects are not damaged and have relatively limited variability in their presentation (since their movement is more restricted in the tube). The challenge is getting the insect into a trap as it is an unnatural structure. If the insect is released from the tube afterwards, there is also a risk of repeat-counting; and
- utilize weather, degree-day information and models to predict and limit possible insect species present.

Imaging of insects can be achieved using sticky and live traps. One of the most common techniques is to obtain images from a camera positioned to observe a sticky trap in the field. Telemetry is used to send images to a base station for processing, or images can be processed automatically, representing a nearly fully automated approach (as discussed later). Alternatively, sticky traps can be collected in the field and brought back to a lab or facility and imaged under consistent lighting conditions and run through a machine learning algorithm, representing a semi-automated system. The advantages of the latter are the likelihood of an improved image by having consistent and/or unique lighting and only needing one camera and processing system. This would replace the need for an experienced (and therefore expensive) human scout, though it would still require labor to maintain and retrieve traps on a regular basis.

An alternative imaging approach is to attract insects, ideally only the target species, or at least a limited number of species, into a trap and image them as they enter. Such traps utilize an attractant/lure and sometimes a funnel structure to encourage the insects to pass through a small tube which is fitted with a camera and optics to image the insect as it passes by. This will provide a relatively consistent image of the insect, albeit with wings closed. However, such a system may be seen as unnatural and therefore avoided by some insects, potentially introducing bias into species detection and enumeration. While not image-based, another class or mode of insect monitoring is sometimes used without imaging as the trap targets specific species with the lure (e.g. sex pheromone) and then simply uses an electronic trigger to count insects as they come through the tube. Some challenges with such systems are non-target insects that might enter are also counted and/or if the insect is able to pass in the reverse direction, it will be counted twice, both yielding inaccurate information. A recent review and explanation of this class of traps are presented by Preti et al. (2021).

A key issue for any system is, as far as possible, trying to ensure relatively consistent positioning and visualization of members of a species entering the system. While this might mean obscuring some features that aid identification, consistency of presentation outweighs the challenges associated with variability due to random presentation.

3 Understanding machine vision image-based insect identification

The term 'machine vision' refers to the technologies and methodologies that allow automated detection and analysis of a target. The cost of the relevant technologies, whether computational capacity/capability and other hardware, such as cameras, continues to decrease. Devices are also decreasing in size while the speed of image capture, quality and resolution are increasing, making the level of detail and amount of data which can be handled ever greater. Image capture and analysis can now be conducted in real-time even where multiple features of many objects (insects) need to be analyzed. Most machine vision technologies involve the following steps (Martineau et al., 2017).

3.1 Image capture (camera, resolution, optics and lighting)

Systems start with a camera plus a light source for optimal imaging. The camera could be an array or a line-scan camera, either of which will produce a two-dimensional digital image. However, a line-scan camera requires the object being imaged to translate in front of the camera to achieve both height and width, i.e. either the camera or the object must move to allow the object to traverse the line of view. The image resolution should be optimized to aid identification, ensuring sufficient details are captured, while minimizing the image file size. If the expected detail of small aspects within the image is required, then high-resolution imaging with an appropriate field of view will be necessary. The optics will determine the field of view.

Artificial lighting should be used to ensure that the object is evenly illuminated and to help limit any false characteristics that can occur with shadows. Lighting and/or spectral filtering of wavelengths can also be selected to enhance potential defining characteristics/features. Imaging the same scene under different wavelengths or wavelength bands can provide additional information about the object of interest. This is termed multi-spectral imaging.

3.2 Image processing

Imaging generally involves digital filtering convolution algorithms, which attempt to enhance an image and make the image more usable, such as running a noise filter over the image to clean up errant pixels or otherwise improving clarity or removing any undesirable artifacts. This should be unnecessary if the physical imaging domain/scene and the hardware used are all optimized for the situation.

3.3 Segmentation

This is the separation or extraction of the target object(s) from the background. There are many methods that can be used to segment, or help segment, objects from their backgrounds. This includes simple color or intensity thresholding, backlighting to first obtain a 'mask' of the object and/or using a highly contrasting background. The task is made much simpler by ensuring the background of the image is uniform, e.g. single color sheets often used in sticky traps provide an ideal background to make masking of the insect easy. For non-uniform backgrounds, if the background is completely stationary, an insect entering the field of view may be accurately masked by subtracting the previous image from the new image. Anything that can be done to enhance accurate segmentation will increase identification accuracy and efficiency within the identification process.

3.4 Feature extraction

This is the process of determining 'measurable' discerning characteristics of the object (insect), which can be 'extracted' from the digital image and used to uniquely identify the object. These can include measuring shapes or angles or overall size/area; counting or measuring a sub-aspect/feature; color; the relationship between phenological parts, etc. Features might be 'global' such as measuring the overall length and/or width or more detailed such as wing tip shape or angle. Features may be identified, developed or selected based on human perception of potential characteristics for discerning between species of insects, i.e. what a trained expert might identify. Established techniques for feature extraction include decision trees or support vector machine techniques (Hassan et al., 2014).

Another means of feature extraction is via more advanced AI machine learning methods. The decision process is developed by the algorithm, either via 'training' the program with tagged data (supervised training), e.g. by showing the system high numbers of labeled insect images, or by unsupervised algorithms. The latter rely on statistical probability to pattern match from large datasets; while they may take considerable time to develop, they can be automated and require less time from entomologists to label the data (Rustia et al., 2019).

Manual or AI approaches have been developed for many image (and other) data analysis settings with unique 'patterns' to differentiate (Junior et al., 2020). Such programs require many training images if the differences between classes are subtle and/or many classes exist. Image analysis can be hugely complicated by variables such as illumination conditions, insect pose, insect density (where individuals are congregated together) or background variables such as plant type and growth stage. If a feature value is variable, due to some of these and other variables (discussed earlier), such as how the insect might 'present' itself in an image, training requirements are greater.

To help address these challenges, image augmentation processes can be used to automatically add in variation to a dataset. These methods manipulate an image dataset, expanding the dataset by modifying copies of existing images (e.g. geometric transformations, color transformations, filters and mixing or removing parts of images). The effect of this provides a larger and more varied dataset of the target object to help the machine learning algorithm cope with variation. A recent survey of insect image classification has highlighted continuing gaps in existing image databases, e.g. in the number of insects covered, images taken in natural conditions (rather than artificially presented to optimize identification) and sufficient coverage to distinguish morphologically similar species (Amarathunga et al., 2021).

3.5 Classification

Classification is the process of combining and 'mapping' the numerical values measured as features to a multidimensional decision matrix space. Within this matrix, each object (insect) ideally has a unique volume or 'vector'. Often, especially in complicated situations and where objects are quite similar, the spaces for a class, such as species of insect, will overlap with others due to fundamental similarity or variability resulting from the presentation. In such cases, misclassifications occur due to false positives and false negatives. Classifiers can be built or programmed manually through several statistically based approaches or can be part of the AI approach noted earlier which extracts features and patterns and builds its own classification using neural networks and pattern matching. Many resources on various classifiers already exist covering a wide range of domains and applications, which are beyond the scope of this chapter but which readers can consult (e.g. Martineau et al., 2017; Rustia et al., 2019; Hansen et al., 2019; Amarathunga et al., 2021).

4 Automated and semi-automated image-based insect identification technologies

Studies to advance automated image-based insect identification have increased greatly in the past 15 years with results often demonstrating identification success above 80% and many studies in the 95+ % range. Many studies are documented in a review by Lima et al. (2020), which describes systems targeting a broad range of insects including Lepidoptera (such as moths), sucking pests (such as whiteflies and aphids), fruit flies and borer insects (such as weevils). While results from these systems are often excellent and demonstrate strong potential, many of the reported studies describe research or prototype devices and are therefore conducted under structured or controlled conditions rather than real-world scenarios. In structured and controlled research environments, the multiple challenges highlighted earlier are much reduced. For example, insects may be carefully manually presented for imaging and/or images are collected in ideal or consistent lighting environments. Many studies demonstrating high levels of identification success are generally capturing insects in a non-active, non-field situation. Additionally, many of the studies demonstrating very high identification accuracy are limited in the domain of insects included and/or they are victim to over-training or over-fitting classifiers making them very specific to the dataset used and limited in robustness. Thorough validation must be conducted on the classification algorithm, e.g. using some form of cross-validation.

Capturing real-world representative images can be more challenging than programming the machine learning process to identify the target. If supervised machine learning is used, correctly labeling all the images is also a significant and time-consuming task requiring trained and experienced entomologists. Any errors at this step will negatively impact the final classification algorithm.

Controlling variations in perspective and depth by ensuring a constant distance between the camera and the target is maintained will improve classification accuracy. For insects that generally crawl rather than fly, this task is easier. For flying insects, the design of the imaging device in the field should be such that the insects' flight is limited or inhibited. A simple and often used solution is to use lethal-type traps (e.g. sticky traps). If lethal traps are not used, then the design of the imaging device should force the flying insect to walk at the point where images are captured.

The use of uniform, sufficiently bright lighting will improve image quality and thus classification accuracy. Good lighting allows a smaller aperture and faster shutter speed which improve image quality. A smaller aperture results in a sharper image with a larger depth of field (i.e. more of the image is sharply in focus), and a faster shutter speed reduces blurring (particularly if the target is moving).

Automated identification of many moth species is made challenging due to their forewings covering their hind wings when in a resting position. Differentiation between similar species is easier with both pairs of wings visible, which show more colors and shape patterns. For many moth species, capturing images of the insect resting in a natural setting with open wings is extremely unlikely. Capturing the insect on a sticky trap can result in partial wing exposure, but due to attempts to escape from the sticky surface, there will be a wide range of body, wing and limb positions, including damaged parts, which adds further difficulty to any classification algorithms. This issue can affect any insect, regardless of whether they have wings or not. An imaging system can reduce some of these problems by either imaging frequently and/or having a trigger sensor on the trap to indicate new insect presence, thus capturing an image before the insect has time to damage itself in the sticky glue.

5 Commercially available systems

There are currently many commercially available image-based insect detection 'systems' of varying degrees of autonomy. Examples of commercially available systems (Table 1) are as follows:

- Crop Vue;
- Diopsis;
- iSCOUT;
- Semios;
- SightTrap;
- Spotta;
- TrapView; and
- TruePest.

Currently, because it is designed for broad-ranging ecological surveys, only the Diopsis device claims to be able to distinguish and identify multiple different species at the same time. Others target specific pests of concern to farmers. However, the same hardware and software can potentially be used to monitor other species although this will incur some development or setup changes to the software and the attractant being used. As development continues and technology advances, more devices capable of multiple target insect monitoring will be available. This could significantly reduce the costs of deploying devices in the field, as one device could monitor multiple species.

Table 1 Examples of commercially available products using image-based insect detection and some of their specifications

Device (company)	Target species	Purpose	Mode of attraction	Imaging device	Detection system	Connectivity	Autonomy	Power
Diopsis (Faunabit)	Any	Ecological survey	Yellow screen + UV layer	8 MP RGB camera	Supervised deep learning to identify and count insects	4G to server	Photograph every 10 s	Solar + battery
TruePest (Russell IPM)	*Trialeurodes vaporariorum*, *Drosophila suzukii*	Pest identification and monitoring	White/yellow sticky sheet (*T. vaporariorum*); kairomone lures (*D. suzukii*)	Compatible mobile phones	Machine learning to identify and count insects	Mobile phone and WiFi networks to server	Human operator to take photo	Mobile phone
Pest pods (Spotta)	*Hylobius abietis*, *Rhynchophorus ferrugineus*	Pest monitoring & action advice	Pheromone + kairomone lures (automated dispenser)	RGB camera	Not disclosed	Mobile phone networks, LoRa, independent WiFi	Fully autonomous for 1 year	Battery
TrapView (Trapview Solutions)	*Cydia pomonella* and many more	Pest monitoring & action advice	Pheromone lure + sticky sheet	RGB camera	Insect detection and verification by human entomologist	Mobile phone networks	Automatic photograph	Solar + battery
SightTrap (Insects Limited)	*Plodia interpunctella*, *Lasioderma serricorne*, *Tribolium* spp.	Pest monitoring	Pheromone lure	5 MP RGB camera	Human entomologist	WiFi	Automated daily photograph	Battery

(Continued)

Table 1 (*Continued*)

Device (company)	Target species	Purpose	Mode of attraction	Imaging device	Detection system	Connectivity	Autonomy	Power
iSCOUT (Pessl Instruments)	Multiple lepidoptera species, e.g. *Agrotis segetum, Cydia pomonella, Helicoverpa armigera*	Pest monitoring	Pheromone and kairomone lures, colored sticky sheet	10 MP RGB camera	Deep learning object detector, with insect classifier to species (for the target) and order/family for non-targets	4G/3G/2G*to server* reduced resolution for 2G	Automated	Battery
Pest Trap Sensor (CropVue)	Lepidoptera (e.g. *Cydia pomonella*)	Pest monitoring & action advice	Pheromone lure + sticky sheet	5 MP RGB camera	Machine learning to count target species	Mobile phone network to cloud server	Autonomous with daily photograph	Solar + battery

Image-based automated or semi-automated insect detection traps can operate at various levels of increasing complexity, such as the following:

1 A sticky or otherwise trap is used to attract and capture insects and the trap is transported at various intervals (days most likely) back to a central and controlled imaging station for automated image analysis and insect identification, avoiding the need for a trained human scout.
2 A camera affixed to and/or viewing an insect immobilization trap, such as a color or lure attracting sticky trap, and either imaged on trigger or at set intervals. The image data are manually downloaded periodically from the trap location by memory card transfer or other manual means.
3 As 2, but images are remotely transferred in either real-time or at set intervals with wireless technology (e.g. GSM or Wi-Fi) to a central processing computer, thereby reducing labor but increasing system cost. This adds the advantage of near-real-time monitoring, and thus reacting, in (near) real-time.
4 As 2 and 3, a live 'trapping' mechanism is used to lure a targeted insect and funnel it (walking/crawling) past a camera system where an image of each individual insect can be captured with high resolution, versus an entire area imaged as in the earlier approaches. Generally, the imaging is triggered by the interruption of a light beam as the insect passes. Collected images could be held for intermittent manual data retrieval or set up with wireless transfer.
5 Full automation with images collected under either the sticky trap or some form of live trap systems would be fully analyzed in real-time at each trap and only the results wirelessly transmitted.

A challenge and limitation for Levels 2-5 is power requirements. This is an added complexity and expense. However, the power demand can be reduced by implementing triggered or interval imaging and/or data transfer so the components are only powered for a small portion of the time; solar/wind electricity generation can also be incorporated.

Most commercial systems are based on conventional trap structures (e.g. 'Delta' traps) which house an attractant and sticky sheet. The trap structure must be modified to contain the camera and other hardware. Some systems can be described as 'self-cleaning' either by incorporating a scrolling sticky surface or by being non-lethal in allowing the target insect to exit the trap.

As noted earlier, traps can use attractants, often pheromone or kairomone semiochemicals or a visual attractant such as colored targets. Using an attractant to bring the insect to the camera allows control over the position and distance of the target relative to the camera and enables a uniform background and lighting of the images. The use of highly species-specific pheromones (e.g. sex pheromones) means that only the target insect species are attracted to the trap and imaged, thus greatly reducing (or perhaps eliminating) the need for computationally complex species identification steps. Overall, these systems are more accurate because they greatly limit the variability and number of unknowns that exist within the images.

Several of the field devices include a simple segmentation step to filter raw images prior to transfer to the server. Once these cropped images are created, they are transferred to a server for further analysis. The segmentation and cropping step greatly reduces the amount of data being transmitted. After the cropped images are transferred to a server, target detection, identification and counting are done either by AI software or human operators.

Many of the companies developing these systems market a monitoring 'service' rather than the devices themselves. The monitoring services (often accessed via a web-browser user interface) provide farms with detailed information on the local population of the target pest, its geographic distribution (if multiple traps are placed around the farm) and sometimes action threshold information. Some of the more advanced services incorporate weather data and pest development prediction models to warn farmers in advance when the pest is likely to become a risk. Some also offer automated messaging services to initiate pest intervention actions (e.g. application of sprays). Selling a service on a subscription-based model rather than the actual devices means the technology companies maintain ownership of the devices, can provide software/hardware updates as required and are responsible for trap maintenance. The service offered provides insights and advice to the farm based on data from the devices. For many farmers, this is more useful than using and maintaining the devices themselves.

This more common service approach is partially driven by the cost and relative complexity of the systems and partially by the ongoing technical support and continued development of the detection systems and user interfaces. The present-day cost of such a service appears to be in the US$30–50 range or more per month, per trap, compared to purchasing smart traps outright which can cost between $1000 and $2500 per trap. This means the trap density required to provide sufficient catch data is an important factor when calculating the total cost. While these costs may seem high when compared to standard non-smart non-IOT (Internet of Things) connected traps, these systems can provide farms with significant savings. These savings come from multiple sources such as reduction in farm labor to check traps, reducing unnecessary spraying (due to the better decision-making based on action thresholds) and near-real-time knowledge of pest populations leading to improved timing of control applications and thus reductions in lost yield.

The exact technology behind commercial systems is generally not publicly shared; however, they are reported to be highly accurate, often at nearly 100% in identifying a particular target insect species. As noted earlier, these figures need to be treated with some degree of caution. Though highly specific to a given insect at any one time, they can be 'tuned' to identify multiple insects by changing the attractant. Multiple traps each 'tuned' to different pests can also be set up for monitoring a site.

Although system performance in field conditions can be variable, these systems are still useful tools in providing automated, real-time time monitoring at a scale and regularity, greater than manual scouting can achieve (Gaston et al., 2004). More robust commercially available image-based insect identification and monitoring systems will evolve since the computing power and technology in 'intelligent' image analysis is already well developed, using AI and training of systems with large databases of insect images. However, the sheer diversity of insect species and range of morphological features within species, together with inherent difficulties in measuring these features in field conditions, makes developing such databases challenging and does not negate the need for trained taxonomists.

6 Conclusion

Obtaining a good quality image, showing as many features of the target insect in sufficient detail as possible, is the first challenge in achieving a successful and robust automated image-based insect detection system. Image processing and classification are relatively simple and well-developed through many applications in other domains. The most efficient approach couples multiple features and decision/classification approaches to converge on the identification of the target species. This can include combining sensor (image capture and processing) information plus *a priori* information such as weather/environmental information, insect activity patterns (e.g. diurnal, nocturnal, crepuscular) or filtering information such as the attractant used (which narrows down likely candidates), to increase the accuracy of the detection model. Using multiple information input sources (a process known as data fusion) narrows the possible options and increases the accuracy of the AI detection model.

Another factor affecting the potential of such systems is the required accuracy. A system which is only 60% accurate in identifying insects may be useful if this level still results in successful identification at the point where action might be needed. A range of studies based on automated identification of target pest species from sticky traps in field conditions suggest accuracy rates of up to 80% are sufficient to enable early and more effective IPM interventions, although some are based on semi-controlled greenhouse rather than field conditions (Guarnieri et al., 2011; Xia et al., 2015; Dey et al., 2016; Sun et al., 2017; Li et al., 2022). Expanded review and detail of automated insect identification concepts and study, including image-based and beyond, can be found in Cardim Ferreira et al. (2020).

Overall, image-based automated insect monitoring has excellent potential to support IPM practices. Automating pest identification helps reduce costs directly by limiting the need for expert manual scouting. However, an even greater potential cost saving is that identification can occur in near real-time in a way that human scouting can struggle to match. The ability to detect population changes of a pest insect immediately rather than knowing 5 or more days later via manual methods can greatly reduce mitigation/control costs. In addition, some of the more advanced systems include insect population modeling algorithms which can predict the risk of the pest in advance, providing the farm with extremely valuable foresight.

Success and widespread use of these systems is pending due to improved reliability as well as feasible cost of deployment and demonstrated value (return of investment) to the farm or other end user, particularly through integration into other systems. An example of such integration might be linking automated pest detection systems with automated messaging services to spray operators so that they know exactly when and where control measures need to be applied. This will minimize, or perhaps remove entirely, unnecessary spraying and ensure that control measures are applied at precisely the correct time to ensure maximum efficiency The use of monitoring data in population modeling to forecast pest risk and action thresholds will also allow timelier and more targeted IPM interventions.

7 References

Amarathunga, D. C., Grundy, J., Parry, H. and Dorin, A. 2021. Methods of insect image capture and classification: A systematic literature review. *Smart Agric. Technol.* 1: 100023.

Cardim Ferreira Lima, M., Damascena de Almeida Leandro, M. E., Valero, C., Pereira Coronel, L. C. and Gonçalves Bazzo, C. O. 2020. Automatic detection and monitoring of insect pests - A review. *Agriculture* 10(5): 161. https://doi.org/10.3390/agriculture10050161.

Dey, A., Bhoumik, D. and Dey, K. N. 2016. Automatic detection of whitefly pests using statistical feature extraction and image classification methods. *Int. Res. J. Eng. Technol.* 3: 950-959.

Gaston, K. J. and O'Neill, M. A. 2004. Automated species identification: Why not? *Philos. Trans. R. Soc. B Biol. Sci.* 359(1444): 655-667.

Guarnieri, A., Maini, S., Molari, G. and Rondelli, V. 2011. Automatic trap for moth detection in integrated pest management. *Bull. Insectol* 64: 247-251.

Hansen, O., Svenning, J. C., Olsen, K., Dupont, S., Garner, B. H., Iosifidis, A., Price, B. W. and Høye, T. T. 2019. Species-level image classification with convolutional neural networks enables insect identification. *Ecol. Evol.* 10(2): 737-747.

Hassan, S. A., Rahman, N. N. S. A. and Zaw, Z. 2014. Vision-based entomology: A survey. *Int. J. Comp. Sci.* 5(1): 19–32.

Hoye, T. T., Ärje, J., Bjerge, K., Hansen, O. L., Iosifidis, A., Leese, F., Mann, H. M., Meissner, K., Melvad, C. and Raitoharju, J. 2020. Deep learning and computer vision will transform entomology. *Proc. Natl Acad. Sci. (PNAS)* 1118(2): e2002545117. https://doi.org/10.1073/pnas.2002545117.

Junior, T. D. C. and Rieder, R. 2020. Automatic identification of insects from digital images: A survey. *Comput. Electron.* 178: 105784.

Li, W., Yang, Z., Lv, J., Zheng, T., Li, M. and Sun, C. 2022. Detection of small-sized insects in sticky trapping images using spectral residual models and machine learning. *Front. Plant Sci.* 13: 915543.

Martineau, M., Conte, D., Raveaux, R., Arnault, I., Munier, D. and Venturini, G. 2017. A survey of image-based insect classification. *Pattern Recognit.* 65: 273–284.

Preti, M., Verheggen, F. and Angeli, S. 2021. Insect pest monitoring with camera-equipped traps: Strengths and limitations. *J. Pest Sci.* 94(2): 203–217. https://doi.org/10.1007/s10340-020-01309-4.

Rustia, D. J. A., Chao, J. J., Chung, J. Y. and Lin, T. T. 2019. An online unsupervised deep learning approach for an automated pest monitoring system. ASABE International Meeting, Boston, MA, USA. https://doi:10.13031/aim.201900477.

Sun, Y., Cheng, H., Cheng, Q., Zhou, H., Li, M., Fan, Y., Shan, G., Damerow, L., Lammers, P. S. and Jones, S. B. 2017. A smart-vision algorithm for counting whiteflies and thrips on stick traps. *Biosys. Eng.* 153: 82–88.

Xia, C., Chon, T., Ren, Z. and Lee, J. 2015. Automatic identification and counting of small-size pests in greenhouse conditions. *Ecol. Inform.* 29: 139–146.

Xiao, Z., Yin, K., Geng, L., Wu, J., Zhang, F. and Liu, Y. 2022. Pest identification via hyperspectral image and deep learning. *SIViP* 16(4): 873–880. https://doi.org/10.1007/s11760-021-02029-7.

Chapter 7

Advances in insect pest monitoring using pest population growth and geospatial data for pest risk assessment

Michael J. Brewer, Texas A&M AgriLife Research, USA; Isaac L. Esquivel, North Florida Research and Education Center, University of Florida, USA; and John W. Gordy, Syngenta Crop Protection, USA

1 Introduction

Insect pest monitoring is central to integrated pest management (IPM), which is designed to encourage the use of compatible and ecologically and economically sound combinations of tactics to prevent economic damage to crops from insect pests. Remedial IPM tactics include insecticide use, the augmentative release of biocontrol agents, and other forms of direct pest disruption that results in near-term pest risk reduction. Preventative IPM tactics include manipulating planting dates, soil and vegetation management, conservation of natural enemies and classical biocontrol, and planting varieties that disrupt pest biology or increase the ability of the crop to resist or tolerate pest presence. The effects of remedial IPM tactics are typically short lived and used for near-term response to an increasing pest threat that is forecast to cause economic harm to the crop. Commonly, the economic threshold concept is used to trigger the use of such remedial IPM tactics (Stern et al., 1959; Pedigo, 1999). In contrast, preventative IPM tactics typically have delayed and long-term effects that are often inserted into regular crop management processes. When they are available, their deployment is appropriate where crop risk to specific

http://dx.doi.org/10.19103/AS.2022.0113.09

insect pests is persistent. Deployed area-wide, they may provide regional pest management and agroecosystem health benefits (Elliott et al., 2008; Brewer and Goodell, 2012). A regional assessment of pest risk is derived, in part, from aggregating crop pest monitoring data from individual fields across a production region, coupled with other regional pest risk assessment tools as available (e.g. Isard et al., 2015).

Insect pest monitoring has two core applications: detecting crop pests that are new to or re-infesting a field or region and evaluating current and forecasting future pest severity. Pest detection focuses on a methodology able to detect very low population levels, such as those associated with the establishment and spread of an invasive pest species (see other chapters in this book) and those that seasonally move across habitats and regions. Morphological and molecular techniques to identify species and species-specific trapping methods (e.g. pheromone traps) fall within the realm of crop pest monitoring. Interested readers are directed to additional chapters in this book and recent reviews on advances in molecular-based identification of insects and trapping to detect pest introduction or movement (e.g. Piper et al., 2019; Martinez et al., 2020). This chapter focuses on insect pest monitoring serving to evaluate and forecast pest severity over time and space with the aim to time the use of near-term remedial IPM tactics and provide information to plan the best use of longer-term preventative IPM tactics. Special attention is given to acquiring and estimating pest population levels and affiliated variables that may influence the trajectory of pest severity and distribution within and across crops and other habitats. The chapter reviews approaches and provides case studies of current advances that are in development or have been applied in demonstration or operational pest management systems. Several of the case studies discuss advances in insect pest monitoring in the context of balancing the benefits of preventative IPM tactics to address near-term pest threats in a specific field with benefits of strategically using available preventative IPM tactics regionally that may accrue benefits over a longer time span and larger scale (Elliott et al., 2008).

As traditionally applied to individual fields, insect monitoring data guide insecticide applications as preferably triggered by an economic threshold or other objective decision-making criteria (Pedigo, 1999). An economic threshold is set at the pest population level above which justifies the use of an insecticide or other remedial IPM tactic to prevent pest increase above a critical level that will cause economic damage to the crop (the economic injury level) (Pedigo, 1999). Insect pest monitoring protocols to support the use of economic thresholds are applied to individual fields, and sections of large fields may be monitored as a separate unit for pest management decision-making purposes (Brewer and Goodell, 2012). To increase time efficiency, albeit with loss of information precision, fields prone to pest infestations may be used as sentinels to gauge when to increase insect monitoring activities across a growing area or to serve as an indicator of need for insecticide use in multiple fields. Introduction of concepts and reviews of efficiencies in insect pest monitoring and IPM decision-making are available (Binns and Nyrop, 1992; Pedigo and Buntin, 1994). The same concept applies at other scales beyond individual fields (e.g. farm-scale as noted above, region-wide as presented elsewhere in this chapter) using insect monitoring and other spatial data. Regionally, economic thresholds should be augmented by other forms of pest risk evaluation such as ecological modeling and evaluation of multiple data streams that affect pest biology and management (Elliott et al., 2008; Brewer et al., 2019).

This chapter first reviews principles and advanced practices in insect pest monitoring at the field scale by incorporating knowledge of expected pest population growth and effects of preventative measures that naturally occur (e.g. resident natural enemies) or that are incorporated into regular field activities (e.g. planting cultivars that are partially pest resistant). The chapter then turns to principles and advancements in using spatial data, including spatially referenced pest, crop, and remote sensing data. Spatial data linked to insect pest monitoring may improve the strategic use of remedial IPM tactics at the field level and use of preventative IPM tactics at larger scales. Additional chapters in this book address pest detection and monitoring beyond individual fields using technologies such as unmanned aerial systems (UAS) and remote sensing, as well as molecular and biochemical techniques applied to trapping techniques. These methods add to information from aggregating insect pest monitoring data across fields in a region. Combining data sources within a pest information network may help guide the use of regional applications or preventative IPM tactics.

2 Pest monitoring basics and economic threshold adjustments

Periodic monitoring of crop insect pests provides pest managers with valuable information ranging from first detection of a pest moving into a field or growing region to estimation of pest density that can be used to decide when to initiate a remedial IPM tactic within an individual field. Depending on the crop-insect pest dynamic, monitoring may be for a certain crop growth period or pest development stage, such as for *Stenodiplosis sorghicola* Coquillett (Diptera: Cecidomyiidae) (sorghum midge) during sorghum bloom (Knutson et al., 2018), or a full production season such as for *Plutella xylostella* L. (Lepidoptera: Plutellidae) (diamondback moth) on cabbage (Talekar and Shelton, 1993). Periodic observations of pest populations are important to capture whether a pest population is increasing or declining in time and space. Such information can improve forecasting the potential for a pest population to exceed an economic threshold (Pedigo, 1999, Sections 2.1 and 2.3 of this chapter).

Many methods of insect sampling have been developed and deployed: soil sampling, sweep nets, beat sheets and buckets, traps with various attractants, and direct visual counts of an insect on a plant or plant part (Pedigo and Buntin, 1994; Pedigo, 1999). The biology and life stage of the pest and the form of crop injury guide the selection of the sampling method. Soil sampling may be needed to monitor insects that feed on roots. The sweep net, beat sheet, and beat bucket are typically used for monitoring insects that are more mobile or difficult to locate such as adult hemipterans and cryptic leaf-feeding lepidopteran and coleopteran larvae. Direct visual counts are often used for less mobile insects that are readily detected by eye, such as aphids (Hemiptera: Aphididae), whiteflies (Hemiptera: Aleyrodidae), and less mobile hemipterans on leaves and lepidopteran and coleopteran larvae feeding on new growth and the external wall of developing fruiting bodies. Use of traps that incorporate pheromone, light, and color attractants can be used for detection of pest arrival and tracking relative pest abundance over time. Pheromone traps have proven successful for monitoring species with multiple generations in a growing season and with annual migratory patterns including *Spodoptera frugiperda* J.E. Smith (fall armyworm) and *Helicoverpa zea* Boddie (corn earworm) (Lepidoptera: Noctuidae) in corn and cotton (Jackson et al., 2008; Cruz et al., 2012) and *Acrobasis nuxvorella* Neunzig (Lepidoptera: Pyralidae) (pecan nut casebearer) in pecan (Knutson and Ree, 2019). Light traps have also been used for decades in monitoring *Nilaparvata lugens* Stål (Hemiptera: Delphacidae) (rice brown planthopper) in rice in Asia (Hirao, 1979; Prasannakumar and Chander, 2014). Indirect measures, such as assessing plant injury caused by the pest, may be used as indicators of pest presence and severity. Some examples include defoliation and chlorosis of leaves, and injury and deformation of flowers, fruit, or roots (e.g. Teetes, 1994; Haile et al., 1998). One consideration of plant injury measurements is their additive nature across time compared with estimation of insect density that may grow or decline over time. Plant injury measurements cannot take advantage of information gained from field estimates of pest population growth.

From the perspective of an IPM practitioner, few metrics are as important as estimating or classifying the population density and comparing it to an economic threshold. Estimating population levels spatially and temporally are key to evaluating the potential for the population to approach or exceed the economic threshold. Once exceeded, a remedial IPM tactic can be implemented to reduce the likelihood that the population causes economic injury. Various forms of sampling plans exist to use specific monitoring methods which vary in cost, time and effort, and complexity (Binns and Nyrop, 1992; Pedigo and Buntin, 1994). The classical form compares estimates of pest density to an economic threshold based on the same sample unit (e.g. number of insects per plant or plant part). Alternative methods include insect population level classification schemes, such as using tally thresholds and sequential sampling plans (Rogers et al., 1994; Hall et al., 2007; Lindenmayer et al., 2020). Use of tally thresholds can be more resistant to the effects of unusual observations, especially at low density levels (Pedigo and Buntin, 1994). There can be time-saving benefits using classification approaches, such as a 50% reduction in time sampling when using a tally-based economic threshold instead of a density-based economic threshold for *Melanaphis sorghi* (Theobald) (= *Melanaphis sacchari* Zehntner) (Hemiptera: Aphididae) (sugarcane aphid) on sorghum (Gordy et al., 2021b).

2.1 Incorporating pest population growth into economic thresholds

Reliability of field estimates of pest population growth is affected by its life history. Furthermore, biotic and abiotic factors (e.g. natural enemies, weather) can affect the rate of pest population growth directly or through effects on the crop host. For example, pests that persist in an area with overlapping generations and continual reproduction (e.g. aphids [Hemiptera: Aphididae] and whitefly [Hemiptera: Aleyrodidae]) may be more easily monitored across time, allowing pest density estimates to be compared with density-based economic thresholds that include projections of pest population growth. Forecasts of population increase of highly mobile pests are challenging and contingent on first detection of movement into a field, such as arrival of long-distance migrating individuals into an area or more local and seasonal movement of pests into a field from alternative hosts and habitat surrounding a field.

Our focus in this section is on insect crop pests that can be reasonably monitored over time in individual fields and across fields in a growing area. For these insects, population rate of increase is relevant to IPM decision-making applied to individual fields. Pest population growth may vary as noted above, while in-field pest populations are particularly sensitive to plant susceptibility and pest regulation by natural enemies, as has been described for *Bemisia tabaci* Gennadius (Hemiptera: Aleyrodidae) (silverleaf whitefly) in cotton and *Aphis glycines* Matsamura (Hemiptera: Aphididae) (soybean aphid) in soybean (Horowitz, 1986; Ragsdale et al., 2007). Along with knowledge of previously determined population growth rates taken from controlled temperature experiments, continued monitoring and regular estimation of pest population growth can help to inform pest managers of when to deploy IPM tactics. This IPM decision-making process is anchored by insect pest monitoring techniques linked to economic thresholds (ETs) and ideally designed to be sensitive to pest population growth potential as discussed in the following sections.

The ET can be defined as a level of insect density at which an IPM tactic is initiated to prevent the population from reaching the economic injury level (EIL) (Pedigo, 1999). The economic injury level is a function of plant susceptibility, crop value, and management cost. It is the lowest number of insects (or other comparable metric) that will cause economic damage (Pedigo, 1999). Within this approach, fixed economic thresholds (a static value some level below the economic injury level) and descriptive economic thresholds (a dynamic value dependent on pest population growth and time needed to initiate an IPM tactic) can be calculated. A fixed economic threshold is often used in operational IPM programs as it requires less detailed knowledge of the potential for insect population growth (Pedigo, 1999). While a descriptive economic threshold requires more information, it can be more useful by better forecasting the future growth potential for the pest population and accounting for additional time to implement a remedial IPM tactic. For purposes of including population growth in calculating a descriptive economic threshold, Pedigo (1999) provided the formula: $ET = EIL/C^x$, where C is an estimate of pest population increase per unit of time, and x is the time needed to implement a management tactic. The daily rate of increase, which is a function of the population growth rate (r), is often used as an estimate of C. In practice, the more time needed to implement a management tactic, the lower the economic threshold will be, while the economic injury level remains unchanged. Likewise, as pest population growth rate increases, the descriptive economic threshold decreases. The economic injury level is not sensitive to changes in pest growth rate or time lag in applying an IPM tactic.

2.2 Consideration of crops exhibiting plant resistance

Plant cultivars that exhibit host plant resistance and tolerance traits to a pest can be an important component of IPM. Host plant resistance is a result of plant traits that reduce the extent of injury caused by a pest insect. Host plant tolerance is a result of plant traits or processes that reduce the amount of plant damage caused per unit injury (Stout, 2013). The cumulative effects of host plant resistance may directly affect the pest, resulting in lower populations, but do not affect plant susceptibility to the pest (Stout, 2013; Smith, 2005). Host plant tolerance does not affect pest biology but results in retaining yield potential at pest densities that would cause yield loss in susceptible cultivars (Smith, 2005).

Adjustment of an economic threshold and economic injury level for a resistant or tolerant plant may be appropriate when there is reduced potential for pest population growth or increased ability of the plant to withstand pest feeding. Simply put, the potential for economic harm from the pest is lowered but not necessarily eliminated. There are practical implications of host plant resistance when calculating economic injury levels and thresholds. Reduced pest population growth rate on resistant cultivars may be seen, while the susceptibility of the plant is not changed. In this instance, an economic injury level does not change but the corresponding descriptive economic threshold will increase due to the reduced pest population growth (i.e. a change in pest population increase per unit time, *C*, of the descriptive economic threshold equation). Alternatively, the time needed to initiate a remedial IPM tactic may be lengthened given the reduced pest population growth potential. This action may functionally leave the economic threshold unchanged or lessen the increase depending on the change in *C* and the desired change in *x* of the equation. For tolerant plants in which plant susceptibility to the pest declines, the economic injury level will increase because of the effect on the yield-insect density relationship used to calculate the economic injury level. This will result in an increase in the economic threshold. When tolerance is part of crop resistance, the slope of the yield-insect density regression used to calculate the economic injury level will decrease and may even approach zero in highly tolerant crops. In such cases, the economic threshold concept does not apply, at least in the range of insect densities experienced in the field (Gordy et al., 2021a). The basic pest status is greatly diminished although it is advisable to periodically inspect for unexpected plant injury as an indicator of potential change in pest virulence (i.e. development of pest biotypes that can overcome plant resistance [Smith, 2005]).

2.3 Case study: sugarcane aphid as a pest of sorghum

Gordy et al. (2019) estimated the yield-aphid density relationship and field-based population growth rates of sugarcane aphid for sorghum hybrids susceptible to sugarcane aphid. Using this information, they estimated the economic injury level and proposed a common descriptive economic threshold useful across a range of conditions found in the large sorghum agroecosystem of the United States. They also estimated field-based population growth rates on sorghum susceptible and partially resistant to sugarcane aphid (Gordy et al., 2021a). Here, we used the data to demonstrate how population growth can be used to inform IPM decision-making based on appropriate adjustments to the economic threshold. For simplification, a single economic injury level of 67 sugarcane aphids per leaf for a susceptible hybrid was used (Gordy et al., 2019).

Gordy et al. (2021a) calculated population growth rates for sugarcane aphid at several locations across several years for both susceptible and resistant sorghum hybrids. While there was considerable variation, the mean doubling time for aphid populations on resistant hybrids was more than double than for populations on susceptible hybrids (13.4 days and 6.2 days, respectively). The decreased rate of population growth observed on resistant hybrids was a likely indicator of host plant resistance which has been confirmed through controlled experimentation with some sorghums (Paudyal et al., 2019). Using this information, a higher economic threshold was appropriate due to the increase in doubling time given the reduction of *C* of the descriptive economic threshold equation. Using the density-based economic injury level of 67 aphids per leaf and daily rates of increase of 1.17 and 1.05 for populations on susceptible and resistant hybrids, respectively, density-based economic thresholds can be calculated for a three- and five-day lag in deploying a remedial IPM tactic (Fig. 1, see left axis scale for density measure of aphids per leaf). For those using a tally-based threshold for sugarcane aphid (Gordy et al., 2021b), infestation proportions of aphids exceeding 50 aphids per leaf replace the density-based information (Fig. 1, see right-axis scale for infestation proportions). For sorghum hybrids with tolerance as part of its sugarcane aphid resistance profile, the calculation of the EIL and ET will be affected, resulting in a higher EIL and ET (Gordy et al., 2021a), as noted in Section 2.2 above.

Consideration of pest population growth on different sorghum hybrids provided three benefits. The first was the direct benefit of utilizing host plant resistance. Monitoring sugarcane aphid field populations over time, Gordy et al. (2021a) proposed host plant resistance mechanisms were involved because of the reduced aphid population growth under field conditions, while some

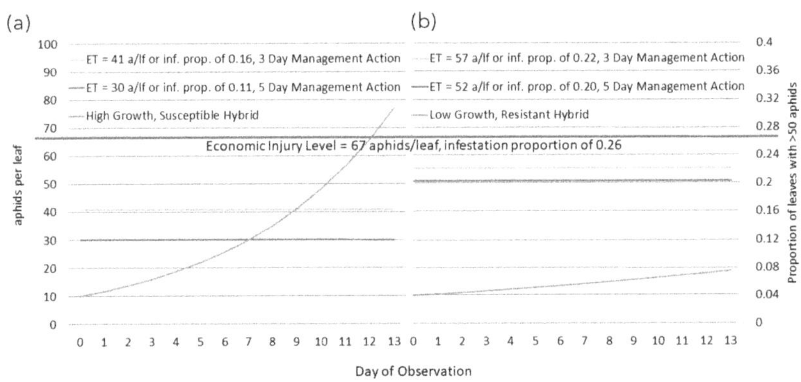

Figure 1 Comparison of the effect of pest population growth on a descriptive economic threshold. Economic injury level is the same for both plots (red line representing 67 aphids per leaf [left axis] or 0.26 infestation proportion of leaves infested with >50 aphids [right axis]). Plot (a) represents a population on a susceptible sorghum hybrid with a high growth potential and daily rate of increase of 1.17 (doubling time of 4.4 days) and corresponding density-based economic thresholds of 41 and 30 aphids/leaf (a/lf) (tally-based infestation proportions [inf. prop.] of 0.16 and 0.11) for a 3- and 5-day lag in deploying an insecticide, respectively. Plot (b) represents a population on a resistant sorghum hybrid with a low growth potential and a daily rate of increase of 1.05 (doubling time of 13.4 days) and corresponding density-based economic thresholds of 57 and 52 aphids/leaf (a/lf) (tally-based infestation proportions [inf. prop.] of 0.22 and 0.20) for a 3- and 5-day lag in deploying an insecticide, respectively.

hybrids may also be partially tolerant. Specifically using field observations of aphid growth on a group of partially aphid-resistant sorghum hybrids, the aphid population growth was reduced for most hybrids, which was appealing to IPM practitioners. The second benefit was incorporating pest population growth into the economic threshold (see exercise in the preceding paragraph and Fig. 1). When a remedial IPM tactic was scheduled within 3 days, the economic threshold increased from 41 aphids to 57 aphids per leaf, and for a tactic implemented within 5 days it increased from 30 aphids to 52 aphids per leaf for susceptible and resistant hybrids, respectively (Fig. 1, left axis). The third benefit was the ability to delay use of a remedial IPM tactic using higher economic thresholds associated with a partially resistant sorghum hybrid. Extending the time period to implement a management tactic may provide opportunity for other aphid regulatory factors (e.g. predator or parasitoid populations, entomopathogenic fungi, weather) to stabilize or reduce aphid populations. This positive benefit of natural enemies and epizootics has been observed for sugarcane aphid on resistant sorghum hybrids (Haar et al., 2018; Faris et al., 2022) and other pests (Ragsdale et al., 2007).

3 Geospatial data for pest risk assessment

Spatial data management and analysis to support pest management can be enabled by geographic information systems (GIS) and remote sensing technologies (Dminić et al., 2010). A geographic information system is a computer tool (including software and hardware) that stores, organizes, manages, analyzes, and displays geospatial data such as geo-referenced insect and plant data, remote sensing data, cropland/land-use data, and weather data. Geospatial data are referenced to a coordinate system and linked to data and information relevant to that space (Birt et al., 2012). For example, a geo-referenced space may be pixel(s) of a raster image that conveys a reflectance, thermal, or other remotely sensed variable value. In a vector image, a geo-referenced space may be an insect sampling site (a point), field edge (a line), or portion or entire crop field (a polygon). The linked information is a variable value such as pest species identity, pest counts or density estimates, crop identity and growth information, and other information relevant to the space and task. Multiple data layers in a common coordinate system may exist, such as a classification layer of soil type and cropland/land use, a weather layer, and/or an insect and crop data layer (USDA NASS, 2021; USDA FSA, 2021). The ability to overlay, manage, and analyze multiple geospatial data layers allows users to explore the spatial aspects of insect populations, consider influences of features portrayed in the data layers, and ultimately aid in pest risk assessment and IPM decision-making (see examples and references in the following text).

Many of these tools have found use in IPM at local and regional scales. Examples include predicting the spatio-temporal distribution of corn earworm (Moral García, 2006), creating GIS-based maps to guide spatially variable insecticide applications (McKinion et al., 2009), using UAS to carry remote sensing equipment for detection of plant stress caused by insects and release of biocontrol agents (Moses-Gonzales and Brewer, 2021), and evaluating natural enemies as biological control agents (Bianchi et al., 2006). It is becoming more feasible to use spatial pest and associated information to improve pest risk assessment and IPM decision-making given the improvements, automation, and linkage of GIS to archived data layers. Weather products are available from automated weather stations (Weather Underground, 2021); cropland-use products (Cropscape) are available from the National Agriculture Statistics Service (USDA NASS, 2021); and roadway and waterway products are available from the geospatial gateway (USDA FSA, 2021). GIS facilitates the importation and manipulation of these data layers and field-collected data layers such as insect monitoring results and UAS imagery (Dminić et al., 2010). This section reviews studies that seek to understand the spatial variation of pest species within and between fields using GIS tools, and how some studies used spatial data managed in a GIS for pest management and risk assessment.

3.1 Spatial variation of insect pests within and between fields

A recent review (Karp et al., 2018) concluded that pest management may be influenced by the landscape, from change in pest and natural enemy abundance to crop sensitivity to pest injury. However, significant variation in the influence of landscape variables on pest management was seen across the studies reviewed. Although a challenging task, Karp et al. (2018) stated that efforts to develop and best deploy pest management tools would benefit from understanding how landscape features influence pest biology and associated crop functions. Spatial and temporal patterns of pest herbivores are affected by their interactions with natural enemies, the suitability of the crop and the surrounding plant community (inclusive of alternative hosts), and the influence of other environmental factors. The spatial size, arrangement, diversity, connectivity, scale, and quality of habitats are relevant in understanding pest-natural enemy-plant interactions (Bianchi et al., 2006). For example, population structure and spatial patterns of highly mobile insects (e.g. lepidopteran crop pests) and their interaction with natural enemies may significantly vary across a region that is diverse in crop and habitat composition and configuration. Whereas more sessile insects like aphids after initial colonization may be influenced by the immediate surroundings of available crop hosts, natural enemies, non-crop habitat, and other abiotic and biotic environmental factors. The affected processes and species traits include pest population growth, pest-natural enemy interactions, pest herbivore movement, and pest affinity within and between crops and other habitats.

Cotton agroecosystems provide good examples of how GIS has been incorporated into studies to investigate the spatial variability of insect pests. The use of broad-spectrum insecticides in cotton has been dramatically reduced after the success of transgenic Bt (*Bacillus thuringiensis*) cotton in many countries (Luttrell et al., 2015). This success has been attributed to the release of a sucking bug complex, consisting of plant bugs and stink bugs (Miridae and Pentatomidae, respectively), from indirect insecticide control (Lu et al., 2010). The adult stage of these pests is highly mobile, which contributes to variable spatial patterns within and between fields. Spatial variation of crop insect pests presents challenges to the traditional use of the economic threshold concept in managing pest insects within individual fields (Brewer and Goodell, 2012), but it also presents opportunities if spatial variation can be tracked and useful information extracted.

To investigate *Lygus hesperus* (Knight) (Hemiptera: Miridae) (western tarnished plant bug) populations in cotton, Carrière et al. (2006) used GIS and spatial statistics to assess the maximum distance in which forage and seed alfalfa, weedy fallow fields, and cotton fields affected western tarnished plant bug populations. They found that forage and seed alfalfa and fallow fields with weeds acted as sources of this plant bug. However, these source effects did not extend beyond 375 m, 500 m, and 1500 m for forage alfalfa, weedy fallow fields, and seed alfalfa, respectively. In contrast, cotton fields in proximity to these fields acted as western tarnished plant bug sinks, but this sink effect did not extend further than 750 m. They concluded that planting cotton fields at a distance greater than 375 m, 500 m, and 1500 m from forage alfalfa, weedy fallow fields, and seed alfalfa, respectively, reduced movements of the pest into cotton during the primary fruiting cycle when cotton is most susceptible to plant bug feeding.

Tillman et al. (2009) investigated the spatial patterns and dispersal of *Nezara viridula* (L.) (southern green stink bug) and *Euschistus servus* (Say) (brown stink bug) (Hemiptera: Pentatomidae) in a peanut-cotton agroecosystem in the southeastern United States. Using GIS and spatial statistics (Spatial Analyses by Distance IndicEs [SADIE]), they interpolated stink bug population maps over time to examine the spatial aggregation of both stink bugs and their association with cotton and peanuts. Nymphs and adults of both species that developed in peanuts dispersed into cotton. Further, when they dispersed into cotton, they tended to aggregate at the interface of the two crops displaying a significant edge effect dependent on landscape structure (cotton adjacent to peanut). Through understanding pest spatial variability, more targeted pest sampling and more strategic use of insecticides or releasing biocontrol agents along field edges in this agroecosystem can be achieved.

In another study in southeastern U.S. cotton agroecosystems, Pilkay et al. (2014) sampled three stink bug species, *E. servus* (brown stink bug), *N. viridula* (southern green stink bug), and *Chinavia hilaris* (Say) (common green stink bug) in wheat, corn, soybean, peanut, and cotton. They used SADIE to analyze spatial data across three years. Populations of the different stink bug pests varied across the years, but spatial analyses indicated aggregated patterns of stink bugs in 18% of year-end totals and 42% of weekly insect counts. Adults were significantly aggregated in a high majority of cases (43 of 56 comparisons), but species differences were detected: adults of brown stink bug were aggregated 39.5% of the time while aggregated distributions occurred less often for southern green stink bug (11.6%), and common green stink bug (9.3%) adults. Most aggregated distributions were found in soybean (32.6%), followed by cotton (20.9%), corn (18.6%), fallow (4.7%), and peanut (2.3%). Stink bug densities were greater in cotton fields adjacent to soybean or peanut fields, and soybean fields adjacent to cotton or peanut fields compared with other crop/adjacent crop combinations. In this study, the patterns of spatial and temporal variation in seasonal dispersion and crop host use by multiple species of stink bugs were important in considering insect monitoring efforts and deploying a remedial IPM tactic.

In cereal crops, consideration of landscape influences within an IPM framework has had some success by incorporating natural enemies in IPM decision-making protocols. One strategy has been incorporating natural enemy occurrence and estimated pest suppression into economic thresholds, as done when monitoring *Schizaphis graminum* Rodani (Hemiptera: Aphididae) (greenbug) and its natural enemy *Lysiphlebus testaceipes* (Cresson) (Hymenoptera: Braconidae) in cereal crops (Royer et al., 2015). In Oklahoma where *L. testaceipes* overwinters, it has been incorporated into management recommendations in winter wheat with success in keeping greenbug population below economic thresholds. Royer et al. (2015) proposed that augmentative releases may suppress greenbug below economic thresholds in areas where *L. testaceipes* does not overwinter.

Simulation of local and regional dynamics of pest species such as sugarcane aphid may aid efficient allocation of local pest management resources and contribute to area-wide pest management (Elliott et al., 2008). For example, Wang et al. (2019) used a spatially explicit, individual-based model to simulate local sugarcane aphid infestations based on temperature-dependent sugarcane aphid growth on sorghum, while an atmospheric model was used to simulate regional sugarcane aphid migration at a large spatial scale (i.e. county level in the central U.S.). Combined, both models were used to forecast the likelihood of winged sugarcane aphid re-infesting fields north of where it overwinters in the far southern Great Plains (Koralewski et al., 2020a). Linking these modeling products with local aphid monitoring results and weather reporting may improve strategies to deploy preventative IPM tactics in regions where pest risk is persistent (Brewer et al., 2019; Koralewski et al., 2020b). Direct monitoring of airborne movement of crop insect pests is considered in additional chapters in this book.

Within-field spatial aggregation and variability of insect crop pests and associated plant injury have been observed. Reay-Jones et al. (2016) used GIS to manage remote sensing data for the aggregation of stink bugs and boll injury in cotton and the spatial association of stink bugs and boll injury. They used an index derived from remotely sensed optical data (i.e. normalized vegetation index [NDVI]) and insect data acquired by field sampling within a grid system. From their spatial analyses, stink bugs were less frequently aggregated (17% in adults and 4% in nymphs) than boll injury (40%). NDVI values were also significantly aggregated within fields in 40% of analyses. Spatial distributions of stink bugs were less stable than NDVI values and the boll injury caused by stink bug feeding. NDVI values were positively associated with boll injury in 11 out of 22 analyses with no negative associations. Other studies have also begun to investigate the use of remote sensing technologies in detecting insect-induced stress in cotton. For example, Reisig and Godfrey (2006) were successful in detecting insect-induced stress using NDVI values for *Aphis gossypii* (cotton aphid) and spider mites (Acari: Tetranychidae). They found a negative association between pest abundance and damage with decreasing NDVI values. Further, Esquivel et al. (2021) found inconsistent trends of densities of *Psuedatomoscelis seriatus* (Reuter) (Hemiptera: Miridae) (cotton fleahopper) regressed on NDVI values (5 of 12 significant regressions but all positive slopes). These studies indicated that more research was needed to utilize remote sensing to detect the type and extent of insect stress in cotton.

Notably, the acquisition of remote sensing data relevant to pest risk assessment has been enabled by recent advances in UAS that carry miniaturized optical sensors and are integrated with flight control automation, post-flight data processing, and GIS technologies (Moses-Gonzales and Brewer, 2021). Research and technology using UAS and remote sensing to support pest management have increased in recent years and may stimulate additional advancements in insect pest monitoring by incorporating spatially variable pest risk into information systems. Advances in pest identification and tracking using UAS and remote sensing technologies are also presented in additional chapters in this book.

3.2 Case studies

3.2.1 Use of GIS to create pest risk maps for in-field use

In order for IPM decision-making in individual fields to receive the maximum benefit of technological advances, pest monitoring observations should be used in near real-time (Isard et al., 2015). Deleon et al. (2017) used a GIS- and GPS-enabled hand-held device to create pest risk maps to aid in pest management decisions for multiple pest species in a cotton–sorghum agroecosystem. Pests included cotton fleahopper and verde plant bug, *Creontiades signatus* (Distant) (Hemiptera: Miridae) in cotton, sugarcane aphid, headworms (corn earworm and fall armyworm), and *Oebalus pugnax* F. (Hemiptera: Pentatomidae) (rice stink bug) in sorghum. GIS software (ArcMap [ESRI, 2021]) was used to create shapefiles to store spatially referenced data obtained from standard cotton and sorghum pest monitoring activities. The sampling sites were projected as mapped points on a base map of cotton and sorghum fields of a demonstration area for display and in-field sampling purposes. The base map was created in the GIS by the digitization of an orthorectified image of the area taken by the National Agriculture Imagery Program and downloaded from the Geospatial Data Gateway (USDA FSA, 2021). Utilizing a GPS-enabled hand-held device (Juno 3B, Trimble [2021]) running ArcPad (ESRI, 2021), insect samplers tracked their position in the field in real-time on the base map with the location and field image seen on the screen of the hand-held device. They recorded insect sampling data at pre-determined geo-referenced sampling points, random points, or other points of interest in a cotton or sorghum field. The hand-held devices were pre-loaded with shapefiles containing the digitized base map and data files for pest monitoring-specific data inputs (empty shapefiles). Data inputs consisted of the number of plants or plant parts sampled (e.g. sorghum leaves, seed heads, cotton bolls inspected), the number of nymphs and adults for cotton fleahopper and verde plant bug on cotton, and the number of sugarcane aphids, stinkbugs, and headworms on sorghum. At each of the sampling sites, insect samplers selected the desired shapefile for the crop and insect pest and then proceeded with data entry.

After returning from the field, shapefiles with the collected data were transferred to a desktop computer running ArcMap to create pest risk maps. ArcMap facilitated adding symbology and labels onto the maps to aid simple visual interpretation. A five-color scheme was used to portray pest densities (i.e. number of insects per sample unit of leaves, bolls, or other plant parts) found at each sampling site relative to economic thresholds that were pre-selected based on the literature and using the same sampling unit as used in the field. The color code consisted of two shades of green for levels below the economic threshold (one shade indicating no pests detected), yellow for densities approaching the pre-selected economic threshold, and orange and red for densities at and above the economic threshold, respectively. Working with the grower cooperators and referring to the literature on economic thresholds for these pests (Brewer et al., 2013; Gordy et al., 2019), the density range of the yellow color code represented the variation associated with the economic threshold given the range of crop value and insecticide application cost. The well-recognized green to red color scheme was used to convey pest population growth and crop risk over time (Fig. 2). When pest populations were high, a more detailed within-field sampling effort was conducted for specific fields (Fig. 3).

The pest maps were delivered as pdf images to grower cooperators and other IPM practitioners within the same day of sampling via text and email. The pest risk maps facilitated snapshot assessments of pest status (Fig. 2) and closer examination of spatial pattern of the pest insect within a field (Fig. 3). Appended directly below the pest risk maps were attribute tables containing all data associated with each sampling point on the map. Based on grower feedback, maps were primarily used to guide whole-field insecticide use decisions (i.e. traditional use of pest monitoring data [Brewer and Goodell, 2012]). These data were occasionally used to consider follow-up insecticide use targeting specific field sections where pest pressure persisted (i.e. a form of within-field precision pest management) (Deleon et al., 2017).

Expanding on this experience, a scenario using insecticides applied to within-field pest management zones guided by the spatial variation of insect densities (within-field zone approach) was compared with a scenario using insecticides applied to whole fields guided by the whole field insect density average derived from the same data (whole field approach). For the within-field zone approach, the area to be sprayed was the proportion of the field represented by corresponding sampling sites that exceeded the economic threshold on a sampling date. If the whole field average was above the economic threshold, then the entire field was projected to be sprayed. The recommended area of insecticide use guided by the two methods was compared using the empirical data from the demonstration project. Within-field variation in insect density reduced the projected area sprayed significantly, between 40% and 70% depending on the pest, when sprays were limited to pest densities of within-field zones exceeding the economic threshold (Table 1). Related, McKinion et al. (2009) demonstrated that spatially referenced field data managed in a GIS can guide spatially variable insecticide applications. They developed an automated methodology to process and analyze multispectral images with pest population data layers to construct maps to guide insecticide application. When using this method, they estimated that insecticide applications can be reduced by 60% compared with traditional whole-field treatments. These studies demonstrated the use of GIS and GPS tools in making pest risk maps that were appealing to cooperating growers and IPM practitioners. Further, it also bridged information on spatial variability of insect pests managed by a GIS to a precision agriculture practice (i.e. within-field zones) by using otherwise standard pest monitoring activities that were placed in the spatial realm of a GIS.

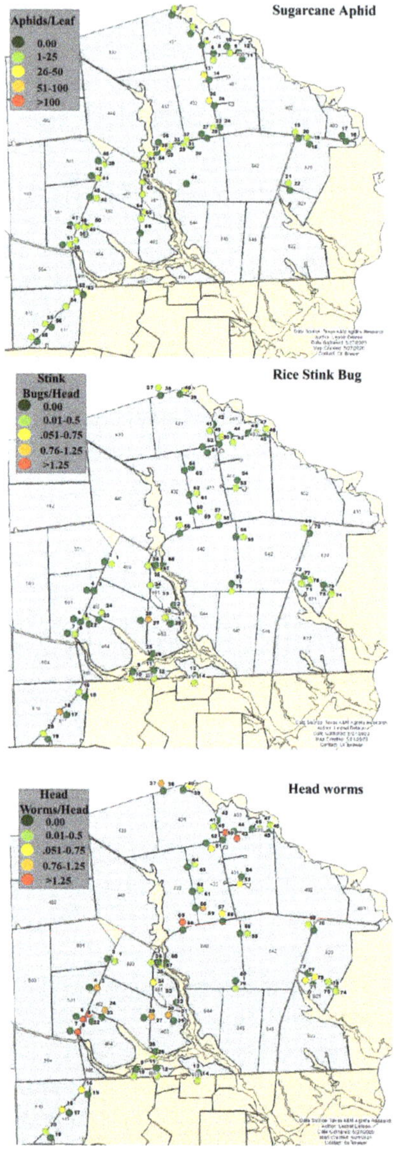

Figure 2 Example of pest risk maps delivered to grower cooperators after completing a pest monitoring event. A pest risk map is made for each species monitored: sugarcane aphid (top), rice stink bug (middle), and headworms (corn earworm and fall armyworm) (bottom). A five-color scheme legend was used to reference the observed pest density estimates to pre-selected economic thresholds: two shades of green represent pest densities below an economic threshold (one indicating no pests were found), yellow represents pest densities approaching the economic threshold, orange and red represent pest densities at and above the economic threshold. Sites were numbered in the order they were visited and were linked to a corresponding attribute table appended below the pest risk map. The maps were sent to grower cooperators and IPM practitioners as pdf files via text and email.

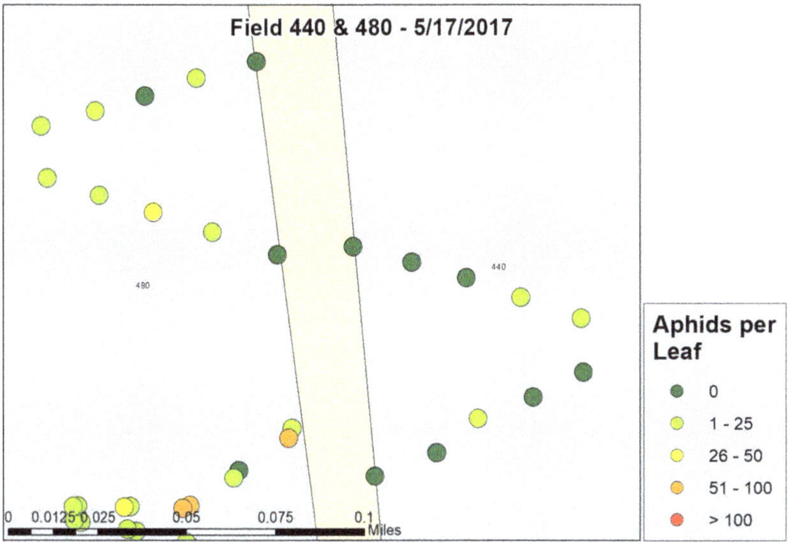

Figure 3 Example of a pest risk map for sugarcane aphid on sorghum showing multiple sampling sites across two adjacent sorghum fields This intensive monitoring effort was stimulated when sugarcane aphid at one field edge exceeded the pre-selected economic threshold. The color code scheme was the same as in Fig. 2.

3.2.2 Spatial variability of cotton pests to improve pest monitoring and management within and between fields

Esquivel et al. (2020) examined the change in pest density and related cotton injury of two cotton pest species from the edge of cotton fields toward field interiors (potential edge effects) and as further influenced by adjacent field type (potential neighbor field effect). Data came from a classical insect monitoring routine with all data placed in the spatial realm with the aid of a commercial GPS-enabled hand-held device (Trimble, 2021) and commercial GIS software (ESRI, 2021) running on a desktop computer. Two species of plant bugs were the predominant insect pests of the South Texas cotton system where this project was conducted: cotton fleahopper and verde plant bug (Brewer et al., 2013). This study was conducted in a large commercial farming operation (1200 ha). Cotton and grain sorghum were rotated annually at an approximate 1:1 ratio and were juxtaposed on the natural gulf prairie habitat that consisted of shrubland and waterways that drained into the Gulf of Mexico. Insect sampling was conducted at multiple field sites across two years. Random point sampling was used to evaluate the extent of edge and neighbor field influences. The random selection of sites was facilitated by spatial randomization using GIS tools applied to a GIS rendering of the fields. Using the GPS-enabled hand-held device, insect samplers were able to track their movement in real-time to the pre-selected sampling points, which eliminated the need for hand measuring

Table 1 Comparison of projected insecticide application frequency and land area sprayed with insecticide for pests of sorghum (sugarcane aphid) and cotton (verde plant bug and cotton fleahopper) using two management scenarios: traditional whole-field approach and a within-field zone approach

				Whole field		Within field		
Pest	Field[a]	Zone[a]	Area[a]	Spray area[b]	% Fields sprayed[b]	Spray area[b]	% Zones sprayed[b]	% Spray reduction[c]
Sugarcane aphid	40	158	12 465	2129	15.0	1250	8.23	41.2
Verde plant bug	43	417	11 245	3852	39.5	1021	15.1	73.5
Cotton fleahopper	58	495	18 634	2574	25.9	1131	16.0	56.0

[a] Number of field visits (Fields) and sample site visits in the fields (Zones). Land area is presented in hectares.
[b] Area in hectares (Spray area) and Percent of fields or zones projected to be sprayed with insecticide of the total number of fields and zones, respectively.
[c] Percent reduction of area projected to be sprayed with insecticide when using the within-field zone approach.

distances, such as would occur for more traditional transect sampling to evaluate edge effects on pest insects. The randomized sampling routine also eliminated sampling bias that may occur due to systematic sampling that occurs when sampling along transects (Esquivel et al., 2020).

Based on these data, insect densities and boll injury significantly declined at a constant rate from cotton field edge to field interior (200–300 m from field edge). For cotton fleahopper, which has weedy hosts along the waterways of this agroecosystem, this edge effect was also influenced by the adjacent field type. For example, cotton fleahopper densities at the interface with semi-natural habitat were higher but declined at a greater rate into the cotton field interior than densities seen at the interfaces with sorghum or another cotton field. For both species, insect densities and boll injury more than 100 m into the field were up to 70% less than at field edges. Cotton fleahopper populations only approached or exceeded economic thresholds in areas less than 100 m from field edge and primarily in areas of a cotton field adjacent to semi-natural habitat (Fig. 4). From a pest management perspective, insect monitoring activities and remedial IPM tactics can be more strategically deployed by

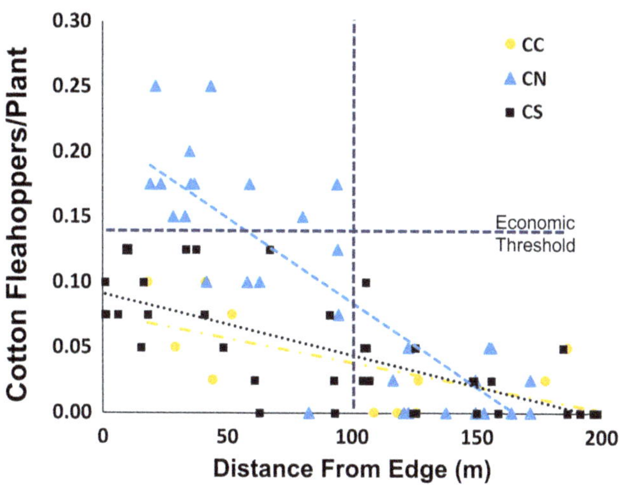

Figure 4 Distribution of cotton fleahopper densities taken from randomized sampling points from distance from field edge visualized by a scatterplot. See legend for interface-type color coding for cotton-cotton (CC), cotton-natural habitat (CN), and cotton-sorghum (CS). The economic threshold, approximately 0.14 cotton fleahopper per plant in the region, was added to the figure (horizontal line). Cotton fleahopper showed a strong edge effect and was influenced by adjacent field types such that populations only approached or exceeded the economic threshold within-field locations that were 100 m or less from the field boundary, and primarily in areas adjacent to natural habitat. In some cases, this represented pest densities that were approximately 70% less in the field interior compared to the edge of a cotton field.

considering influence of field edges and adjacent crops and vegetation such as found along waterways.

3.2.3 Spatial pest risk assessment across a cotton production region

Brewer et al. (2020) used GIS-based mapping and geo-referenced captures of boll weevil from standard baited monitoring traps along cotton field edges to investigate if landscape features were associated with spatially variable boll weevil detections. Seven years of trap data were used, principally from the lower Rio Grande Valley where an increase in boll weevil detections occurred in 2018 after a period of low incidence likely related to the boll weevil eradication effort in the region. Boll weevil trap captures were correlated with data extracted from data layers of cropland classifications, transportation roadways and corridors, and hydrological features. In 14 of 24 correlations, the number of boll weevil captures was negatively associated with the distance of the trap to selected landscape features. Boll weevil captures decreased linearly as the distance of the traps from water bodies and watermelon fields increased in 2010 and to a lesser extent in 2014. For boll weevil, these analyses may guide structured comparisons of trap deployment strategies across landscape features, such as vegetation associated with water bodies and the Rio Grande River, with the aim to best place traps that support the boll weevil eradication effort of this cotton production region (Brewer et al., 2020).

3.2.4 Pest information networks for regional pest risk assessment

More elaborate IPM practitioner-oriented information networks have been created with the aim to expand the usefulness and integration of information for local to regional pest risk assessment. A public-supported example is the Integrated Pest Information Platform for Extension and Education (iPiPE), which was initiated in 2010 and expanded in 2015 (Isard et al., 2015). The intent was to improve prevention, early detection, rapid diagnosis, containment, mitigation, and recovery from new, foreign, and emerging crop pests. An important iPiPE program goal was to catalyze and support a national network of similar internet-based networks operated by various organizations, with a focus on providing information to enhance IPM and food security. Within the iPiPE structure, several information networks for specific pests and crops were supported. The pests included plant pathogens primarily (e.g. southern corn rust, and soybean rust) and a few insect pests (e.g. pecan nut casebearer) (Isard et al., 2015; IPM IPIPE, 2021). Public (U.S. Department of Agriculture) support for these programs has ended, while specific databases and tools are still accessible with a focus on county-level pest incidence reporting and forecasting tools for selected pests (IPM IPIPE, 2021). This program showcased that an information network can manage large spatially referenced data sets across large scales for use in pest risk assessment. It is unclear if the iPiPE program catalyzed sustainable internet-based networks in the private sector or in private-public sector ventures, but certainly capabilities and structure were demonstrated.

Another research-based concept of an information network (Koralewski et al., 2020a) used an integrated ecological model to simulate regional infestation dynamics of the invasive sugarcane aphid on sorghum in the southern and central Great Plains of the U.S. (Texas to Kansas). A spatially explicit, individual-based model was used to simulate local aphid population dynamics and was integrated with an atmospheric model that computes inert air particle dispersion and deposition to simulate regional sugarcane aphid wind-aided long-distance movement. Their findings indicated that initial sugarcane aphid infestations in the most southern part of Texas affected the spatial and temporal patterns farther north into north Texas and Oklahoma. Models predicted that annual sugarcane aphid migration into the central Great Plains, where expansive plantings of sorghum were located, appeared driven by capacity of sugarcane aphid to move long distances via seasonal wind-aided long-distance flights (Koralewski et al., 2020a). This prediction was visually displayed and recently validated with four years of field monitoring of sugarcane aphids from Texas to southern Kansas (Koralewski et al., 2022). This model can be inserted as a forecasting tool in an information network. The forecasts can help guide regional strategies in timing sugarcane aphid monitoring and planting partially aphid-resistant sorghums where infestation probabilities are high (Brewer et al., 2019; Koralewski et al., 2020b).

4 Conclusion and future trends in research

It is useful from an applications standpoint to advance insect pest monitoring systems by building from the traditional use of economic thresholds to guide the implementation of remedial IPM tactics at the individual field level (Stern et al., 1959; Brewer and Goodell, 2012). About 50 years of research and development have put this concept into practice, and many of the tools developed have been adopted or adapted by the crop management community and IPM practitioners. To this end, field-derived estimates of pest population growth were useful to adjust economic thresholds calculated by the descriptive economic threshold equation (Pedigo, 1999) (Section 2.1). A descriptive economic threshold (and economic injury level as appropriate) may be further adjusted upward resulting from the use of (partially) resistant host plants (Section 2.2). These adjustments were dependent on field-derived data from pest monitoring activities conducted over time and supplemented by monitoring plant health and other pest population regulatory factors, such as natural enemies and weather. Controlled experimentation, for example host plant mechanism studies, can provide additional data and confidence in the adjustments. These advances were demonstrated in the case study of the invasive sugarcane aphid-threatening cultivated sorghum throughout North America (Section 2.3). We acknowledge that calculating descriptive economic thresholds was counter to the common IPM practitioner's question 'what is the threshold' implying preference for one static threshold, but descriptive economic thresholds have provided benefits for in-field use that merit consideration (Section 2.3).

To move the conversation toward the relevance of pest, plant, and landscape variation in IPM decision-making, spatial variation of insect pests within and between fields was shown to improve crop pest monitoring systems and IPM outcomes in several case studies (Section 3). The first case study introduced the field use of a GPS-enabled hand-held device linked to a GIS that together facilitated data entry, storage, management, analysis, and presentation in the spatial realm. The resulting pest risk maps linked to pre-selected economic thresholds were incorporated into traditional IPM decision-making for individual cotton and sorghum fields (Section 3.2.1). This same process, but with attention to more robust and random sampling, was used to characterize the spatial variation of cotton insect pests within and between fields (Section 3.2.2). Cooperating growers and IPM practitioners have taken steps to use this more detailed information to modify remedial IPM tactics within fields (i.e. insecticide use in this case but also relevant to the augmentative release of biocontrol agents). These case studies demonstrated advances in moving known concepts, processes, and technologies into operational IPM systems using commercial GIS technologies and mapping to visualize pest risk within an otherwise traditional crop insect pest monitoring procedure.

Further case studies demonstrated the use of geospatial data for pest risk assessment beyond individual fields with the aim to improve insect pest monitoring and IPM outcomes regionally. Strategies for spatial deployment of an insect trapping network across a cotton production region were considered using archived data that was originally placed in a GIS to support data reporting (Section 3.2.3). The iPiPE network provides pest mapping and forecasting tools in the public domain, although the sustainability of the effort may be in question with a lapse in public support (Section 3.2.4). In a proposed pest information network, several ecological models have been linked to simulate regional infestation dynamics of the invasive sugarcane aphid pest. The model outputs have been validated and in concept may be updated with field-derived insect monitoring activities in near real-time within an information network (Section 3.2.4).

Throughout this chapter, technologies (e.g. descriptive economic thresholds, GIS, UAS, ecological modeling, and information networks) were presented as tools to improve insect pest monitoring strategies and IPM decision-making for individual fields as well as for regional applications. Some were operational and some were in research and proof-of-concept phases. Implied was the availability of technology products that can be adopted 'off-shelf' or adapted based on research for specific pest management applications. For the future, we see technology advances increasing at a more rapid rate, presenting opportunities for further adoption and adaptions to improve insect pest monitoring and IPM systems. We acknowledge and appreciate that the existing crop IPM structure focuses on individual field pest risk assessment, but we see opportunities to expand and extend the system by considering pest risk assessment across fields and regionally that may further improve pest management.

5 Where to look for further information

Throughout this chapter, literature was cited that provided reviews and specific research of the concepts and case studies pertaining to pest monitoring basics, economic threshold adjustments, and geospatial data for pest risk assessment. Below is a selection with a brief description of the content.

5.1 *Pest monitoring basics and economic threshold adjustments*

- Pedigo (1999) provides background information on sampling methods, detailed information on economic threshold calculation, and consideration of pest insect population growth.
- Gordy et al. (2019, 2021a) provide experimental details for the sugarcane aphid case study of Section 2.3, including calculation of economic thresholds and considering pest insect population growth.

5.2 *Geospatial data for pest risk assessment*

- Karp et al. (2018) use meta-analysis of data from 31 countries, 132 studies, and 6759 sampling sites to investigate the classic agroecological paradigm that non-crop habitat enhances pest control and represents a win-win opportunity to conserve biodiversity and increase yield.
- Moses-Gonzales and Brewer (2021) introduce a series of articles archived in a special collection that covers research and technology focusing on UAS (drone) pest management services including pest risk assessment using sensing drones and the release of biocontrol agents by actuation drones.
- Dminić et al. (2010) review studies across the world that show the importance of combining geostatistical analysis of agronomic and climatic

parameters with insect distribution patterns in improving pest monitoring and management at different spatial scales.

6 Acknowledgements

We thank Stephanie Klock (Texas A&M AgriLife Research) for preparing the References section. This writing project was conducted in part under the umbrella of USDA NIFA Hatch project TEX0-2-9394.

7 References

Bianchi, F. J. J. A., Booij, C. J. H. and Tscharntke, T. (2006). Sustainable pest regulation in agricultural landscapes: a review on landscape composition, biodiversity and natural pest control. *Proc. Biol. Sci.* 273(1595): 1715–1727.

Binns, M. R. and Nyrop, J. P. (1992). Sampling insect populations for the purpose of IPM decision making. *Annu. Rev. Entomol.* 37(1): 427–453.

Birt, A. G., Calixto, A., Tchakerian, M., Dean, A., Coulson, R. N. and Harris, M. K. (2012). Harnessing information technology (IT) for use in production agriculture. *J. Integr. Pest Manag.* 3(1): D1–D8. doi: 10.1603/IPM11008.

Brewer, M. J., Anderson, D. J. and Armstrong, J. S. (2013). Plant growth stage-specific injury and economic injury level for verde plant bug, *Creontiades signatus* (Hemiptera: Miridae), on cotton: effect of bloom period of infestation. *J. Econ. Entomol.* 106(5): 2077–2083.

Brewer, M. J., Deleon, L. and Esquivel, I. L. (2020). Geographic information system (GIS)-based mapping and spatial analyses applied to risk assessment and resource allocation for boll weevil (Coleoptera: Curculionidae) detection. *Ann. Entomol. Soc. Am.* 113(2): 71–78.

Brewer, M. J. and Goodell, P. B. (2012). Approaches and incentives to implement integrated pest management that address regional and environmental issues. *Annu. Rev. Entomol.* 57: 41–59.

Brewer, M. J., Peairs, F. B. and Elliott, N. C. (2019). Invasive cereal aphids of North America: ecology and pest management. *Annu. Rev. Entomol.* 64: 73–93.

Carrière, Y., Ellsworth, P. C., Dutilleul, P., Ellers-Kirk, C., Barkley, V. and Antilla, L. (2006). A GIS-based approach for areawide pest management: the scales of *Lygus hesperus* movements to cotton from alfalfa, weeds, and cotton. *Entomol. Exp. Appl.* 118(3): 203–210.

Cruz, I., de Lourdes Correa Figueiredo, M., Braga de Silva, R., da Silva, I. F., deSouza Paula, C. and Foster, J. E. (2012). Using sex pheromone traps in the decision-making process for pesticide application against fall armyworm (*Spodoptera frugiperda* [Smith][Lepidoptera: Noctuidae]) larvae in maize. *Int. J. Pest Manag.* 58(1): 83–90.

Deleon, L., Brewer, M. J., Esquivel, I. L. and Halcomb, J. (2017). Use of a geographic information system to produce pest monitoring maps for South Texas cotton and sorghum land managers. *Crop Prot.* 101: 50–57.

Dminić, I., Kozina, A., Bažok, R. and Barčić, J. I. (2010). Geographic information systems (GIS) and entomological research: a review. *J. Food Agric. Environ.* 8: 1193–1198.

Elliott, N. C., Onstad, D. W. and Brewer, M. J. (2008). History and ecological basis for areawide pest management, pp. 15–33. In: Koul, O., Cuperus, G. W. and Elliott, N.

C. (eds), *Area-Wide Pest Management: Theory to Implementation*. CABI Publishing, Oxfordshire, UK.

Esquivel, I. L., Brewer, M. J. and Coulson, R. N. (2020). Field edge and field-to-field ecotone-type influences on two cotton herbivores: cotton fleahopper, *Pseudatomoscelis seriatus* (Hemiptera: Miridae), and verde plant bug, *Creontiades signatus*. *J. Econ. Entomol.* 113(5): 2213-2222.

Esquivel, I. L., Starek, M. J., Popescu, S., Brewer, M. J. and Coulson, R. N. (2021). Preliminary approach in detecting cotton fleahopper induced damage via unmanned aerial systems and normalized difference vegetation indices. *J. Cotton Sci.* 25(2): 79-90.

ESRI (2021). ArcMap, ArcPad, and ArcCatalog online documentation. Available at: http://www.esri.com/software/arcgis.

Faris, A. M., Elliott, N. C. and Brewer, M. J. (2022). Suppression of the sugarcane aphid, *Melanaphis sacchari* Zehntner (Hemiptera: Aphididae), by resident natural enemies on susceptible and resistant sorghum hybrids. *Environ. Entomol.* 51(2): 332-339. doi: 10.1093/ee/nvab147.

Gordy, J. W., Brewer, M. J., Bowling, R. D., Buntin, G. D., Seiter, N. J., Kerns, D. L., Reay-Jones, F. P. F. and Way, M. O. (2019). Development of economic thresholds for sugarcane aphid (Hemiptera: Aphididae) in susceptible grain sorghum hybrids. *J. Econ. Entomol.* 112(3): 1251-1259.

Gordy, J. W., Seiter, N. J., Kerns, D. L., Reay-Jones, F. P. F., Bowling, R. D., Way, M. O. and Brewer, M. J. (2021a). Field assessment of aphid doubling time and yield of sorghum susceptible and partially resistant to sugarcane aphid (Hemiptera: Aphididae). *J. Econ. Entomol.* 114(5): 2076-2087.

Gordy, J. W., Way, M. O., Bowling, R. D. and Brewer, M. J. (2021b). Tally-based thresholds as an alternative to density-based thresholds for sugarcane aphid, *Melanaphis sacchari* (Hemiptera: Aphididae) in grain sorghum. *Crop Prot.* 148: 105749. doi: 10.1016/j.cropro.2021.105749.

Haar, P. J., Bowling, R., Gardner, W. A. and Buntin, G. D. (2018). Epizootics of the entomopathogenic fungus *Lecanicillium lecani* (Hypocreales: Clavicipitaceae) in sugarcane aphid (Hemiptera: Aphididae) populations infesting grain sorghum in Georgia and Texas 1. *J. Entomol. Sci.* 53(1): 104-106.

Haile, F. J., Higley, L. G. and Specht, J. E. (1998). Soybean cultivars and insect defoliation: yield loss and economic injury levels. *Agron. J.* 90(3): 344-352.

Hall, D. G., Childers, C. C. and Eger, J. E. (2007). Binomial sampling to estimate rust mite (Acari: Eriophyidae) densities on orange fruit. *J. Econ. Entomol.* 100(1): 233-240.

Hirao, J. (1979). Forecasting brown planthopper outbreaks in Japan, pp. 101-112. In: *Brown Planthopper: Threat to Rice Production in Asia*. International Rice Research Institute, Los Banos, Philippines.

Horowitz, A. R. (1986). Population dynamics of *Bemisia tabaci* (Gennadius): with special emphasis on cotton fields. *Agric. Ecosyst. Environ.* 17(1-2): 37-47.

Integrated Pest Information Platform for Extension and Education (IPM IPIPE). (2021). ipmPIPE mission & vision. U.S. Department of Agriculture, Southern IPM Center. Available at: https://www.ipmpipe.org/

Isard, S. A., Russo, J. M., Magarey, R. D., Golod, J. and VanKirk, J. R. (2015). Integrated pest information platform for extension and education (iPiPE): progress through sharing. *J. Integr. Pest Manag.* 6(1): 15. doi: 10.1093/jipm/pmv013.

Jackson, R. E., Bradley, J. R., Van Duyn, J. O., Leonard, B. R., Allen, K. C., Luttrell, R., Ruberson, J., Adamczyk, J., Gore, J., Hardee, D. D., Voth, R., Sivasupramaniam, S., Mullins, J. W.

and Head, G. (2008). Regional assessment of *Helicoverpa zea* populations on cotton and non-cotton crop hosts. *Entomol. Exp. Appl.* 126(2): 89-106.

Karp, D. S., Chaplin-Kramer, R., Meehan, T. D., Martin, E. A., DeClerck, F., Grab, H., Gratton, C., Hunt, L., Larsen, A. E., Martínez-Salinas, A., O'Rourke, M. E., Rusch, A., Poveda, K., Jonsson, M., Rosenheim, J. A., Schellhorn, N. A., Tscharntke, T., Wratten, S. D., Zhang, W., Iverson, A. L., Adler, L. S., Albrecht, M., Alignier, A., Angelella, G. M., Zubair Anjum, M., Avelino, J., Batáry, P., Baveco, J. M., Bianchi, F. J. J. A., Birkhofer, K., Bohnenblust, E. W., Bommarco, R., Brewer, M. J., Caballero-López, B., Carrière, Y., Carvalheiro, L. G., Cayuela, L., Centrella, M., Ćetković, A., Henri, D. C., Chabert, A., Costamagna, A. C., De la Mora, A., de Kraker, J., Desneux, N., Diehl, E., Diekötter, T., Dormann, C. F., Eckberg, J. O., Entling, M. H., Fiedler, D., Franck, P., Frank van Veen, F. J., Frank, T., Gagic, V., Garratt, M. P. D., Getachew, A., Gonthier, D. J., Goodell, P. B., Graziosi, I., Groves, R. L., Gurr, G. M., Hajian-Forooshani, Z., Heimpel, G. E., Herrmann, J. D., Huseth, A. S., Inclán, D. J., Ingrao, A. J., Iv, P., Jacot, K., Johnson, G. A., Jones, L., Kaiser, M., Kaser, J. M., Keasar, T., Kim, T. N., Kishinevsky, M., Landis, D. A., Lavandero, B., Lavigne, C., Le Ralec, A., Lemessa, D., Letourneau, D. K., Liere, H., Lu, Y., Lubin, Y., Luttermoser, T., Maas, B., Mace, K., Madeira, F., Mader, V., Cortesero, A. M., Marini, L., Martinez, E., Martinson, H. M., Menozzi, P., Mitchell, M. G. E., Miyashita, T., Molina, G. A. R., Molina-Montenegro, M. A., O'Neal, M. E., Opatovsky, I., Ortiz-Martinez, S., Nash, M., Östman, Ö, Ouin, A., Pak, D., Paredes, D., Parsa, S., Parry, H., Perez-Alvarez, R., Perović, D. J., Peterson, J. A., Petit, S., Philpott, S. M., Plantegenest, M., Plećaš, M., Pluess, T., Pons, X., Potts, S. G., Pywell, R. F., Ragsdale, D. W., Rand, T. A., Raymond, L., Ricci, B., Sargent, C., Sarthou, J. P., Saulais, J., Schäckermann, J., Schmidt, N. P., Schneider, G., Schüepp, C., Sivakoff, F. S., Smith, H. G., Stack Whitney, K., Stutz, S., Szendrei, Z., Takada, M. B., Taki, H., Tamburini, G., Thomson, L. J., Tricault, Y., Tsafack, N., Tschumi, M., Valantin-Morison, M., Van Trinh, M., van der Werf, W., Vierling, K. T., Werling, B. P., Wickens, J. B., Wickens, V. J., Woodcock, B. A., Wyckhuys, K., Xiao, H., Yasuda, M., Yoshioka, A. and Zou, Y. (2018). Crop pests and predators exhibit inconsistent responses to surrounding landscape composition. *Proc. Natl. Acad. Sci. U. S. A.* 115(33): E7863-E7870. doi: 10.1073/pnas.1800042115.

Knutson, A., Bynum, E., Kerns, D., Porter, P., Biles, S. and Reed, B. (2018). *Managing Insect and Mite Pests of Texas Sorghum, ENTO-085, Texas A&M AgriLife Extension*. College Station, TX.

Knutson, A. E. and Ree, B. (2019). Biology and management of the pecan nut casebearer (Lepidoptera: Pyralidae). *J. Integr. Pest Manag.* 10(1): 1-6. doi: 10.1093/jipm/pmz028.

Koralewski, T. E., Wang, H.-H., Grant, W. E., Brewer, M. J. and Elliott, N. C. (2022). Evaluation of areawide forecasts of wind-borne crop pests: sugarcane aphid (Hemiptera: Aphididae) infestations of sorghum in the Great Plains of North America. *J. Econ. Entomol.* 115(3): 863-868.

Koralewski, T. E., Wang, H.-H., Grant, W. E., Brewer, M. J., Elliott, N. C., Westbrook, J. K., Szczepaniec, A., Knutson, A., Giles, K. L. and Michaud, J. P. (2020a). Integrating models of atmospheric dispersion and crop-pest dynamics: linking detection of local aphid infestations to forecasts of region-wide invasion of cereal crops. *Ann. Entomol. Soc. Am.* 113(2): 79-87.

Koralewski, T. E., Wang, H.-H., Grant, W. E., LaForest, J. H., Brewer, M. J., Elliott, N. C. and Westbrook, J. K. (2020b). Toward near-real-time forecasts of airborne crop pests:

aphid invasions of cereal grains in North America. *Comput. Electron. Agric.* 179: 105861, doi: 10.1016/j.compag.2020.105861.

Lindenmayer, J. C., Giles, K. L., Elliott, N. C., Knutson, A. E., Bowling, R., Brewer, M. J., Seiter, N. J., McCornack, B., Brown, S. A., Catchot, A. L. and Royer, T. A. (2020). Development of binomial sequential sampling plans for sugarcane aphid (Hemiptera: Aphididae) in commercial grain sorghum. *J. Econ. Entomol.* 113(4): 1990–1998.

Lu, Y., Wu, K., Jiang, Y., Xia, B., Li, P., Feng, H., Wyckhuys, K. A. and Guo, Y. (2010). Mirid bug outbreaks in multiple crops correlated with wide-scale adoption of Bt cotton in China. *Science* 328(5982): 1151–1154.

Luttrell, R. G., Teague, T. G. and Brewer, M. J. (2015). Cotton insect pest management, pp. 509–546. In: Fang, D. D. and Percy, R. G. (Eds). *Cotton*, 2nd ed., Monograph 57. Madison, WI.

Martinez, B., Reaser, J. K., Dehgan, A., Zamft, B., Baisch, D., McCormick, C., Giordano, A. J., Aicher, R. and Selbe, S. (2020). Technology innovation: advancing capacities for the early detection of and rapid response to invasive species. *Biol. Invas.* 22: 75–100.

McKinion, J. M., Jenkins, J. N., Willers, J. L. and Zumanis, A. (2009). Spatially variable insecticide applications for early season control of cotton insect pests. *Comput. Electron. Agric.* 67(1-2): 71–79.

Moral García, F. J. (2006). Analysis of the spatio-temporal distribution of *Helicoverpa armigera* Hb. in a tomato field using a stochastic approach. *Biosyst. Eng.* 93(3): 253–259.

Moses-Gonzales, N. and Brewer, M. J. (2021). A special collection: drones to improve insect pest management. *J. Econ. Entomol.* 114(5): 1853–1856.

Paudyal, S., Armstrong, J. S., Giles, K. L., Payton, M. E., Opit, G. P. and Limaje, A. (2019). Categories of resistance to sugarcane aphid (Hemiptera: Aphididae) among sorghum genotypes. *J. Econ. Entomol.* 112(4): 1932–1940.

Pedigo, L. P. (1999). *Entomology and Pest Management* (3rd edn.). Prentice-Hall Publishing, Upper Saddle River, NJ.

Pedigo, L. P. and Buntin, G. D. (1994). *Handbook of Sampling Methods for Arthropods in Agriculture*. CRC Press, Boca Raton, FL.

Pilkay, G. L., Reay-Jones, F. P. F., Toews, M. D., Greene, J. K. and Bridges, W. C. (2014). Spatial and temporal dynamics of stink bugs in southeastern farmscapes. *J. Insect Sci.* 15(1): 23. doi: 10.1093/jisesa/iev006.

Piper, A. M., Batovskad, J., Cogan, N. O. I., Weiss, J., Cunningham, J. P., Rodoni, B. C., Blacket, M. J. (2019). Prospects and challenges of implementing DNA metabarcoding for high-throughput insect surveillance. *GigaScience* 8(8): 1-22. doi: 10.1093/gigascience/giz092.

Prasannakumar, N. R. and Chander, S. (2014). Weather-based brown planthopper prediction model at Mandya, Karnataka. *J. Agrometeorol.* 16(1): 126–129.

Ragsdale, D. W., McCornack, B. P., Venette, R. C., Potter, B. D., MacRae, I. V., Hodgson, E. W., O'Neal, M. E., Johnson, K. D., O'Neil, R. J., DiFonzo, C. D., Hunt, T. E., Glogoza, P. A. and Cullen, E. M. (2007). Economic threshold for soybean aphid (Hemiptera: Aphididae). *J. Econ. Entomol.* 100(4): 1258–1267.

Reay-Jones, F. P. F., Greene, J. K. and Bauer, P. J. (2016). Stability of spatial distributions of stink bugs, Boll Injury, and NDVI in Cotton. *Environ. Entomol.* 45(5): 1243–1254.

Reisig, D. and Godfrey, L. (2006). Remote sensing for detection of cotton aphid (Homoptera: Aphididae) and spider mite (Acari: Tetranychidae) infested cotton in the San Joaquin Valley. *Environ. Entomol.* 35(6): 1635–1646.

Rogers, J. S., McCoy, C. W. and Manners, M. M. (1994). Standardized visual comparison keys for rapid estimations of citrus rust mite (Acari: Eriophyidae) populations. *J. Econ. Entomol.* 87(6): 1507–1512.

Royer, T. A., Pendleton, B. B., Elliott, N. C. and Giles, K. L. (2015). Greenbug (Hemiptera: Aphididae) biology, ecology, and management in wheat and sorghum. *J. Integr. Pest Manag.* 6(1): 19–19. doi: 10.1093/jipm/pmv018.

Smith, C. M. (2005). *Plant Resistance to Arthropods: Molecular and Conventional Approaches*. Springer, Dordrecht.

Stern, V. M., Smith, R. F., van den Bosch, R. and Hagen, K. S. (1959). The integration of chemical and biological control of the spotted alfalfa aphid: the integrated control concept. *Hilgardia* 29(2): 81–101.

Stout, M. J. (2013). Reevaluating the conceptual framework for applied research on host-plant resistance. *Insect Sci.* 20(3): 263–272.

Talekar, N. S. and Shelton, A. M. (1993). Biology, ecology, and management of the diamondback moth. *Annu. Rev. Entomol.* 38(1): 275–301.

Teetes, G. L. (1994). Adjusting crop management recommendations for insect-resistant crop varieties. *J. Agric. Entomol.* 11: 191–200.

Tillman, P. G., Northfield, T. D., Mizell, R. F. and Riddle, T. C. (2009). Spatiotemporal patterns and dispersal of stink bugs (Heteroptera: Pentatomidae) in peanut-cotton farmscapes. *Environ. Entomol.* 38(4): 1038–1052.

Trimble (2021). GIS data collection & integration. Timble Geospatial. Available at: https://geospatial.trimble.com/GIS.

U.S. Department of Agriculture-Farm Service (USDA FSA). (2021). *Geospatial Data Gateway. NAIP Imagery*. Available at: https://www.fsa.usda.gov/programs-and-services/aerial-photography/imagery-programs/naip-imagery/.

U.S. Department of Agriculture-National Agricultural Statistics Service (USDA FSA). (2021). *Cropscape-Cropland Data Layer*. Available at: https://nassgeodata.gmu.edu/CropScape/.

Wang, H.-H., Grant, W. E., Elliott, N. C., Brewer, M. J., Koralewski, T. E., Westbrook, J. K., Alves, T. M. and Sword, G. A. (2019). Integrated modelling of the life cycle and aeroecology of wind-borne pests in temporally variable spatially heterogeneous environment. *Ecol. Modell.* 399: 23–38.

Weather Underground (2021). *Weather Underground*. Available at: https://www.wunderground.com.

Chapter 8

Advances in pest risk assessment techniques focusing on invertebrate pests of European outdoor crops

Mark W. Ramsden, Samuel Telling, Daniel J. Leybourne, Natasha Alonso and Sacha White, RSK ADAS Ltd, UK; and Nikos Georgantzis, Burgundy School of Business, France

1 Introduction

Every year outdoor crops are infested with a diverse range of invertebrate pests (any invertebrate capable of causing economic injury to commercial crops). When widespread and abundant, a pest infestation (during which a pest arrives, feeds, and reproduces on the crop in a manner likely to lead to crop damage) can result in significant crop injury and economic losses (Barzman et al., 2015). Management of these pests is best achieved through a combination of mitigating measures that reduce the likelihood of infestation and/or reduce the subsequent injury that it causes. This integrated approach to pest management (integrated pest management, IPM) has been increasingly implemented since the turn of the twenty-first century and was made a requirement in European countries when the Sustainable Use Directive was introduced in 2009 (SUD

http://dx.doi.org/10.19103/AS.2022.0113.10

2009/128/EC). Despite the availability of various IPM tactics (i.e. individual measures put in place as part of an overall IPM strategy), as well as legislative, political, and public pressure for the implementation of IPM, pest management in outdoor crops remains largely reliant on synthetic chemical pesticides. Overuse of these has resulted in the emergence of resistant pest populations, which in the absence of alternative control strategies can result in crop failure (Hillocks, 2012; Sparks and Nauen, 2015). Excessive pesticide application has also resulted in reduced biodiversity and associated ecosystem services in agricultural landscapes (Barzman et al., 2015; Sanchez-Bayo and Wyckhuys, 2021). Synthetic pesticides are, however, readily available and promise a simple solution to pest management. The relatively low price of insecticide products means that if applied when the pest risk is low, the cost penalty is also low, whereas the cost of failing to manage a pest infestation can be substantial. However, if pesticides are always applied irrespective of actual pest risk, it is impossible to determine if the treatment was cost-effective; the outcome is always the same. As a result, the decision to apply pesticides is reinforced each year; breaking out from this cycle is difficult. Instigating IPM strategies can be disruptive to the farm, expensive, and require a greater understanding of the pests and their interactions with the crop and wider landscape. Farmers and their advisors need to learn and practice new agronomic skills, and increased monitoring is needed, resulting in higher labour costs. Ultimately, there are risks associated with changing pest control strategies, but there are also risks from not changing. This is exhibited in a rise in the development of pesticide resistance in key pests, loss of active ingredients in pesticides actives, pressure to reduce chemical residues in food and water, and recognition that pesticides can affect important non-target species (e.g. pollinators and natural enemies), which can impact productivity (Lamichhane et al., 2015).

This chapter outlines current approaches for pest risk assessments (PRAs), their interpretation, and future direction. We focus on invertebrate pests of outdoor field crops in Europe as examples, but the approaches are transferable to other pests/sectors. We first outline the current application of PRAs and the current availability of decision support systems (DSSs) that support the assessment of risk. We then discuss the factors limiting the application of PRAs and then consider how the varying perceptions of risk can affect the success of assessments. Finally, we provide an outline of steps that can be followed to make an effective PRA, and how these might be refined by future research.

Pest control strategies must be continuously reviewed and updated to reflect changing conditions. Climate change may alter the spatial and temporal distribution of pests or their life cycle, producing new threats or enhancing current ones (Grünig et al., 2020). New crop species and varieties may be vulnerable to pests or damage (Ratnadass et al., 2012), and legislation changes often dictate the IPM tactics, products, or chemicals that must, can, or cannot be used (Barzman et al., 2015). IPM strategies are widely employed by many farmers (Creissen et al., 2019), often quantified as a reduction in Treatment Frequency Index (TFI); the number of recommended doses applied to a crop area (Lechenet et al., 2017). This is an effective tool for assessing the impact of IPM strategies at a regional level; however, it is less informative for individual farmers and advisors for evaluating and improving their own IPM strategy. At a farm level, TFI can vary year on year reflecting the crop rotation, and so TFI use as a measure of implementation of IPM is limited and does not include any measure of the effective management of pests. The practice of self-improvement through benchmarking is established in other aspects of crop production, for example, the Yield Enhancement Network (Lacoste et al., 2022). However, there are few tools that enable farmers to evaluate the success (or otherwise) of their IPM strategy. Facilitated networks, such as the Dephy network in France (Lechenet et al., 2017), promote the sharing and discussion making of observations of pest abundance and the impact of various IPM tactics, enabling some level of benchmarking against which an individual can assess their own strategy.

PRAs provide a simple evidence-based evaluation of the risk posed by one or more pests to crops (Jeger et al., 2018). This is different from pest risk analysis, which is undertaken at a regional level, and which takes a broad view of the risks posed by specific pests to wider industry objectives. The PRAs use a combination of the likelihood of an infestation occurring, and the hazard this represents to the crop and wider farm management. These two elements can be scored individually and multiplied together to give an overall risk score (Fig. 1). This allows pests to be compared and ranked and allows for an assessment of the impact of preventative and mitigation measures, giving farmers and advisors more confidence in deciding not to apply pesticides. While undertaking a PRA is, in principle, straightforward, the initial process requires consideration of many factors, though subsequent review and updates can be relatively quick. Assessments should be completed before the start of the cropping season and reviewed throughout. Where possible, quantitative information should be used to score the relative likelihood and hazard, using data collected on the farm, from remote data, using DSS or analysis of historic data. DSSs are interactive systems (usually software-based) that help users identify and solve problems and make decisions as part of developing an IPM strategy.

2 Assessing the likelihood of pest infestations

The likelihood of pest infestations is largely dependent on specific pest ecology relating to biotic and abiotic conditions. The phenology of most agricultural pests is underpinned by climatic conditions, where specific development stages are related directly or indirectly to factors such as temperature or rainfall (Luquet et al., 2019). As these factors can be forecast, it is possible to predict pest activity,

		Hazard Score Upon infestation: 1. Very unlikely to lead to economic injury 2. Unlikely to lead to economic injury 3. May lead to some economic injury 4. Likely to lead to notable economic injury 5. Very likely to lead to substantial economic injury				
		1	2	3	4	5
Likelihood Score 1. Very unlikely to infest crop 2. Unlikely to infest crop 3. May infest crop 4. Likely to infest crop 5. Very likely to infest crop	1	1	2	3	4	5
	2	2	4	6	8	10
	3	3	6	9	12	15
	4	4	8	12	16	20
	5	5	10	15	20	25

Total Risk Score	Interpretation
1-4	Very low risk
5-8	Low risk
9-12	Moderate risk
13-16	High risk
17+	Very high risk

Figure 1 Calculation of pest risk assessment scores for a given pest/crop combination.

including when they are likely to be active in the crop (Gözüaçik et al., 2016). Other factors can also be used to make predictions about the likelihood of future infestations, such as the abundance of pre-damaging life stages in and around the crops. Where forecasts or other factors are not available, observation networks can be invaluable to growers, providing insight into the current infection risk. To make an accurate assessment of the risk of pest infestation, farmers and advisors use a combination of in-field monitoring, forecasting systems, observation networks, and remote sensing technologies. Each approach has benefits and disadvantages specific to the pest being assessed.

2.1 Likelihood of infestation based on field monitoring

One or more life stages of a pest may be completed within fields without any direct impact on the crop, and assessment of these stages can be used to predict future infestation and determine the potential hazard. Potato cyst nematodes (PCNs), for example, remain in the soil after potato crops have been harvested (Jones, 1970). The two species most responsible for the damage are *Globodera pallida* and *G. rostochiensis*. Second-stage juveniles hatch when the host plant begins extending roots into the soil. The nematodes enter the roots near the root tip and migrate to the inner cortex layers, and the resulting damage is estimated to cause total losses of about 9% to worldwide production (Varandas, Egas and Conceicao, 2020). PCNs are difficult to control because they can survive for many years in the soil in the absence of their host, and there are few rapid control treatments available (Bairwa et al., 2021). Statutory regulations require soil sampling for PCN in fields selected for potato seed production and are recommended for ware production to avoid substantial losses. Soil sampling can be used to assess both the species and current abundance of PCN in a field. Soil sampling takes place between September and March, prior to potato cropping, with at least 30 cores collected across no more than 4 ha in a field, to a depth of 10-15 cm. The soil cores are extracted, combined, and examined for PCN under a microscope. The PCN infestation is expressed as number of eggs per species per gram of soil. The farmer and agronomist can then consider the best action to take, based on available thresholds and considering potential varietal resistance/tolerance, field history, PCN species and abundance, and how these factors contribute to the likelihood of an economic injury and associated losses in the next crop (Contina et al., 2019).

Such assessments lend themselves to pests that can be assessed *in situ* during a non-damaging life stage, or prior to the crop being grown at that location. Similar soil analysis can also be done for wireworms, the soil-dwelling larvae of click beetles (*Agriotes* spp.), as part of best practice for risk assessment (Poggi et al., 2021). In these cases, the time needed to collect, process, analyse, and report on pest abundance can be completed before management decisions need to be made. The main advantage is that this process gives information on actual abundance at the location of interest; however, the analysis is time-consuming and costly, limiting the frequency at which it is done. Alternatively, proactive trapping of pests can identify current activity in the field. This assessment can give a better overview of potential pest abundance but normally needs to be done just before or during periods of high risk of infestation, meaning that decisions on management may need to be made at short notice.

Semiochemical-baited traps can be used to detect the presence of pests and keep track of their development to determine when to apply pesticides (Witzgall et al., 2010). Baited traps are installed on the edges or within the crop, for example, to assess the abundance of Bruchid beetles (*Bruchus rufimanus*, Boheman, 1833) in field beans to inform treatment decisions. The baited traps contain floral attractants and are placed at the same height as the bean flowers (Bruce et al., 2011). This type of trapping can be expensive and time-consuming and is often used to evaluate the actual presence and abundance of pests following alerts raised by remote systems or pest forecasts.

2.2 Likelihood of infestation based on forecast models

Methods for assessing pre-infestation or infesting life stages are not available for all pests, and baited trapping requires regular field assessments with associated direct and indirect costs. Even where a monitoring method has been defined, it is often time-consuming and cannot practically be undertaken in all fields across every farm, for every pest (Ramsden et al., 2017). Forecast systems can be used to prioritise which pests to monitor in the field and when. These DSSs consist of algorithms predicting the arrival/emergence, abundance, and/or activity of pests, based on previously collected data and experiments. This is largely climate based, but often with some additional agronomic input required.

Slugs are important pests of many outdoor crops, killing seeds and seedlings, destroying stems and growing points, and reducing plant leaf area. The grey field slug, *Deroceras reticulatum,* for example, is widespread in Europe and present in crops all year round. Population activity and growth are influenced by various factors, including climatic conditions – especially daily air temperature and rainfall (Choi et al., 2004). Based on field data, predictive models can be used to identify when periods of high activity can be expected, during which more targeted monitoring and/or treatment may be appropriate (Young, Port and Green, 1993; Petrovskii et al., 2022). Colorado potato beetle (*Leptinotarsa decemlineata*) is the most destructive potato pest in Europe. In Germany, the SIMLEP1-Start model predicts the first appearance of the beetle in spring after winter diapause, as well as subsequent emergence events later in the season (Jorg et al., 2007). Both adults and larvae feed on host plants, and often cause complete defoliation, with yield losses of up to 50%. Emergence is calculated based on the accumulated soil temperature at 20 cm depth (above a base of 8°C) from 1 March each year. When 95% of emergent beetles are predicted, the user should check the crops for egg clutches. This calculated date can also be used as the 'first egg found' input in a second model; SIMLEP3 (Racca et al., 2014). This model calculates the population dynamics of the beetle from air temperature and the first appearance of eggs to the appearance of adult larvae. It is a tool for determining optimal dates for decision assessments (maximum number of egg clutches) and control measures (maximum number of young larvae). These two models are currently made available to growers through the Information System for Integrated Plant Production, run by participating federal states in Germany and coordinated by Zentralstelle der Länder für EDV-gestützte Entscheidungshilfen und Programme im Pflanzenschutz (ZEPP)[1].

Creating reliable predictive models for pests requires several years of monitoring and research, to identify and model the key parameters that determine pest population dynamics. The resulting algorithm then needs to be integrated into a suitable user interface, in which outputs can be easily interpreted. The

[1] https://www.isip.de/isip/servlet/isip-de.

user should also be provided with additional supporting information, such as assumptions and conditions of use (Damos, 2015). Reliable forecast systems enable farmers to update their PRA and prioritise management activities. If low levels of pest activity are predicted, this can give farmers the confidence not to apply treatments and tailor monitoring efforts accordingly. Where high levels of activity are predicted, this helps the farmer target monitoring and make informed decisions about pesticide applications. DSS should only ever be used to support pest management decisions and will never replace location and crop-specific experience based on the user's own crop monitoring and technical knowledge. DSS should help growers make effective decisions amid the increasing complexity of crop protection. In an industry with growing regulation and scrutiny of the use of pesticides and other inputs both by the government and processors, DSS can play a vital role in effective decision-making.

DSS can make substantial contributions to improving pest management by enabling better targeting of treatments for infestations; however, uptake is relatively poor (Rossi et al., 2012). This is largely due to the regional fragmentation of DSS development, both nationally and internationally, a lack of consistent methodology for quantifying and communicating the benefits of DSS, and inadequate validation of DSS in general but especially in regions outside those in which they were developed.

2.3 Likelihood of infestation based on observation networks

Not all pests are present in or around crops during their entire life cycle and may represent an occasional, but significant risk to production. The silver-Y moth, *Autographa gamma*, and diamond-back moth, *Plutella xylostella*, are important migratory pests of several vegetable and brassica crops, respectively, in the UK. Large migration events across Europe can occur for both species, however the speed of subsequent infestation can make in field monitoring in response to in field observation too slow to inform management. In these cases, pest observation networks are highly effective in providing early warning of potential infestations (Wainwright et al., 2020). The diamond-back moth monitoring network has shown that the use of citizen science data is an effective approach for alerting farmers to the risk of infestation. While a network of pheromone traps provides crop-specific information, the traps require frequent, regular attention to match the data provided by public reporting of sightings on social media (Wainwright et al., 2020). In the UK, observations of both diamond-back moth and silver-Y moth are collated on the Warwick Crop Centre website, along with observations of several other important pests[2]. Similar observation networks have been established elsewhere in Europe, such as for carrot fly, *Chamaepsila rosae*, in Finland (Collier et al., 2020), and the

[2] https://warwick.ac.uk/fac/sci/lifesci/wcc/research/pests/activity2022/.

Rothamsted Insect Survey in the UK (Bell et al., 2015). These networks alert growers to an impending infestation, enabling them to monitor their crops more proactively and consider potential mitigation/treatment options should a large pest infestation occur. When collated across regions, farm monitoring data can be used to improve understanding of a pest's threat to production and help the development of forecasting tools (e.g. development of the wheat bulb fly forecast, Leybourne et al., 2022).

3 Assessing the hazard of pest infestations

Once a farmer has established that an infestation is likely to occur, they then need to assess the hazard this represents to the crop, i.e. will there be sufficient damage to justify treatment? Here again, there are DSS available to help farmers. In particular, accurate predictions of the timing of infestations, predictions of population growth rates, and economic thresholds (ETs) and associated economic injury levels.

The arrival of a pest does not necessarily mean the crop is at risk of economic injury. The emergence of adult orange wheat blossom midge (OWBM), *Sitodiplosis mosellana*, for example, is related to temperature and rainfall events. These meteorological parameters trigger larvae dormant in the soil to migrate to the soil surface, pupate, and subsequently emerge as adults that move into wheat crops (Jacquemin et al., 2014). Adult OWBM lay their eggs on the spikelets, and larvae feed within the grains causing direct damage and facilitating fungal infestation (Jacquemin et al., 2014). The crop is only vulnerable to infestation during mid-ear emergence (growth stage 53-59). If OWBM emergence does not coincide with these growth stages, the infestation, however large, represents no hazard to the crop. The emergence of OWBM can be modelled (Jacquemin et al., 2014) and compared to crop growth stage to assess the hazard, or farmers can use pheromone and yellow sticky trapping to monitor actual presence/abundance in vulnerable crops (Bruce et al., 2007).

ETs are a valuable method of assessing whether control measures, such as insecticides, are necessary to reduce the risk of economic losses. They are usually defined in terms of the number of a pest per unit area, per plant, or per part of plant, above which action should be taken. While farmers recognise the value of ETs (Ramsden et al., 2022), there is little evidence that farmers regularly use them to support insecticide application decisions (Ramsden et al., 2017). In part, this is due to a lack of trust in over-simplified thresholds that do not effectively account for the ability of crops to resist or tolerate pest damage. More recent ETs, such as for pollen beetle, *Meligethes* spp., in oilseed rape, are better designed to reflect variable tolerance but can require increased monitoring effort and interpretation from the farmer or advisor. For pollen beetle, the crop is only vulnerable to damage during the green bud stage; however, the plants can tolerate some loss of buds due to pollen beetle damage without any impact on yield. The number of buds that can be lost is related to the number of branches, and so to crop density (Ellis and Berry, 2012). Other ETs also capture changing tolerance in their design, such as the ETs for cereal aphids which change with the growth stage (Oakley and Walters, 1994). In this case, it is the frequency of infestation that dictates the hazard to crop performance, citing the proportion of tillers infested as the critical factor for assessment.

Infestations of invertebrate pests are usually initially patchy, becoming increasingly homogenous across the crop as infestations spread (Fievet et al., 2007). In-field assessment of infestations is therefore always necessary to make effective decisions on the next management steps. Where pests are difficult to monitor, or correct identification is challenging, advances in remote sensing and image-recognition technology can help farmers and growers assess their risk with greater confidence.

3.1 Remote sensing and image-recognition technology to assess pest infestation

Recent advances in remote sensing technology mean that it is possible to receive and interpret large-scale, multi-level information on the Earth's surface, including environmental data and multispectral imaging data. Remote sensing technology can use this data, often collected through aviation (e.g. planes and drones) or via satellite feeds, at a variety of spatial and temporal scales to monitor pest migration, damage, and outbreak patterns (Abd El-Ghany et al., 2020; Sishodia et al., 2020) which can be used by farmers to assess their hazard of potential pest infestation in near real-time. Two examples that showcase the use of remote sensing technology to monitor pest hazards include the use of multispectral normalised difference vegetation index (NDVI) data to detect Hessian fly infestation in the USA (Bhattarai et al., 2019) and the use of remote sensing techniques to identify aphid presence and damage on wheat (Luo et al., 2013).

One important aspect of successful IPM is the correct identification of pest species, which supports farmers in assessing their overall hazard potential and, if required, helps with the identification of the selection and implementation of appropriate management and treatment options. Accurate identification of insect pests is challenging for most farmers, and low confidence in their ability to correctly identify pests usually results in the application of insurance sprays. One avenue for improving this decision-making process is the use of Artificial Intelligence (AI) algorithms that use image-analysis processes to support on-farm insect identification and prescriptive treatment. At the fundamental level, AI-based image-recognition processes work through a workflow in a similar way a human taxonomist would work through a taxonomic key. By developing, training, and validating AI algorithms, researchers can develop mobile applications that can support end-users with pest identification, and therefore hazard determination (Liu et al., 2019). Future developments will increase end-user uptake in these technologies as user confidence increases, and eventually, these will provide farmers and growers access to a suite of tools that will help them when determining their pest hazard potential.

4 Reducing risk

Once a farmer has identified a high risk to their crop, they can consider appropriate actions to manage the infestation.

4.1 Preventing infestations

Preventing a pest from infesting crops is challenging and relies largely on good cultural management of the overall farming landscape. Using wide crop rotations with a diversity of crops and varieties and practising good hygiene between fields and crops will minimise the likelihood of repeated infestations; however, it cannot entirely prevent pests from arriving. The use of resistant varieties can drastically reduce the risk of pest damage. Crop resistance to OWBM, for example, was developed in the 1990s to minimise infestation by utilising a gene that produces a physical barrier to larvae entering the grain (Kassa et al., 2016).

4.2 Managing infestations

Various strategies are available to reduce the risk of infestation reaching damaging levels, and most successful approaches require the integration of these to achieve effective control (i.e. IPM). One approach that can be used is to promote the natural enemies of crop pests across the landscape, by providing additional resources that enhance populations of predators and parasitoids, in so-called Conservation Biological Control (CBC). Both ground-dwelling and flying natural enemies can reduce pest population growth rates during periods when the crop is vulnerable or delay the timing of peak abundance to a point where the crop is no longer vulnerable (Holland et al., 2012). By promoting earlier and more abundant populations of natural enemies, farmers can increase their impact on pest populations (Ramsden et al., 2016). Unlike most outdoor crop pests, which complete their lifecycle within the host crop or other widely available habitats (e.g. grass field margins), most natural enemies use a diverse range of habitats over the course of their life cycle, and their populations may be bottlenecked by a lack of one or more key resources (Ramsden et al., 2015). To increase the size of the natural enemy population at an early stage of pest infestation, targeted resource provision is required within agro-ecosystems throughout the year. Providing targeted additional resources across agro-ecosystems will support a robust and diverse natural enemy population, which in turn will contribute to the suppression of pest populations (Wäckers and van Rijn, 2012, Fountain, 2022).

5 Risk versus the perception of risk

Pre-existing bias can affect pest management decisions, resulting in inferior outcomes for both the decision makers and wider society (Akaka et al., 2023), with consequences for farmers, consumers, and the environment. To achieve productive agriculture with low pesticide inputs requires farmers and agronomists to identify an appropriate combination of alternative pest control methods (Jorgensen et al., 2017). Decision-making is complicated by the multiplicity of IPM measures, the development of pesticide resistance, and uncertainty caused by the effects of weather on pest epidemics. The characteristics of IPM, and the economic and social circumstances under which decisions are made, create many possibilities for behavioural biases. For example, as pest outbreaks are highly dependent on specific weather conditions, there is often a high probability of small losses in yield and a low probability of large losses. Probability weighting and loss aversion may bias farmers' risk assessment, driving the overuse of pesticides.

A variety of behavioural biases may explain why growers do not adopt the quantitative outputs from risk assessments and associated DSS, even where they are demonstrated to be beneficial to the farmer and/or wider society/environment (Abedullah et al., 2015; Villano et al., 2015). Risk aversion, for example, makes us hesitate to adopt, or over-commit to, new methods or innovative decision-support tools (Barham et al., 2014; Barham et al., 2015). A distorted perception of the probabilities of pest infestations occurring often leads to the risk being overestimated rather than underestimated. We may choose to avoid losses more than chasing gains of the same magnitude (relating to the concept of loss aversion as in Kahneman et al. (1979), especially in the presence of a long-standing status quo.

Determining a reference point (current earnings, aspiration level, subsistence level, etc.), makes it possible to frame the consequences of decisions as gains or losses, depending on the reference point adopted. Such individual behavioural biases need to be considered when DSSs are developed, as well as social and economic behavioural biases. The social context and effect of engagement of farmers within relevant learning networks have an impact on the decision to adopt innovative practices. Access to social networks, gender, age, and farm size are among the key socioeconomic influences on decision-making (Feder and Umali, 1993).

Decisions are rarely made alone, and farmers work closely with their advisors and peers to assess pest risks and select appropriate actions. The behavioural, social, and economic influences can have varying impacts on all those involved, alongside individual motivations. The decision to not apply pesticides, for example, has multiple considerations and outcomes. The farmer and advisor will prioritise short-term productivity, but the farmer will balance this against the long-term management of the farm and its environmental health. However, they may also have separate motivations, influenced by company policies, professional reputation (i.e. consistent avoidance of pest damage), and wider outlook across multiple farms in a region (i.e. effective early control on one farm may help avoid the need to treat on a neighbouring farm). These occasionally conflicting motivations and perceptions can lead to a lack of confidence in DSS, particularly where the systems lack transparency about the underlying models and assumptions behind the outputs. Conversely, transparent systems facilitate an improved understanding of the pest and associated risks, driving informed discussion around specific decisions and improving overall IPM strategies.

6 Summary steps to pest risk assessment

Given the complexity of managing multiple pests across a dynamic agricultural landscape, it is understandable that farmers default to responsive pest management. These results in IPM tactics being used within existing management strategies rather than seeking to proactively reduce the risks associated with pest infestations earlier in the season. Proactive efforts, including undertaking PRA throughout the season, will improve the overall management of crops. This begins at the very start of crop planning when, based on field history (i.e. past crop rotation and historic pest infestations), a list of priority pests to monitor should be drawn up. For each pest on this list, the farmer can then plan their monitoring approach and select any DSS they intend to use to make more detailed pest assessments, relating this to the crop variety and expected performance in the upcoming season. The DSS can be used to inform monitoring efforts, and so maintain a continual assessment of the likelihood of these priority pests infesting the crop. Where the risk of infestation is high, and the crop is vulnerable, the farmer can then assess the hazard this infestation represents and act accordingly.

Five steps to PRA are as follows:

1 identify priority pests for your crop;

2 assess the likelihood of these pests infesting your crop(s);
3 assess the hazard an infestation represents to crop(s) performance;
4 take action to prevent and/or manage the impact of priority pests; and
5 review and plan for following season/crop.

7 Worked example: risk assessment of barley yellow dwarf virus in winter wheat

In this section we provide a summary worked example for undertaking a PRA for barley yellow dwarf virus (BYDV) in winter wheat, using the steps outlined earlier.

7.1 Pest priority: the need for integrated pest management of barley yellow dwarf virus in winter wheat

BYDV is one of the most important diseases of winter wheat in the UK. It is transmitted by aphids in the autumn/winter and can lead to losses of up to 2.1 t/ha in winter wheat, 80% of winter barley yield, and 2.7 t/ha in winter oats (McKirdy et al., 2002; Dedryver et al., 2010). There are several strains of BYDV, of which BYDV-PAV is most common in western Europe (Henry et al., 1993; Plumb, 1995). BYDV is transmitted primarily by two species of aphid, the bird cherry-oat aphid, *Rhopalosiphum padi*, and grain aphid, *Sitobion avenae* (Miller & Rasochova, 1997). Rose-grain aphid, *Metopolophium dirhodum*, although a vector for BYDV, is considered low risk to winter cereals in the UK as it is absent from crops in the autumn although it can have a yield impact on spring sown cereals (Mann et al., 1997). The impact on yield and relatively low cost of pyrethroid insecticides has led to extensive treatment for this pest.

Bird cherry-oat aphid and grain aphid migrate from grasses into cereals in the autumn. A mild autumn and winter increase BYDV risk by lengthening the aphid migration period and allowing rapid reproduction and spread within the crop (Lowles et al., 1996). Reducing the secondary infections of BYDV within the crop is the most important factor in minimising yield loss. Traditionally, this has been based on detecting aphids within the crop and applying foliar insecticides such as synthetic pyrethroids. However, as the number of aphids migrating is relatively low, infested plants are difficult to find. This has often led to the prophylactic application of sprays even when aphids have not been observed as insurance to prevent yield loss. In order to improve monitoring, using observation networks such as the Rothamsted suction traps to identify when the migration of the winged aphids is occurring can inform monitoring timing (Harrington et al., 1999). Then using in-field traps (e.g. water or sticky traps) can improve the detection of pest presence on a field basis. This can be used to determine whether the primary infection (the initial arrival of winged aphids) has occurred.

7.2 Assessing the likelihood of infestation

The secondary spread of aphids begins when the second-generation offspring is produced, so understanding the phenological and environmental conditions that influence the speed of development is crucial for effective BYDV management. The appearance of the second-generation offspring can be forecast based on the accumulation of approximately 170 degree days above 3°C (the baseline temperature below which there is no aphid reproduction, Brabec et al. (2014)). The Agriculture and Horticulture Development Board in the UK developed a tool to manage BYDV more effectively which incorporates the day degrees model to help time sprays in the autumn.[3] This DSS can be used to identify periods during which aphids are likely to start moving within the crop and spreading BYDV. In addition to forecasts and in-field monitoring, national observation networks such as the UK aphid monitoring network can be used to assess the likelihood of infestations.[4]

7.3 Assess the hazard an infestation represents to crop performance

Having established that second-generation aphid populations are likely to be present, the farmer needs to assess the actual hazard of transition of BYDV in the crop. This consists of two factors, (1) is the abundance of aphid colonies within the field currently large enough to justify management, and (2) are populations carrying BYDV?

7.3.1 Assessing abundance

Various methods can be used to assess the spread and abundance of aphids in cereal crops, including water traps, yellow sticky straps, and visual assessments - each with various benefits and disadvantages (Holland et al., 2021). While abundance is a useful metric to assess, for this pest, its presence/absence across the field is a more informative factor for consideration since this will indicate the current and future potential spread of BYDV.

7.3.2 Assessing barley yellow dwarf virus load

Not all cereal aphids carry BYDV, and the proportion of aphids carrying the virus is usually under 10% and varies both spatially and over time. This means that not all infestations represent any hazard to the crop. There is a molecular test that can be used to assess whether aphids carry the virus, but to date,

[3] https://ahdb.org.uk/bydv.
[4] https://ahdb.org.uk/knowledge-library/the-uk-aphid-monitoring-network.

the logistics and costs generally preclude individual farmers from using this to inform their decisions. A risk-averse approach is usually taken, whereby it is assumed that aphid infestations carry, and are spreading, BYDV. Improving regional and/or on-farm testing for the presence of BYDV in aphids and public reporting may give farmers and advisors more confidence in leaving aphid populations untreated in low-risk crops.

7.4 Preventing and/or mitigating for barley yellow dwarf virus

Using the scoring system outlined in the introduction, a farmer may initially assess BYDV as very likely to infest their crops (5) and likely to lead to notable economic injury (4), giving a total score of 20: very high risk of infestation and subsequent substantial economic injury. If mitigation actions are enacted, this could enable the farmer to reduce their estimation of both the likelihood and hazard. Having an established wide rotation across the farm, minimising the 'green bridge' (in-field volunteers and weeds which aphids can use to infect the new crop), delaying sowing, and providing resources for natural enemies of aphids could justify a reduced likelihood score to 3 (may infest crops). Using a forecast tool to predict secondary infection in the crop, monitoring the crop to assess the actual presence and spread of aphids, and monitoring wider aphid population dynamics and BYDV load through regional or national observation networks could also justify reducing the hazard score to 3 (may lead to some economic injury) and giving a revised score of 9 (moderate risk). Ultimately, the farmer may decide that based on actual crop conditions, treatment is required; however, this will now be based on a thorough assessment of the risk and consideration of mitigating factors. This process of informed decision-making is essential in moving towards low-input, sustainable farming.

8 Conclusion

PRAs should be undertaken by farmers and their advisors throughout the growing season. Initially, this should focus on identifying the priority pests to monitor and be built into a farm-specific IPM strategy. Decisions to apply pesticides or other remedial tactics should only be made where there is a high risk of infestation and then a high risk of substantial crop damage and associated commercial loss. Farmers and their advisors should review PRAs each year, and ideally through the year, as changes in pest resistance, crop varietal tolerance/resistance, and product availability change annually. Over time, PRAs can form the basis of more ambitions Holistic IPM approaches which manage multiple pests across multiple crops. This can facilitate collaboration with neighbouring farms to coordinate landscape management approaches to pest suppression.

It is difficult to assess the success of individual IPM tactics, such as the implementation of CBC to reduce the risk of pest infestations. This is due to the difficulty in quantifying the direct and indirect impact that CBC interventions have on pest suppression. Conversely, the costs associated with installing and managing non-crop habitats are easily quantified. A holistic, integrated approach to pest management, which recognises the wider use of multiple tactics to target a range of pests across the farm, can spread the costs and benefits across the whole production system, and ultimately lead to a more resilient and sustainable farm (Lechenet et al., 2017).

9 Future trends in research

- As the pressure on farmers and their advisors increases to reduce agricultural inputs while maintaining productivity, greater use of pest management reporting, such as through PRAs, is likely.
- Increased access to IPM DSS, and remote sensing and identification of pests, will enable greater uptake and implementation of IPM, driving improved assessment of the likelihood of pest infestations.
- Improved availability of tactics for mitigating the impact of infestations, and wider demonstration of such tactics in action will help build confidence in IPM strategies.

10 Where to look for further information

10.1 Further reading, journals and conferences

- IPM Decisions Platform, an IPM DSS platform operating a European scale https://platform.ipmdecisions.net/ (accessed on 7 November 2022).
- VIPS, Pest forecasting system in Norway. www.vips-landbruk.no (accessed on 7 November 2022).
- Information System for Integrated Plant Production in Germany (ISIP). https://isip.de/isip/servlet/isip-de (accessed on 7 November 2022).
- Horta srl: a permanent platform in Italy for enhancing results from research in the agroalimentary sector through https://www.horta-srl.it/ (accessed on 7 November 2022).
- IOBC Working groups - https://www.iobc-wprs.org/expert_groups/wg.html (accessed on 7 November 2022).
- Association of Applied Biologists Conference series, https://www.aab.org.uk/conferences (accessed on 7 November 2022).
- Journal of Crop Protection, https://www.journals.elsevier.com/crop-protection (accessed on 7 November 2022).

10.2 Major international research projects

- AFARCLOUD - Aggregate Farming in the Cloud is an H2020 EU-funded project No. 783221; http://www.afarcloud.eu/.
- IPM Decisions is an H2020 EU-funded project No. 817617; https://www.ipmdecisions.net/.
- IPMWorks is an H2020 EU-funded project No. 101000339; https://ipmworks.net/.

11 Acknowledgements

Production of this chapter was supported in part by the IPM Decisions project, which has received funding from the European Research Council (ERC) under the European Union's Horizon 2020 research and innovation programme (grant agreement 817617), and by the Alexander von Humboldt Foundation through a Research Fellowship to DJL.

12 References

Abd El-Ghany, N. M., Abd El-Aziz, S. E. and Marei, S. S. (2020). A review: Application of remote sensing as a promising strategy for insect pests and diseases management. *Environmental Science and Pollution Research International* 27(27), 33503–33515.

Abedullah, K. S., Kouser, S. and Qaim, M. (2015). Bt cotton, pesticide use and environmental efficiency in Pakistan. *Journal of Agricultural Economics* 66(1), 66–86. https://doi.org/10.1111/1477-9552.12072.

Bairwa, A., Venkatasalam, E. P., Priyank, H. M. and Sharma, S. (2021). Introduction of potato cyst nematodes, life cycle and their management through biobased amendments. In: Kaushal, M. and Prasad, R. (Eds.), *Microbial Biotechnology in Crop Protection*. Springer, Singapore. https://doi.org/10.1007/978-981-16-0049-4_3.

Barham, B. L., Chavas, J. P., Fitz, D., Rios-Salas, V. R. and Schechter, L. (2014). The roles of risk and ambiguity in technology adoption. *Journal of Economic Behavior and Organization* 97, 204–218. https://doi.org/10.1016/j.jebo.2013.06.014.

Barham, B. L., Chavas, J. P., Fitz, D., Rios-Salas, V. and Schechter, L. (2015). Risk, learning, and technology adoption. *Agricultural Economics* 46(1), 11–24. https://doi.org/10.1111/agec.12123.

Barzman, M., Bàrberi, P., Birch, A. N. E., Boonekamp, P., Dachbrodt-Saaydeh, S., Graf, B., Hommel, B., Jensen, J. E., Kiss, J., Kudsk, P., Lamichhane, J. R., Messéan, A., Moonen, A., Ratnadass, A., Ricci, P., Sarah, J. and Sattin, M. (2015). Eight principles of integrated pest management. *Agronomy for Sustainable Development* 35(4), 1199–1215. https://doi.org/10.1007/s13593-015-0327-9.

Bell, J. R., Alderson, L., Izera, D., Kruger, T., Parker, S., Pickup, J., Shortall, C. R., Tayloer, M. S., Verrier, P. and Harrintorn, R. (2015). Long-term phenological trends, species accumulation rates, aphid traits and climate: Five decades of change in migrating aphids. *Journal of Animal Ecology* 84(1), 21–34. https://doi.org/10.1111/1365-2656.12282.

Bhattarai, G. P., Schmid, R. B. and McCornack, B. P. (2019). Remote sensing data to detect Hessian fly infestation in commercial wheat fields. *Scientific Reports* 9(1), 6109.
Brabec, M., Honek, A., Pekar, S. and Martinkova, Z. (2014). Population dynamics of aphids on cereals: Digging in the time-series data to reveal population regulation caused by temperature. *PLoS ONE* 9(9), e106228. https://doi.org/10.1371/journal.pone.0106228.
Bruce, T. J., Hooper, A. M., Ireland, L., Jones, O. T., Martin, J. L., Smart, L. E., Oakley, J. and Wadhams, L. J. (2007). Development of a pheromone trap monitoring system for orange wheat blossom midge, *Sitodiplosis mosellana*, in the UK. *Pest Management Science* 63(1), 49–56 https://doi.org/10.1002/ps.1307.
Bruce, T. J., Martin, J. L., Smart, L. E. and Pickett, J. A. (2011). Development of semiochemical attractants for monitoring bean seed beetle, *Bruchus rufimanus*. *Pest Management Science* 67(10), 1303–1308. https://doi.org/10.1002/ps.2186.
Choi, Y. H., Bohan, D. A., Powers, S. J., Wiltshire, C. W., Glen, D. M. and Semenov, M. A. (2004). Modelling *Deroceras reticulatum* (Gastropoda) population dynamics based on daily temperature and rainfall. *Agriculture, Ecosystems and Environment* 103(3), 519–525. https://doi.org/10.1016/j.agee.2003.11.012.
Collier, R., Mazzi, D., Schjoll, A.-F., Schorpp, Q., Thoming, G., Johansen, T. J., Meadow, R., Meyling, N. V., Cortesero, A. M., Vogler, U., Gaffney, M. T. and Hommes, M. (2020). The potential for decision support tools to improve the management of root-feeding fly pests of vegetables in Western Europe. *Insects* 11(6), 369. https://doi.org/10.3390/insects11060369.
Contina, J. B., Dandurand, L. M. and Kndsen, G. R. (2019). A predictive risk model analysis of the potato cyst nematode *Globodera pallida* in Idaho. *Plant Disease* 103(12), 3117-3128. https://doi.org/10.1094/PDIS-04-19-0717-RE.
Creissen, H. E., Jones, P. J., Tranter, R. B., Girling, R. D., Jess, S., Burnett, F. J., Gaffney, M., Thorne, F. S. and Kildea,S. (2019). Measuring the unmeasurable? A method to quantify adoption of integrated pest management practices in temperate arable farming systems. *Pest Management Science* 75(12), 3144–3152. https://doi.org/10.1002/ps.5428.
Damos, P. (2015). Modular structure of web-based decision support systems for integrated pest management: A review. *Agronomy for Sustainable Development* 35(4), 1347–1372.
Dedryver, C.-A., Le Ralec, A. and Fabre, F. (2010). The conflicting relationships between aphids and men: A review of aphid damage and control strategiesLes relations conflictuelles entre les pucerons et les hommes : une revue sur leurs dégâts et les stratégies de lutte. *Comptes Rendus Biologies* 333(6–7), 539–553. ISSN 1631–0691. https://doi.org/10.1016/j.crvi.2010.03.009.
Dewar, A. M., Ferguson, A., Pell, J. K., Nicholls, C. and Watts, J. (2016). A review of pest management needs in cereals and oilseed rape in the UK. *AHDB Research Review* 86.
EFSA Panel on Plant Health (PLH), Jeger, M., Bragard, C., Caffier, D., Candresse, T., Chatzivassiliou, E., Dehnen-Schmutz, K., Grégoire, J. C., Jaques Miret, J. A., MacLeod, A., Navajas Navarro, M., Niere, B., Parnell, S., Potting, R., Rafoss, T., Rossi, V., Urek, G., Van Bruggen, A., Van Der Werf, W., West, J., Winter, S., Hart, A., Schans, J., Schrader, G., Suffert, M., Kertész, V., Kozelska, S., Mannino, M. R., Mosbach-Schulz, O., Pautasso, M., Stancanelli, G., Tramontini, S., Vos, S. and Gilioli, G. (2018). Guidance on quantitative pest risk assessment. *EFSA Journal. European Food Safety Authority* 16(8), e05350. https://doi.org/10.2903/j.efsa.2018.5350.

Ellis, S. A. and Berry, P. M. (2012). *Re-Evaluating Thresholds for Pollen Beetle in Oilseed Rape*. Project Report 495. Home-Grown Cereals Authority, London.

Feder, G. and Umali, D. L. (1993). The adoption of agricultural innovations: A review. *Technological Forecasting and Social Change* 43(3-4), 215-239. https://doi.org/10.1016/0040-1625(93)90053-A.

Ferguson, A. W., Skellern, M. P., Johnen, A., von Richthofen, J. S., Watts, N. P., Bardsley, E., Murray, D. A. and Cook, S. M. (2016). The potential of decision support systems to improve risk assessment for pollen beetle management in winter oilseed rape. *Pest Management Science* 72(3), 609-617. https://doi.org/10.1002/ps.4069.

Fievet, V., Dedryver, C., Plantegenest, M., Simon, J. and Outreman, Y. (2007). Aphid colony turn-over influences the spatial distribution of the grain aphid *Sitobion avenae* over the wheat growing season. *Agricultural and Forest Entomology* 9(2), 125-134. https://doi.org/10.1111/j.1461-9563.2007.00331.x.

Foster, S. P., Paul, V. L., Slater, R., Warren, A., Denholm, I., Field, L. M. and Williamson, M. S. (2014). A mutation (L1014F) in the voltage-gated sodium channel of the grain aphid, *Sitobion avenae*, is associated with resistance to pyrethroid insecticides. *Pest Management Science* 70(8), 1249-1253. https://doi.org/10.1002/ps.3683.

Fountain, M. T. (2022). Impacts of wildflower interventions on beneficial insects in fruit crops: A review. *Insects* 13(3), 304. https://doi.org/10.3390/insects13030304.

Gözüaçik, C., Yiğit, A. and Şimşek, Z. (2016). Predicting the development of critical biological stages of Sunn pest, *Eurygaster integriceps* Put. (Hemiptera: Scutelleridae), by using sum of degree-days for timing its chemical control in wheat. *Turkish Journal of Agriculture and Forestry* 40, 577-582. http://doi.org/10.3906/tar-1511-64.

Grünig, M., Mazzi, D., Calanca, P., Karger, D. N. and Pellissier, L. (2020). Crop and forest pest metawebs shift towards increased linkage and suitability overlap under climate change. *Communications Biology* 3(1), 233. https://doi.org/10.1038/s42003-020-0962-9.

Harrington, R., Mann, J. A., Burgess, A. J., Tones, S. J., Rogers, R., Foster, G. N., Blake, S., Morrison, S. F., Ward, L., Barker, I., Morgan, D. and Walters, K. F. A. (1999). *Development and Validation of Decision Support Methodology for Control of Barley Yellow Dwarf Virus*. HGCA Project Report No. 205. Agriculture and Horticulture Development Board.

Henry, M., George, S., Arnold, G. M., Dedryver, C. A., Kendall, D. A., Robert, Y. and Smith, B. D. (1993). Occurrence of barley yellow dwarf virus (BYDV) isolates in different farmland habitats in western France and south-west England. *Annals of Applied Biology* 123(2), 315-329. https://doi.org/10.1111/j.1744-7348.1993.tb04094.x.

Hillocks, R. J. (2012). Farming with fewer pesticides: EU pesticide review and resulting challenges for UK agriculture. *Crop Protection* 31(1), 85-93. https://doi.org/10.1016/j.cropro.2011.08.008.

Holland, J. M., McHugh, N. M. and Salinari, F. (2021). Field specific monitoring of cereal yellow dwarf virus aphid vectors and factors influencing their immigration within fields. *Pest Management Science* 77(9), 4100-4108. https://doi.org/10.1002/ps.6435.

Holland, J. M., Oaten, H., Moreby, S., Birkett, T., Simper, J., Southway, S. and Smith, B. M. (2012). Agrienvironment scheme enhancing ecosystem services: A demonstration of improved biological control in cereal crops. *Agriculture, Ecosystems and Environment* 155, 147-152. https://doi.org/10.1016/j.agee.2012.04.014.

Jacquemin, G., Chavalle, S. and De Proft, M. (2014). Forecasting the emergence of the adult orange wheat blossom midge, *Sitodiplosis mosellana* (Géhin) (Diptera:

Cecidomyiidae) in Belgium. *Crop Protection* 58, 6–13. https://doi.org/10.1016/j.cropro.2013.12.021.

Jakubowska, M., Bocianowski, J., Nowosad, K. and Kowalska, J. (2020). Decision support system to improve the effectiveness of chemical control against cutworms in sugar beet. *Sugar Tech* 22(5), 911-922. https://doi.org/10.1007/s12355-020-00808-z.

Jones, F. G. W. (1970). The control of the potato cyst-nematode. *Journal of the Royal Society of Arts* 118, 179–199.

Jorg, E., Racca, P., Preiß, U., Butturini, A., Schmiedl, J. and Wójtowicz, A. (2007). Control of Colorado potato beetle with the SIMLEP decision support system. *EPPO Bulletin* 37(2), 353–358. https://doi.org/10.1111/j.1365-2338.2007.01137.x.

Jorgensen, L. N., van den Bosch, F., Oliver, R. P., Heick, T. M. and Paveley, N. D. (2017). Targeting fungicide inputs according to need. *Annual Review of Phytopathology* 55, 181-203. https://doi.org/10.1146/annurev-phyto-080516-035357.

Kahneman, D. and Tversky, A. (1979). Prospect theory: An analysis of decision under risk. *Econometrica* 47(2), 263–291. https://doi.org/10.2307/1914185.

Kassa, M. T., Haas, S., Schliephake, E., Lewis, C., You, F. M., Pozniak, C. J., Krämer, L., Perovic,D., Sharpe, A. G., Fobert, P. R., Koch, P., Wise, I. L., Fenwick, P., Berry, S., Simmonds, J., Hourcade, D., Senellart, P., Duchalais, L., Robert, L., Förster, J., Thomas, J. B., Friedt, W., Ordon, F., Uauy, C. and McCartney, C. A. (2016). A saturated SNP linkage map for the orange wheat blossom midge resistance gene Sm1. *Theoretical and Applied Genetics* 129, 1507–1517. https://doi.org/10.1007/s00122-016-2720-4.

Lacoste, M., Cook, S., McNee, M., Gale, D., Ingram, J., Bellon-Maurel, V., MacMillan, T., Sylvester-Bradley, R., Kindred, D., Bramley, R., Tremblay, N., Longchamps, L., Thompson, L., Ruiz, J., García, F. O., Maxwell, B., Griffin, T., Oberthür, T., Huyghe, C., Zhang, W., McNamara, J. and Hall, A. (2022). On-farm experimentation to transform global agriculture. *Nature Food* 3(1), 11-18. https://doi.org/10.1038/s43016-021-00424-4.

Lamichhane, J. R., Dachbrodt-Saaydeh, S., Kudsk, P. and Messean, A. (2016). Toward a reduced reliance on conventional pesticides in European agriculture. *Plant Disease* 100(1), 10–24. https://doi.org/10.1094/PDIS-05-15-0574-FE.

Lechenet, M., Dessaint, F., Py, G., Makowski, D. and Munier-Jolain, N. (2017). Reducing pesticide use while preserving crop productivity and profitability on arable farms. *Nature Plants* 3, 17008. https://doi.org/10.1038/nplants.2017.8.

Leybourne, D. J., Storer, K. E., Berry, P. and Ellis, S. (2022). Development of a pest threshold decision support system for minimising damage to winter wheat from wheat bulb fly, *Delia coarctata*. *Annals of Applied Biology* 180(1), 118–131. https://doi.org/10.1111/aab.12718.

Liu, L., Wang, R., Xie, C., Yang, P., Wang, F., Sudirman, S. and Liu, W. (2019). PestNet: An end-to-end deep learning approach for large-scale multi-class pest detection and classification. *IEEE Access* 7, 45301–45312 https://doi.org/10.1109/ACCESS.2019.2909522.

Lowles, A. J., Tatchell, G. M., Harrington, R. and Clark, S. J. (1996). The effect of temperature and inoculation access period on the transmission of barley yellow dwarf virus by *Rhopalosiphum padi* (L.) and *Sitobion avenae* (f.). *Annals of Applied Biology* 128(1), 45-53. https://doi.org/10.1111/j.1744-7348.1996.tb07088.x.

Luo, J., Huang, W., Guan, Q., Zhao, J. and Zhang, J. (2013). Hyperspectral image for discriminating aphid and aphid damage region of winter wheat leaf 2013, 3726-3729. https://doi.org/10.1109/IGARSS.2013.6723640.

Luquet, M., Hullé, M., Simon, J. C., Parisey, N., Buchard, C. and Jaloux, B. (2019). Relative importance of long-term changes in climate and land-use on the phenology and abundance of legume crop specialist and generalist aphids. *Insect Science* 26(5), 881–896. https://doi.org/10.1111/1744-7917.12585.

Mann, J. A., Harrington, R., Carter, N. and Plumb, R. T. (1997). Control of aphids and barley yellow dwarf virus in spring-sown cereals. *Crop Protection* 16(1), 81–87, ISSN 0261-2194, https://doi.org/10.1016/S0261-2194(96)00068-3.

McKirdy, S. J., Jones, R. A. C. and Nutter, F. W., Jr. (2002). Quantification of yield losses caused by Barley yellow dwarf virus in wheat and oats. *Plant Disease*. 86: 769-773.

Miller, W. A. and Rsachová, L. (1997). Barley Yellow Dwarf viruses. *Annual Review of Phytopathology*, 35, 167-190.

Oakley, J. N. and Walters, K. F. A. (1994). A field evaluation of different criteria for determining the need to treat winter wheat against the grain aphid *Sitobion avenae* and the rose-grain aphid *Metopolophium dirhodum*. *Annals of Applied Biology* 124(2), 195–211. https://doi.org/10.1111/j.1744-7348.1994.tb04128.x.

Ortega-Ramos, P. A., Coston, D. J., Seimandi-Cordi, G., Mauchline, A. L. and Cook, S. M. (2021). Integrated pest management strategies for cabbage stem flea beetle (*Psylliodes chrysocephala*) in oilseed rape. *GCB Bioenergy* 14, 267–286. https://doi.org/10.1111/gcbb.12918.

Petrovskii, S., Ellis, J., Forbes, E., Petrovskaya, N. and Walters, K. F. A. (2022). A predictive model and a field study on heterogeneous slug distribution in arable fields arising from density dependent movement. *Scientific Reports* 12(1), 2274. https://doi.org/10.1038/s41598-022-05881-w.

Plumb, R. T. (1995). Epidemiology of barley yellow dwarf in Europe. In: D'Arcy, C. J. and Burnett, P. A. (Eds.), *Barley Yellow Dwarf, 40 Years of Progress*. American Phytopathological Society, 107–127.

Poggi, S., Le Cointe, R., Lehmhus, J., Plantegenest, M. and Furlan, L. (2021). Alternative strategies for controlling wireworms in field crops: A review. *Agriculture* 11(5), 436. https://doi.org/10.3390/agriculture11050436.

Racca, P., Tschöpe, B., Falke, K., Kleinhenz, B. and Rossberg, D. (2014). Forecasting of Colorado potato beetle development with computer aided system SIMLEP decision support system. In: Dharam, P. (Ed.), *Integrated Pest Management*. Academic Press, 79–91. https://doi.org/10.1016/B978-0-12-398529-3.00006-3.

Ramsden, M., Furiosi, M., Debenedettis, P., Stojacic, I., Mendes, M., Munier-Jolain, N.. and Caffi, T. (2022). Development of an online pan-European Integrated Pest Management Resource Toolbox [version 1; peer review: Awaiting peer review]. *Open Research Europe* 2, 72. https://doi.org/10.12688/openreseurope.14679.1.

Ramsden, M. W., Kendall, S. L., Ellis, S. A. and Berry, P. M. (2017). A review of economic thresholds for invertebrate pests in UK arable crops. *Crop Protection* 96, 30–43. https://doi.org/10.1016/j.cropro.2017.01.009.

Ramsden, M. W., Menéndez, R., Leather, S. R. and Wäckers, F. (2015). Optimizing field margins for biocontrol services: The relative role of aphid abundance, annual floral resources, and overwinter habitat in enhancing aphid natural enemies. *Agriculture, Ecosystems and Environment* 199, 94–104. https://doi.org/10.1016/j.agee.2014.08.024.

Ramsden, M. W., Menendez, R., Leather, S. R. and Wäckers, F. L. (2017). Do natural enemies really make a difference? Field scale impacts of parasitoid wasps and hoverfly larvae on cereal aphid populations. *Agricultural and Forest Entomology* 19(2), 139–145. https://doi.org/10.1111/afe.12191.

Ratnadass, A., Fernandes, P., Avelion, J. and Habib, R. (2012). Plant species diversity for sustainable management of crop pests and diseases in agroecosystems: A review. *Agronomy for Sustainable Development* 32(1), 273–303. https://doi.org/10.1007/s13593-011-0022-4.

Rossi, V., Caffi, T. and Salinari, F. (2012). Helping farmers face the increasing complexity of decision-making for crop protection. *Phytopathologia Mediterranea* 51, 457–479. Available at: http://www.jstor.org/stable/43872334 (accessed 31 August 2022).

Sanchez-Bayo, F. and Wyckhus, K. A. G. (2021). Further evidence for a global decline of the entomofauna. *Austral Entomology* 60(1), 9–26. https://doi.org/10.1111/aen.12509.

Segers, A., Megido, R. C., Lognay, G. and Francis, F. (2021). Overview of *Bruchus rufimanus* Boheman 1833 (Coleoptera: Chrysomelidae): Biology, chemical ecology and semiochemical opportunities in integrated pest management programs. *Crop Protection* 140, 105411. https://doi.org/10.1016/j.cropro.2020.105411.

Sishodia, R. P., Ray, R. L. and Singh, S. K. (2020). Applications of remote sensing in precision agriculture: A review. *Remote Sensing* 12(19), 3136. https://doi.org/10.3390/rs12193136.

Sparks, T. C. and Nauen, R. (2015). IRAC: Mode of action classification and insecticide resistance management. *Pesticide Biochemistry and Physiology* 121, 122–128.

Van den Eynde, R., Van Leeuwen, T. and Haesaert, G. (2020). Identifying drivers of spatio-temporal dynamics in barley yellow dwarf virus epidemiology as a critical factor in disease control. *Pest Management Science* 76(8), 2548–2556. https://doi.org/10.1002/ps.5851.

Varandas, R., Egas, C. and Conceicao, I. L. (2020). Potato cyst nematodes: New solutions to an old problem. *Crop Protection* 137, 105303. https://doi.org/10.1016/j.cropro.2020.105303.

Villano, R., Braavo-Ereta, B., Solis, D. and Fleming, E. (2015). Modern rice technologies and productivity in the Philippines: Disentangling technology from managerial gaps. *Journal of Agricultural Economics* 66(1), 129–154. https://doi.org/10.1111/1477-9552.12081.

Wäckers, F. L. and van Rijn, P. C. J. (2012). Pick and mix: Selecting flowering plants to meet the requirements of target biological control insects. In: Gurr, S. D., Snyder, W. E. and Read, D. M. Y. (Eds.), *Biodiversity and Insect Pests: Key Issues for Sustainable Management*. John Wiley and Sons Ltd., Chichester, UK, 139–165. https://doi.org/10.1002/9781118231838.ch9.

Wainwright, C., Jenkins, S., Wilson, D., Elliot, M., Jukes, A. and Collier, R. (2020). Phenology of the diamondback moth (*Plutella xylostella*) in the UK and provision of decision support for brassica growers. *Insects* 11(2), 118. https://doi.org/10.3390/insects11020118.

Witzgall, P., Kirsch, P. and Cork, A. (2010). Sex pheromones and their impact on pest management. *Journal of Chemical Ecology* 36(1), 80–100. https://doi.org/10.1007/s10886-009-9737-y.

Young, A. G., Port, G. R. and Green, D. B. (1993). Development of a forecast of slug activity: Validation of models to predict slug activity from meteorological conditions. *Crop Protection* 12(3), 232–236. https://doi.org/10.1016/0261-2194(93)90115-Y.

Part 3

Invasive species

Chapter 9

Assessing the potential economic impact of invasive plant pests

Monique Mourits and Alfons Oude Lansink, Wageningen University, The Netherlands

1 Introduction

Although most introduced, non-native, pest species (including insects and pathogens) are relatively innocuous (Holmes et al., 2009), some could lead to massive economic impacts in the invaded region. Examples of recent invaders with widespread economic impacts are Emerald ash borer (*Agrilus planipennis*), *Phytophtora ramorum*, *Hymenoscyphus fraxineus*, the pine wood nematode (*Bursaphelenchus xylophilus*) and *Xylella fastidiosa* (Aukema et al., 2011; Petucco et al., 2019; Hill et al., 2019; Soliman et al., 2012b; Schneider et al., 2020).

Economic impact assessments form part of a pest risk analysis to determine whether the introduction of a pest will result in unacceptable economic consequences. In cases where consequences are judged to be unacceptable, phytosanitary measures need to be considered to reduce the economic risk to an acceptable level. Hence, economic impact assessments are essential in developing and prioritizing invasive species management (Holmes et al., 2009; Epanchin-Niell, 2017).

Economic impact assessments need to account for the bio-economic characteristics of the invasion (Kovacs et al., 2010; Haight et al., 2011).

http://dx.doi.org/10.19103/AS.2022.0113.12

Information on hosts at risk should be integrated with information on the potential area of establishment and spread, the development of the pest population density over time and space, the biophysical and harvest-related damage caused by the pest and the resulting economic consequences on society. However, due to the dynamic and spatial complexity of a pest invasion, impact estimations can be challenging.

A basic approach often applied in economic impact assessments of invasive pests is centered on the identification of the potential amount of host resource at risk (potential production losses), multiplied by some unit market price (e.g. Pimentel et al., 2005). However, this approach does not account for biological invasions being spatial and dynamic processes where the economic well-being of stakeholders is affected in different ways and intensity (Holmes et al., 2009).

The economic impact of invasive species consists of direct and indirect market impacts. Direct impacts are related to expected losses in biomass production, either in quality and/or quantity (often referred to as yield), resulting from damage to the crops, trees and/or their products. Indirect impacts are related to the consequential market responses to the altered biomass supply as reflected by changes in market prices, producer and consumer responses to price changes and effects on international trade (Soliman et al., 2012b; Petucco et al., 2019; Schneider et al., 2021).

Appropriate techniques to assess the various economic impacts consist of partial budgeting, partial equilibrium models, input-output models and computable general equilibrium models (Soliman et al., 2010; Welsh et al., 2021). These methods differ in the width of economic scope and consequently in their ability to assess the economic effects of a pest introduction. A wider scope generally requires more elaborated data and skills. In this chapter, we evaluate each method in terms of its economic concepts, scope and required data and skills, followed by a general reflection on the question of how to select the most appropriate technique, considering the trade-off between completeness in economic scope and resources required.

The chapter concludes with two case studies of recent pest threats in Europe (*B. xylophilus*, a nematode infecting pine, and *X. fastidiosa*, an insect-vectored bacterial disease in olive trees), illustrating the potential of the most commonly applied methods.

2 Methods to assess the potential economic impact of invasive plant pests

2.1 Partial budgeting

Partial budgeting is a straightforward approach in which the net direct impact of a change in a system, like a plant pest invasion, is obtained by assessing the associated negative effects (i.e. additional costs and reduced returns) and positive effects (i.e. additional returns and reduced costs). In general, direct impacts related to a pest invasion include only negative effects, viz. yield losses and additional production costs (Soliman et al., 2010).

When applying partial budgeting, spatially explicit data on host plants, pest spread and climate need to be combined to determine the endangered area, i.e. the area where the pest can establish at a given time and cause damage. Subsequently, a relationship between the defined endangered area and the likely level of damage (production loss expressed in quantity and/or quality losses) needs to be identified. Ideally, this level of production loss is estimated as a function of pest population density. However, it is often very difficult to reliably model the density of an invasive pest population, and so instead, spread models are more often used to generate spatial data on the presence-absence of the pest, which can only be used to delimit the endangered area (Robinet et al., 2012). In those cases, additional data are needed to serve as a proxy for the missing population data, e.g. climate and/or other ecological features that are highly correlated with the pest density (Kriticos et al., 2013; Soliman et al., 2012a,b).

Direct impacts could be estimated as cumulative sums over time and space from the start of the invasion till a steady-state situation or other arbitrary end point, or as yearly flows in a steady state. To account for the time value of money concept (a euro today is worth more than it would be next year, due to the opportunity to invest and gain interest over a period of a year), costs accrued over time are aggregated by discounting their values to the present value (e.g. Hillier et al., 2020).

Although the method is designed to evaluate the direct impact at the producer level, partial budgeting can also be used at the national or continental level by scaling up the budgetary impacts (Rich et al., 2005). MacLeod et al. (2004) used partial budgeting, for instance, at the national level to assess the impact of an uncontrolled melon thrips, *Thrips palmi*, invasion in England. To this end, the potential damage resulting from yield and quality losses in horticulture, export losses and costs related to additional research requirements and plant health certifications were evaluated over a period of 10 years after invasion.

By comparing the evaluated direct impact to the estimated costs of potential control measures, decisions on mitigation can be supported if the resulting ratio is sufficiently high. In the study of MacLeod et al. (2004), the resulting benefit (= the value of losses prevented) to cost ratio was at least 4:1, justifying the implementation of the evaluated mitigation measures.

Soliman et al. (2012a) used the partial budgeting approach at a continental scale to estimate the direct impacts of an unregulated spread of Potato Spindle Tuber viroid (PSTVd) in Europe. Direct impacts on potato and tomato producers were estimated by accounting for expected yield losses and additional crop protection costs in the expected established steady state. Yield losses were estimated for spatial explicit agri-environmental zones to account for host distribution and temperature-dependent damage levels of the pest. Aggregation of the losses and additional crop protection costs over the spatial zones indicated a direct annual impact on potato growers of 2.1 M€ and on tomato producers of 3.5 M€. Implemented phytosanitary measures to prevent spread within Europe accounted for, approximately, 5 M€ a year, indicating that the implementation of these measures is economically questionable when considering only the direct impacts (Soliman et al., 2012a).

The strength of the partial budgeting approach is its simplicity and transparency. It has little complexity in terms of its resource needs, as it requires a limited amount of data, expertise and time investment (Holland, 2007). However, partial budgeting is not suitable for measuring long-term effects or impacts in other sectors of the economy as it only accounts for the direct impact of a change in an isolated setting, assuming fixed budgets and market prices. Changes in production volume caused by a pest invasion will, however, have long-term effects on total market supply and prices, thereby affecting other producers and other sectors of the economy, such as the processing industry (Petucco et al., 2019), but also consumers. This will especially be the case with forest pests, where an invasion will not only reduce the present availability of the affected tree species but also its future availability (Soliman et al., 2012b; Petucco et al., 2019). The partial budgeting technique is not able to assess these indirect market effects. This shortcoming can be counterbalanced by a complementary use of the partial equilibrium modeling technique.

2.2 Partial equilibrium modeling

Partial equilibrium modeling requires the definition of functional relationships describing the supply and demand of the product of interest in the market to determine the so-called market equilibrium. Market equilibrium occurs when the supply in the market meets the demand in the market, reflecting the combination of price and quantity that maximizes economic welfare. Maximum economic welfare is realized when producers and consumers maximize their profits and utilities (Mas-Colell et al., 1995).

Figure 1 illustrates this concept in more detail by showing a downward-sloping demand curve, reflecting the diminishing marginal utility of an increased consumption, and an upward-sloping supply curve, reflecting increasing marginal costs of production. The market equilibrium, where quantity supplied equals quantity demanded, occurs at an equilibrium price of P_E and quantity Q_E. The difference between P_E and the demand curve represents how much consumers benefit by being able to purchase the product for a price (P_E) that is

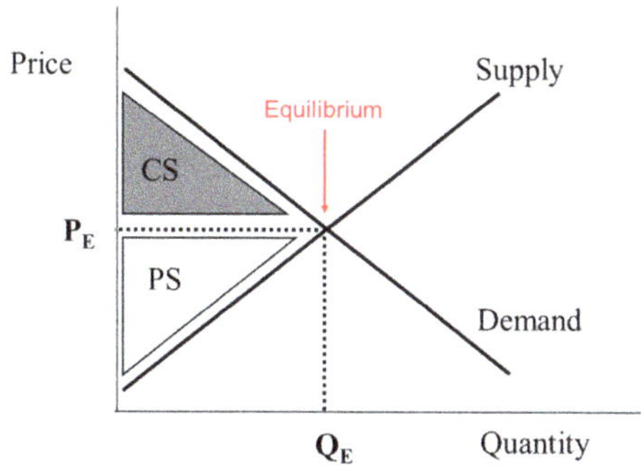

Figure 1 Illustration market equilibrium concept.

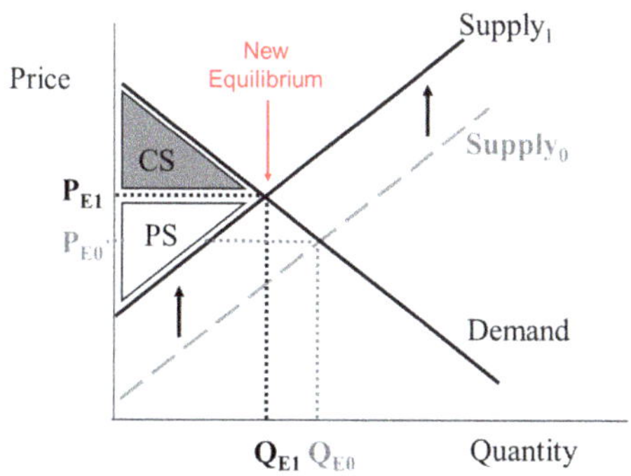

Figure 2 Illustration impact pest invasion on market equilibrium.

less than they would be willing to pay. The total benefit derived by consumers, or consumer surplus, is represented by the triangle CS.

Since the supply curve represents the marginal variable cost of production, the area below the curve equals the total variable costs. The revenues from sales are equal to price (P_E) times quantity (Q_E), which is the area enclosed between the dashed lines. Hence the producer surplus, defined as the difference between total revenue and total variable costs, is reflected by the triangle PS. Total economic welfare is the sum of consumer surplus and producer surplus (Mas-Colell et al., 1995).

At market equilibrium, P_E will not change unless an external factor - like a pest invasion - changes the supply or demand, which results in a disruption of the equilibrium and, subsequently, in a change in economic welfare. By partial equilibrium modeling, these pest-induced changes in economic welfare can be assessed by measuring the differences in equilibrium price and quantity before and after the pest invasion.

Pest invasions may lead to a loss in yield and an increase in production costs, resulting in an upward shift in the supply curve (Fig. 2). This shift in the supply curve alters the equilibrium point, implying a decrease in quantity supplied (from Q_{E0} to Q_{E1}) and an increase in market price (from P_{E0} to P_{E1}). Consumer losses, or the reduction in consumer welfare, that result from the new equilibrium point can be calculated by comparing CS before and after invasion. In the same way, changes in producer welfare can be calculated.

Partial equilibrium modeling can be conducted with respect to one sector (a single-sector model) or multiple sectors (a multi-market model). Multi-market models link related markets and are, therefore, able to capture spillover effects between main markets. Within each partial equilibrium model, main assumptions need to be made to define the behavior of the affected market(s) (e.g. perfect or imperfect competition), the level of homogeneity for products from exogenous markets and the influence of domestic producers on the world market. Data requirements can be substantial (Mas-Colell et al., 1995; Petucco et al., 2019) as data are needed to reflect the affected markets, including data on prices, production and consumption quantities, market prices and international trade, to determine price elasticities of both supply and demand (i.e. the degrees to which supply and demand respond to a change in price).

The strength of the partial equilibrium approach is the consideration of both direct and indirect impacts by analyzing the change in consumer and producer welfare. Results provide insight into the total economic impact, including the extent to which producers can transfer part of the negative direct impact to consumers. Following upon the earlier described evaluation of the direct impacts of an unregulated PSTVd spread in Europe by partial budgeting, Soliman et al. (2012a) also applied a partial equilibrium modeling approach to consider the indirect market effects resulting from the reduced supply and increased production costs. Change in total welfare equaled to a loss of 4.4 M€ for potato producers and consumers and of 5.7 M€ for tomato producers and consumers. Consumers, however, bore 92% of the total welfare loss in the case of affected potato production and 77% in the case of affected tomato production, indicating a considerable transfer of direct impacts from producers (i.e. 2.1 M€ on potato growers and of 3.5 M€ on tomato producers) to consumers. Hence, the economic impact of PSTVd appeared to be important (and therefore the justification for restrictive measures) when accounting for the expected market impact, but questionable if only the direct impact was considered.

Most recent examples of partial equilibrium applications in invasive plant pest economic impact are the analyses performed by Petucco et al. (2019) and Schneider et al. (2021). Petucco et al. (2019) applied the partial equilibrium approach to assessing the impact on economic welfare caused by the ash dieback invasion in France. Schneider et al. (2021) used the approach to evaluate the potential indirect market impact of an uncontrolled *X. fastidiosa* infestation at a European scale (see Section 5).

Despite its suitability for evaluating market effects, partial equilibrium modeling is limited in its ability to account for economy-wide effects. Economic impact assessments using partial equilibrium models are therefore only appropriate when the indirect market effects of the pest are not expected to affect other markets or cause measurable macroeconomic changes (e.g. changes in employment). For applications that require an economy-wide scope, input-output analysis (Section 2.3) or computable general equilibrium modeling (Section 2.4) approaches can be considered.

2.3 Input-output analysis

Input-output (I-O) analyses focus on the interdependencies of sectors in an economy (regional or national), making it an appropriate technique to predict the ripple effects throughout an economy due to a shock within a particular sector (Leontief, 1986). Central to an I-O analysis is the specification of the monetary flows of inputs and outputs among the productive sectors of an economy (Miller and Blair, 1985), reflected by so-called input-output tables. In an I-O table, each sector is represented by a row and a column. The rows specify the distribution of the total output of a specific sector sold to other sectors (intermediate demand) or to final demand (final consumption and exports). The columns refer to the production side of a specific sector, by indicating the value of the inputs of each sector needed to produce output.

Any change in the final demand for the products of an industry has both direct and indirect effects on the economy as a whole. Changes cause primary ripples by directly changing the purchasing patterns of the affected sector. Suppliers to the affected sector must change their purchasing patterns to meet the demand of the sector originally affected by the change in final demand, creating a smaller secondary ripple. Those who meet the needs of the suppliers must in turn change their purchasing patterns to meet the demand of the suppliers of the affected sector and so on.

The relationship between the initial change and the total effects generated by the change is known as the multiplier effect of the sector, or the impact of the sector on the economy. To compute this multiplier effect, I-O tables are mathematically converted into matrices of multipliers that reflect the amount by which production, employment and income would alter because of one-unit change in final demand (Miller and Blair, 1985).

Based on I-O analysis, the impact of a pest invasion on an economy can be evaluated by adjusting the final demand in the affected sector (such as agriculture or forestry) according to the expected shock to demand (such as decrease in the production of agricultural commodities or reduction in exports), multiplied by the multiplier matrices. Examples of I-O applications in pest risk assessment are the analyses of Elliston et al. (2005) and Juliá et al. (2007). Elliston et al. (2005) investigated the regional economic impact of a potential invasion of Karnal bunt (caused by the fungus *Tilletia indica*) in wheat in Queensland. In the scenario of a widespread incursion, the direct effect on the wheat and other grain industries was estimated as an AUD M$89 decline in output and a loss of 400 full-time jobs over a 15-year planning horizon. The indirect effects in all other industries resulted in a decline of AUD M$38 in output and a decline in employment of 200 full-time jobs.

In the I-O analysis of Juliá et al. (2007), direct and indirect economic effects of the invasive weed, yellow starthistle (*Centaurea solstitialis*), in the rangelands of Idaho were determined in relation to its interference with agricultural and non-agricultural benefits (e.g. wildlife recreation expenditure and water quality). Agriculture-related economic impacts accounted for 79% of the total impact on the rangeland economy, and non-agricultural impacts for the remaining 21%.

Apart from the high data requirements, application of an I-O analysis is generally limited by two fundamental assumptions. First, I-O models only account for changes in the economy induced by shifts in demand. Supply is assumed to be perfectly elastic (e.g. there are no restrictions to inputs), which in practice is often not the case. Especially in agriculture and forestry, supply is often limited, which means that an I-O analysis of a pest introduction may not capture all the important economic effects. Second, I-O models are static and cannot account for changes in prices or for changes in the structure of a sector over time. This means that I-O models assume fixed prices, no substitution between inputs and constant returns to scale (e.g. underlying data and relationships are not affected by the economic changes). However, this static assumption can be justified if the I-O technique is used to analyze only short-term impacts.

In conclusion, the I-O approach provides an opportunity to measure short-term, spillover effects across sectors of the economy given a shock in a particular sector that affect the demand side only. For assessments that require the economy-wide scope of I-O models as well as the economic realism of partial equilibrium models (price effects), a computable general equilibrium (CGE) modeling approach would be more appropriate.

2.4 Computable general equilibrium modeling

The CGE modeling approach uses I-O tables to represent all sectors in the economy with the inclusion of functional relationships between these sectors as in a PE model. By the specification of demand relationships, production technologies, relationships between domestic and imported goods, prices, household income and equilibrium conditions, CGE models address questions concerning impacts across sectors as well as price changes and longer-run impacts. This capacity, however, makes CGE models highly complex (Sadoulet and de Janvry, 1995; Dixon and Parmenter, 1996).

Incorporating the whole economy in one model requires a high level of aggregation to keep it practical. High levels of aggregation, however, make it difficult to analyze a change in a subsector of the economy. Many agriculture-oriented CGE models are disaggregated into only two subsectors, such as tradable and non-tradable crops, or food and cash crops (Bourguignon and Pereira da Silva, 2003). The economic consequences of a pest that causes large damages in only a subset of crops might remain unnoticed due to the level of aggregations. Applications of CGE models are, therefore, only appropriate to address large-scale problems that generate measurable macroeconomic impacts. Pest invasions rarely trigger changes in overall employment, income or the inflation rate. Consequently, there are only a few applications of CGE applications in pest risk assessments known.

McDermott et al. (2013) developed a CGE model for the state of Ohio to estimate the economic impact caused by the invasive Emerald ash borer, *A. planipennis*. Annual economic impact of the beetle was estimated at USD M$70. Most of the damage (~80%) was incurred by recreation sectors, households and state government. Recreation sectors had to add the costs of removing infested ash trees to their production costs. The cost of ash removal reduced the disposable income of households, while these expenditures flow to the garden sector due to an increase in the demand for their services. The government had to adjust revenues to account for ash removal, shifting this expenditure to the garden sector.

3 Selection of appropriate level of complexity

3.1 Context setting economic impact assessment

Invading plant pests impose economic consequences that vary across stakeholders (plant growers, forest managers, residents/homeowners, industries or sectors, consumers, government) and regions, and are dynamic over time. Induced disturbances do not only reduce the present availability of affected hosts but also alter their future availability, population structure and distribution. Due to this three-dimensional context (scope, scale, timeline), economic impact assessments require explicit settings of each of these impact dimensions, resulting in a specific demarcation (objective) of the analysis.

- The scope dimension refers to the perspective (level) of the assessment. Is it only accounting for the impacts experienced by specific stakeholders (producers, residents/homeowners, industries or sectors, consumers, government) or is it performed from the perspective of society as a whole? The selected perspective is of primary importance with respect to the identification of the most relevant impacts; an analysis from the

perspective of forest managers will neglect potential impacts on, for instance, residents/homeowners, industries, or consumers, while an analysis from the perspective of a national government ideally considers all main societal impacts.
- The scale dimension refers to the geographical dimension of the analyses. Is the assessment conducted on a local, regional, national or international scale? Data availability is of primary importance in determining the scale to which a quantitative economic impact assessment can be applied.
- The time dimension refers to the time horizon of the assessment, which is key in economic impact assessments. The economic impacts could be estimated as cumulative sums over time and space from the start of the invasion till a steady-state situation or other arbitrary end point, or as yearly flows in a steady state. Modeling the development of the economic impacts over time requires a detailed modeling of spread. With an increase in timespan, uncertainty in the evaluated economic systems increases, primarily due to behavioral responses (adaptation) of stakeholders who attempt to cope with the invasion. Producers can, for instance, decrease negative impacts by adapting their crop mix or crop management. A first action involves switching to resistant varieties of the same crop or to crops that are not vulnerable to the pest. Producers may also change their crop management to decrease the negative consequences of pests and diseases. In fact, it is unrealistic to assume that crop management practices will remain unchanged following a pest invasion; nevertheless, economic impact studies rarely account for such adaptations (Epanchin-Niell, 2017; Petucco et al., 2019).

3.2 Selection of appropriate assessment method

In general, the first step in the assessment of economic impacts in most pest risk analyses consists of a quick qualitative analysis of the expected impacts, based on experts' judgments. For this purpose, plant protection agencies generally apply structured decision support schemes as outlined in the International Standards on Phytosanitary Measures (ISPM) No. 11 (FAO, 2004). These schemes consist of a logical sequence of questions to capture experts' opinions. For each question, experts provide their answer by selecting a score within some predefined ordinal risk classification scheme. In such a scheme, economic consequences are only described in broad qualitative terms, without strict guidelines defining time frame and scope. Economic impacts are assessed by questions considering the expected yield or quality losses, the potential impact on international trade and expected control costs. Although this approach is helpful for classifying pest impacts (acceptable, unacceptable), the assessment is too weak to justify or select the appropriate level of management measures. For this purpose, a quantitative assessment is needed (Soliman et al., 2015).

The four evaluated assessment techniques that can be applied for this purpose differ markedly in terms of the scope of impacts covered and consequently also in terms of data requirements and model complexity (Table 1) (Soliman et al., 2015; Welsh et al., 2021). While partial budgeting is a basic and easily understood technique for assessing direct impacts, its scope does not include indirect effects of pest damage resulting from effects on market prices, supply and demand, nor does it address spillover effects to other sectors of the economy. Partial equilibrium or CGE modeling techniques widen the scope to include indirect effects, in the case of partial equilibrium only for the sector of the affected commodity and in the case of CGE for the economy (society) as a whole. A technique between partial budgeting and equilibrium modeling is I-O analysis, as it accounts for the spillover effects of a reduction in the production of an agricultural commodity to other sectors in the economy without addressing the effects of price changes.

Given the differences between the four techniques (Table 1), each risk analyst faces the question: what is the appropriate method to choose? As the scope and size of impacts can differ for each pest and host (commodity), an 'one size fits all' selection procedure of the most appropriate method is not existing. However, some guidance can be provided based on (1) the defined objective (demarcation) of the analysis, indicating the desired scope of insights into economic impacts, and (2) the available resources in terms of data, skills and time. The remainder of this section reflects on the extent to which each assessment method aligns with these considerations.

Despite its limitations, the partial budgeting approach remains the default method of choice for an assessment of the economic impacts of pests and diseases. This technique provides insight into the direct impacts of the pest, while its results are easily understood and communicated to stakeholders such as policymakers. The required data can often be obtained at a reasonable level of accuracy, either through secondary data collection or, in the absence of

Table 1 Illustration of the extent to which the assessment methods differ in their economic width (scope), resource requirements and output credibility

Method	Economic width	Data and skill requirement	Output credibility
Partial budgeting			
Partial equilibrium modeling			
Input-output analysis			
Computable general equilibrium modeling			

these, through expert elicitation, and the technical expertise needed to apply the method is modest. In addition, the results of partial budgeting evaluations provide insights that can be used to determine whether further analysis is needed using more sophisticated techniques. If the outcome of the partial budgeting shows the direct economic impacts are devastating, then there is no rationale for conducting further analysis in the context of a pest risk analysis (i.e. answering the question of whether the pest needs to be controlled or not).

If the objective goes beyond a first assessment of the costs of pest introduction, or if indirect impacts are expected to substantially alter the outcome of the evaluation, then more sophisticated techniques are warranted. Partial equilibrium modeling is worth considering if the expected changes in production volumes are very large, while demand or supply is inelastic; in such cases, substantial price effects are very likely to occur. Generally, a pest invasion reduces the supply of crops, thereby increasing the price. However, with the occurrence of price effects, part of these invasion costs is transferred from affected producers to consumers who pay a higher price for the consumables, whereas non-affected producers may even benefit (see case studies). These insights are relevant in deciding on the need for supporting policy responses, e.g. compensation policies. Partial equilibrium modeling requires more advanced data and skills (knowledge) compared to partial budgeting. The complexity of the model will vary with the spatial scale of the analysis, the number of sectors considered, the functional specification of the supply and demand functions and the market structure (e.g. monopoly or oligopoly). A more broad-based economic technique like I-O analysis or CGE modeling could be considered if large spillover effects to other sectors of the economy are expected such as transport, trade and tourism. However, these techniques have rarely been used in pest assessment due to the large data and skill requirements as well as the fact that pest invasions rarely generate measurable macroeconomic impacts. Hence, in most cases, a combination of partial budgeting and partial equilibrium modeling can provide a sufficient scope in those cases where both direct and indirect impacts are expected to occur (Rich et al., 2005; Soliman et al., 2015; Welsh et al., 2021).

Generally, it is best to start by means of a partial budget analysis and to incorporate only as much detail and complexity as is needed to achieve the defined objective of the assessment. A fundamental dilemma underlying the choice of developing a partial equilibrium model in addition to a partial budgeting analysis is that it is difficult to know in advance whether the more advanced partial equilibrium modeling technique is justified without applying the method. The results of a partial budgeting analysis do not inform about the occurrence of potential market effects and, hence, about the need of using a partial equilibrium modeling approach. This can only be evaluated by exploring data on the price elasticity of supply and demand. The more inelastic the demand or supply, the greater the likelihood of price effects and thus the likelihood that a partial equilibrium model approach will provide additional insights that could alter the conclusion from the partial budgeting model.

Although the presented techniques provide useful assessment models to obtain insights into the potential economic consequences of invasive pests, it is important to note that these models do not provide exact predictions. Uncertainties arise in different stages of an economic impact assessment as many aspects of both invasion dynamics and affected processes are highly uncertain. Prior to or early in a pest invasion, for example, little may be known about the future pattern of spread and the potential damages that the invader might cause. Usually, economic assessment models account for uncertainty by assuming 'known' distributions of uncertain parameters and subsequently calculate the probability-weighted average across potential outcomes that arise from the uncertainty, viz. the expected outcome (Epanchin-Niell, 2017). In reality, uncertainty about ecological and economic parameters related to plant pest invasions can be severe, which makes sensitivity analyses a required part of the assessment to see how changes in the underlying model assumptions affect the estimated total economic impact.

Generally, the credibility of the outcomes of economic models decreases with their level of sophistication, because the greater sophistication entails making assumptions (i.e. introducing uncertainty) about processes that may work quite differently from how they are modeled. Thus, partial budgeting has a greater potential of giving credible results, particularly in the short run, while confidence is bolstered as anybody can check the assumptions underlying its calculations. Partial equilibrium and CGE techniques give more uncertain results because relationships between prices and supply and demand of affected commodities must be specified in simplified mathematical models, which impose strong assumptions on producer and consumer behavior. This is not to say that any sophisticated model is wrong. The models are theoretically correct, but they are simplifications of economic reality. Therefore, the outcomes of such models should be interpreted as plausible trends, inferred from past behavior, and should be used to complement the results of a partial budgeting rather than replace them. Results of partial equilibrium and CGE modeling should, in particular, be interpreted with caution as to the absolute magnitude of the effects. It is well-known that small changes in price elasticities may have large effects on the magnitude of economic effects.

4 Economic analyses to support pest risk management

Economic impact risk assessments raise awareness about the size of the economic impacts due to a pest invasion but cannot be used directly to support managerial responses. Hence, additional decision analyses are needed to evaluate the costs and benefits associated with potential management strategies. Generally, decision analyses on the mitigation of invading/emerging pest species rely on the concept of optimal resource allocation (Clark, 1990), aiming to maximize social welfare. Existing decision analyses studies directed toward forest biosecurity, for instance, estimated the costs and benefits of specific control strategies (Gatto et al., 2009), evaluated the trade-offs between eradication and slow-the-spread programs (Sharov and Liebhold, 1998; Brockerhoff et al., 2010) or analyzed the cost-effectiveness of surveillance programs (Epanchin-Niell et al., 2014) and control strategies targeted toward particular life stages of an invader (Buhle et al., 2005). Techniques to value the economic costs and benefits are in line with those described in the previous sections.

It is important to note that the management of invading pest incurs various economic consequences. The main effects relate to the extent the implemented management measures slow down the spread of the invading pest. By slowing down the spread, damage to the susceptible hosts is prevented and thus the overall economic impact is reduced. However, implementation of management measures also directly results in control costs (e.g. costs to bio-control releases or removal of hazard trees) and in opportunity costs due to the occurrence of potential profits foregone (e.g. from advancing or delaying harvest or due to movement restrictions out of a demarcated zone). Indirectly, it might also influence trade opportunities of other commodities or services (e.g. due to movement restrictions, increased inspections, cargo treatments) resulting in additional consequential losses (opportunity costs).

Cost-benefit analysis provides an approach for structuring the information on the various types of effects to aid decisions about control management design, by evaluating whether the advantages (benefits) of an invasion mitigation strategy outweigh its disadvantages (costs) (Boardman et al., 2006; Kehlenbeck et al, 2012). By expressing the advantages and disadvantages in monetary terms, the different effects become comparable and can be used to determine whether the evaluated management strategy increases total welfare or not.

5 Case studies: pine wood nematode (*B. xylophilus*) and *X. fastidiosa*

To illustrate the application of an economic impact assessment using partial budgeting and partial equilibrium models, two case studies on pest organisms currently threatening Europe are presented in more detail. The first case study describes the insights obtained from the impact assessment evaluating the consequences of the establishment of pine wood nematode (*B. xylophilus*) in Europe. As pine wood nematode affects forestry stock and future production volume, indirect impacts on the European market lag the moment of actual damage. The second case presents the economic impact assessment of an uncontrolled infestation by *X. fastidiosa* vectored by xylem-feeding spittlebug nymphs (*Philaenus spumarius*) affecting the European olive production. Presented results highlight the distribution of impacts among various European countries and market participants.

5.1 Case study 1: pine wood nematode

PWN (*B. xylophilus*) is recognized worldwide as a major forest pest. The nematode can multiply rapidly at persistently high temperatures (>20°C for at least 8 weeks), creating large populations that disrupt water transport in the infected tree, causing pine wilt disease. Ultimately, pine wilt disease results in the death of the infested trees. PWN is vectored from diseased to healthy trees by bark beetles in the genus *Monochamus*. Pines (*Pinus* spp.) are favored hosts but other genera of conifers are also attacked (Evans et al., 1996).

Since its introduction to Portugal in 2008, PWN has been a serious threat to the European forestry industry. The presented case study is based on Soliman et al. (2012b) who analyzed the risk of PWN in the EU by estimating the decline in forest resources available for wood supply and its resulting impact on the downstream roundwood market over a time span of 22 years (2008–30). Spatial analysis was used to integrate the required information on climate suitability, host distribution, pest spread and value of assets. Information on spread was derived from the output of the process-based PWN spread model of Robinet et al. (2011) indicating the spread potential from the initial infested sites in Portugal to the rest of Europe. Direct economic impacts resulting from wood loss were computed using partial budgeting at the regional scale, while market impacts on the round wood market were computed using partial equilibrium modeling at an EU scale.

Under the assumption of no regulatory control measures, the value of lost forestry stock cumulated over the evaluated period of 22 years was estimated at €22 bn. Substantial direct impacts were expected in Portugal, Spain, Southern France and North-West Italy (Fig. 3). The greatest yearly loss of stock was expected to occur in the period 2014–9, with a peak of €3 bn in 2016, after which it stabilized at 300–800 M€/year (Soliman et al., 2012b).

The results of the PE analysis showed that the reduction in the domestic supply of industrial round wood caused by the PWN invasion led to an increase in the EU market price and a decrease in EU demand. Annual net total welfare (= total impact on affected, non-affected producers and consumers) after 22 years of spread (inducing a shift in wood supply due to an accumulated loss in forestry stock) was reduced by M€ 218 (Table 2). The shortage in EU supply (the gap between supply and demand) was covered by an increase in imports and/or

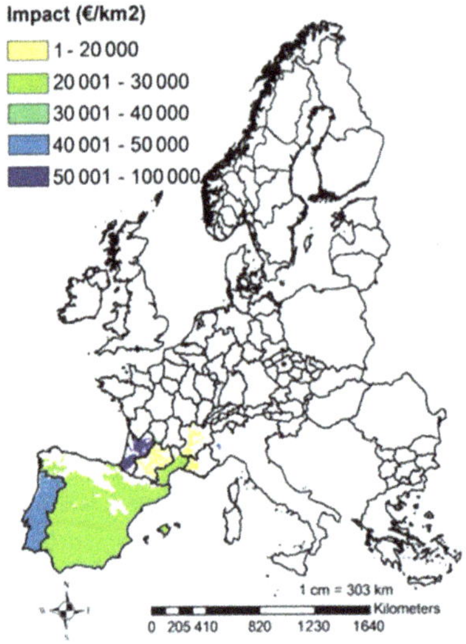

	Infested area (%)	Direct impact (M€)
Portugal	82.4	5164
Spain	67.5	14 649
France	13.2	2532
Italy	0.2	30
Total	10.6	22 375

Figure 3 Cumulative direct impact over 22 years of uncontrolled PWN spread in the EU (estimated by partial budgeting).

Table 2 Annual impact of a PWN invasion in the EU after 22 years of uncontrolled spread (estimated by partial equilibrium modeling)

	Absolute change	% change
Supply (Mm^3)	−1.89	−0.8
Demand (Mm^3)	−0.77	−0.3
Price (€/m^3)	1.44	2.8
Net trade (Mm^3)	−1	−14.6
Consumer surplus (M€)	−357	−2.5
Producer surplus (M€)	139	2.0
Total welfare (M€)	−218	−1.0

decrease in exports (i.e. change in net trade between the EU and the world market), triggering an increase in the world price for round wood. The increase in market prices caused an increase in supply leading to a new equilibrium in the industrial round wood market. Due to the higher prices, consumers suffered a reduction in their surplus of M€357. Non-affected producers experienced a positive net impact as they benefited from the higher market price in the new equilibrium situation without suffering losses in the stock. Affected producers experienced a loss as the reduction in production volume was not fully compensated by the price increase. Overall, the total producer surplus increased by M€139.

Figure 3 and Table 2 indicate a large difference in the magnitude of the estimated direct and indirect impacts. This is because the direct impacts represent the reduction in the value of standing forestry stock (future production potential), while the indirect impacts refer to the changes in the yearly flow of wood to the round wood market. These flows represented only 2% of the standing stock. Therefore, losses in wood flow accumulated slowly and over a long period of time after the standing stock was lost due to PWD. This means that if the spread of PWD had stopped, the annual direct loss would have disappeared, while the effect on the market would have continued to increase until mitigation efforts became effective (i.e. replanted [resistant] trees flow into the round wood market).

5.2 Case study 2: X. fastidiosa

X. fastidiosa is one of the most dangerous plant pathogenic bacteria worldwide causing a variety of diseases, with huge economic impact on agriculture, public gardens and the environment (EFSA, 2018). Multiplication of the bacteria within susceptible host plants results in clogging of the vascular system, decreasing water and nutrient flows. Consequently, large parts of the crown or foliage dry out and die off. Yields start to decline and fruit quality is reduced, and eventually the host plant dies. The bacterium is naturally transmitted by insect vectors (spittlebug nymphs), which feed on the xylem of host plants.

Since 2013, *X. fastidiosa* has been reported to have infected several hosts in different areas of Europe. To date, most damage has been inflicted on the olive orchards of southern Apulia (Italy), where infection with *X. fastidiosa* subspecies *pauca* has led to the death of millions of trees. To reduce the spread of the pathogen and the associated olive quick decline syndrome (OQDS), various control measures have been implemented, including containment measures such as tree felling and vector control and the establishment of demarcated areas. Unfortunately, the size of the area affected and the hidden reservoir of symptomless, but infectious, host plants have hindered disease eradication (Morelli et al., 2021). Worse, the dense network of olive groves in Apulia (Italy) might serve as a European reservoir of *X. fastidiosa* subspecies *pauca* (Strona et al., 2017).

This second economic impact case study is based on the research of Schneider et al. (2020; 2021) who evaluated the potential economic impact of *X. fastidiosa* on olives in the EU. Via a combination of a climatic suitability map to delineate the area of potential spread, a radial range expansion model (Robinet et al., 2012) to estimate the spread and an economic impact assessment module, a bio-economic model was developed to simulate the annual EU olive production area at risk within a time horizon of 50 years, under the assumptions of an uncontrolled infestation and the Apulia region as a point of entry (Fig. 4). The production area at risk was represented by the

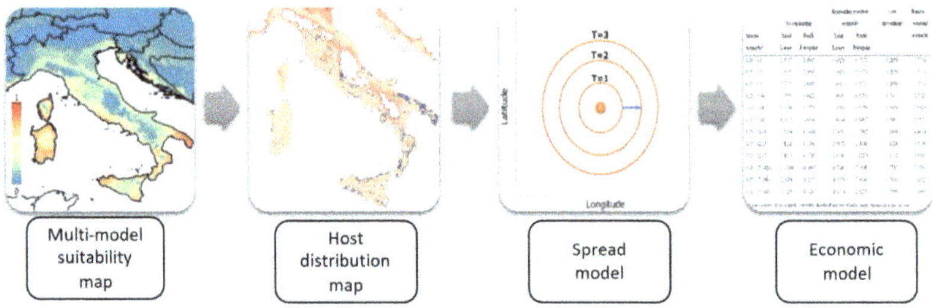

Figure 4 Illustration of bio-economic modeling framework to estimate the economic impact of *X. fastidiosa* in European olive orchards.

olive production sites of Italy, Greece and Spain, accounting for around 95% of European olive oil production.

The net profit flows of the following scenarios were evaluated using the modeling framework:

1) A reference situation in which production is not affected by *X. fastidiosa* to estimate the unaffected profits and replanting investments during the modeled planning horizon.
2) A situation in which infected hectares remain out of production after host death by *X. fastidiosa* for the remainder of the planning horizon (resulting in long-term yield and investment losses).
3) A situation in which infested hectares are replanted after host death by *X. fastidiosa*-resistant varieties, which - once full grown - produce the same yield and quality as susceptible equivalents (resulting in short/midterm yield and investment losses).

The comparison of situations 1 and 2 provides insights into the range of potential economic impacts from a *X. fastidiosa* epidemic, while the comparison of situations 2 and 3 provides insight into the effectiveness of replanting with resistant varieties as an adaptation measure. To account for the uncertainty and data scarcity on *X. fastidiosa* spread, scenario analyses were conducted by utilizing the 5th, 50th and 95th percentiles of the EFSA elicited range distribution to project the epidemic beyond Apulia, the area of entry (Bragard et al., 2019). These settings corresponded to radial dispersal rates varying between 1.10 km/year and 12.35 km/year, with 5.18 km/year as the median value (Fig. 5).

An economic module computed the direct impact by accounting for discounted foregone profits (e.g. yield losses) and losses in investments (e.g. tree dieback) on the olive orchard level using a partial budgeting approach (Schneider et al., 2020).

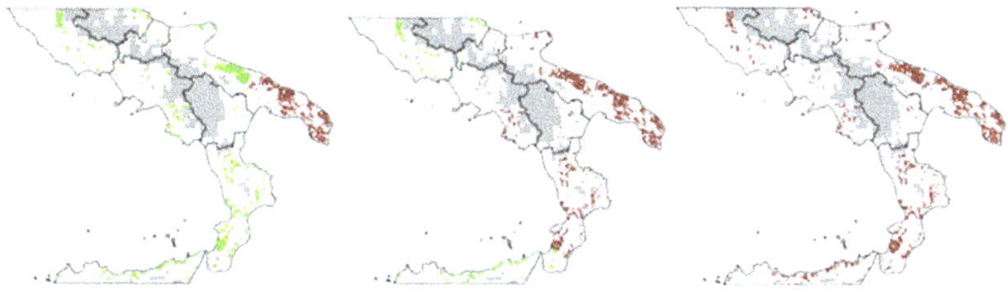

Figure 5 Sensitivity of proportion of invaded olive production area after 50 years to the rate of radial expansion (19–88% of the Italian area of production invaded at t = 50 years). Unsuitable area in gray, not infested olive cells in green, infested olive cells in red.

Table 3 Estimated direct economic impact due to *X. fastidiosa* in olives over 50 years

	Direct economic impact (€ bn)		
Radial dispersal rate (km/year)	No replanting (situation 1-2)	Resistant replanting (situation 1-3)	Benefit replanting (situation 3-2)
1.10	2.38	0.80	1.58
5.18	4.02	1.47	2.55
12.35	7.42	2.91	4.51

Across the evaluated rates of radial range expansion, the potential direct impact over a 50-year period ranged from €2.38 bn to €7.42 bn if replanting with resistant cultivars was not feasible (Table 3). If replanting with resistant cultivars was feasible, the impact ranged from €0.80 bn to €2.91 bn. Moreover, results indicate that, depending on whether replanting was feasible, €0.67-€1.63 bn could be saved over the course of 50 years if the spread was reduced from 5.18 km to 1.1 km per year (50th and 5th percentiles of the elicited spread rate distribution). Based on these results, it was concluded that, irrespective of the uncertainty in the rate of dispersal, the direct impacts over a 50-year planning horizon were sizable and hence warranted actions from regulators and the public. Moreover, results indicated clear economic savings that could be secured if resistant varieties are breed and/or the spread rate is reduced.

For the evaluation of the indirect impacts, a partial equilibrium model was defined on the European olive oil market to estimate changes to consumer and producer welfare from the reduction in supply and the associated increase in price. Within the defined model, Italy, Greece, Spain and the rest of EU (RoEU) and the rest of the world (ROW) were included as separate markets. Spread simulations were used to compute the newly infected share of the production areas in the four included EU countries (Table 4); their outcomes were used to

Table 4 Shares of olive production areas affected after 50 years of uncontrolled spread in dependence on the rate of radial range expansion

Radial dispersal rate (km/year)	Italy (%)	Greece (%)	Spain[a](%)
1.10	19	0	0
5.18	52	5	0
12.35	88	53	0

[a]Independent of the dispersal rate, Spain remained unaffected during the simulated time horizon.

predict the change in supply due to reduced yields and changes in operational costs.

Prices across countries were assumed to be coupled to the world market price but differed in trade costs/tariffs by a constant wedge, which was calibrated on historical market data. Changes in price due to reduced supply were determined by country-specific elasticities of supply and demand. The total indirect impact was computed as the sum of total discounted future welfare losses due to the spread of the disease and associated changes in the equilibrium price and quantity (for more model details, see Schneider et al., 2021).

Table 5 depicts the indirect economic impact for all countries computed by aggregating the difference in welfare between the baseline scenario without *X. fastidiosa* and the different spread rate scenarios over the time horizon of 50 years. Negative values indicate economic losses, while positive values depict gains. Impacts on total welfare (TW) arise through changes in the consumer surplus (CS) and total producer surplus (PS), where the latter consists of changes in the producer surplus for non-affected farmers (PSn) and affected farmers (PSa).

Over the course of 50 years, the median rate of radial expansion of 5.18 km/year resulted in an indirect impact of −€9.91 bn in Italy. This outcome was mostly driven by negative changes in the producer surplus (€8.95 bn) but also due to a decrease in the Italian consumer surplus (€2.71 bn). For Greece, the net result suggested economic benefits due to the epidemic of around €0.50 bn over the course of 50 years. This outcome was driven by a higher producer surplus in Greece (€1.31 bn) due to the increase in prices over the time horizon. Greek consumers were adversely affected by this price increase with their surplus decreasing by around €0.86 bn. Similarly, Spain experienced economic benefits of around €4.70 bn over the course of 50 years through a sizable increase in producer surplus (€7.74 bn) which outweighed the adverse effects on consumers (€3.04 bn). However, Europe-wide the total welfare loss following a median rate of radial range expansion of the epidemic was €5.26 bn. This loss was driven by the aforementioned adverse impacts on consumers (€7.78 bn), which outweighed the benefits that arise to producers due to higher prices (€2.52 bn).

Table 5 Indirect economic impact over 50 years for all countries in € bn given the evaluated spread rate scenarios (RoEU = rest of EU)

Country	CS	PS	PSn	PSa	TS
Radial expansion rate 1.10 km/year					
Italy	−1.54	−4.07	−4.94	0.87	−5.61
Greece	−0.49	0.84	0.84	0.00	0.35
Spain	−1.73	4.34	4.34	0.00	2.61
RoEU	−0.67	0.35	0.35	0.00	−0.31
EU total	**−4.43**	**1.47**	**0.60**	**0.87**	**−2.96**
Radial expansion rate 5.18 km/year					
Italy	−2.71	−7.20	−8.95	1.75	−9.91
Greece	−0.86	1.35	1.31	0.05	0.50
Spain	−3.04	7.74	7.74	0.00	4.70
RoEU	−1.18	0.63	0.63	0.00	−0.55
EU total	**−7.78**	**2.52**	**0.73**	**1.80**	**−5.26**
Radial expansion rate 12.35 km/year					
Italy	−5.35	−12.66	−15.76	3.10	−18.01
Greece	−1.67	0.97	0.31	0.66	−0.70
Spain	−6.06	16.01	16.01	0.00	9.95
RoEU	−2.38	1.33	1.33	0.00	−1.05
EU total	**−15.46**	**5.65**	**1.89**	**3.76**	**−9.81**

The tendencies described earlier were also found for the other two spread rates (i.e. the 1.10 km/year and 15.35 km/year). In general, all European consumers and producers in Italy were adversely affected, whereas Greek, Spanish and other European producers benefited from the reduction in supply and the associated increase in prices. As high rates of radial range expansion resulted in larger reductions in supply, the magnitude of impact increases depending on the rate of radial range expansion.

Around €2.3 bn in welfare could be saved if the spread was reduced from 5.18 km to 1.1 km per year. A reduction of the rate of radial range expansion from 12.35 km to 5.18 km per year would save around €4.55 bn in welfare. The indirect benefits consumers derived from this reduction in the spread were even larger than the aforementioned net benefits. Around €3.35 bn and €7.68 bn in consumer surplus could be saved by reducing the rate of radial range expansion from 5.18 km/year to 1.10 km/year and from 12.35 km/year to 5.18 km/year, respectively.

The analysis of indirect economic impacts extends the insights obtained from the direct impact assessment by highlighting the distribution of impacts among countries and market participants. Results indicated that most of the indirect economic impacts is suffered by Italian producers and consumers.

While consumers in other EU countries also incurred sizable indirect economic impacts due to higher prices for olive oil, producers in non-affected EU territories were expected to benefit from the reduced supply in the EU market and the associated increase in prices. The results showed that spread reduction of *X. fastidiosa* not only benefited Italian producers but more so the European consumers. Utilizing public funds to enact regulatory measures or provide economic incentives to reduce the rate of spread should find general approval by regulators and the public given the sizable economic welfare that can be saved for the European consumers/taxpayers.

6 Conclusion and future trends

Economic research on invasive pests has expanded rapidly in recent decades, focusing on impact assessments to determine whether the introduction of a pest results in unacceptable economic consequences and - in case they are judged to be unacceptable - to identify cost-effective prevention and control measures. Hence, economic impact assessments have been essential in developing and prioritizing mitigation options.

On conducting an economic impact assessment, risk analysts must make choices regarding the most appropriate technique to apply. Depending on the objective of the study (scope, scale, timeline), the appropriate technique needs to provide an acceptable estimate of the economic impacts, while minimizing uncertainty with respect to conclusion. In addition, the technique should use the minimum possible resources in terms of data, skills and time.

Partial budgeting is often considered the default method of choice. This technique estimates the producer-level impacts with low uncertainty and requires modest number of resources. However, its economic scope is limited and does not include indirect effects of pest damage that follow from changes in market prices, supply and demand. In those cases where a significant change in social welfare is expected due to the occurrence of indirect effects, partial equilibrium modeling is the method of choice. Resource requirements for partial equilibrium modeling are, however, higher than requirements for partial budgeting. The I-O and CGE modeling techniques are rarely used in pest impact assessments; they are resource-intensive and their scope usually exceeds what is required for an impact assessment.

Independent of the method applied, a critical limitation in economic assessment modeling remains the ability to model feedback mechanisms between the biological and economic systems. Behavioral responses (adaptation) of stakeholders with respect to the spread of the pest and the political and institutional environment of the endangered area can influence the degree of social acceptability and technical feasibility of disease control measures, which strongly determines the course of the invasion (Fenichel et al., 2011). These inherent feedback mechanisms, therefore, require further research attention (Pacilly et al., 2016).

As invasive plant pests are expanding their ranges due to climate change, new challenges are created with respect to invasive pest management. Early detection could tackle some early invasions, but limited resources make it impossible to monitor every range-shifting pest species. Prioritizing range-shifting invasive pests provides a unique opportunity for early detection and rapid tackling of future problems of pest species before they can establish and spread. Economic impact assessments will be indispensable in establishing these targeted priority pest monitoring lists.

7 Where to look for further information

7.1 Further reading

- A good review on the concepts of economic impact assessments is provided by Epanchin-Niell (2017), Soliman et al. (2015) and Welsh et al. (2021).
- The book New Approaches to the Economics of Plant Health by Oude Lansink (2007) focuses on a number of scientific developments regarding the economic analysis of the impacts of invading pest organisms.
- International standards for conducting a pest risk assessment, including an economic analysis, are defined by ISPM11 'Pest risk analysis for quarantine pests' and ISPM21 'Pest risk analysis for regulated non quarantine pests', which can be inspected by: https://www.ippc.int/en/core-activities/standards-setting/ispms/.

7.2 Key journals

- Journals covering economic impact assessments related to pest invasions include Crop protection, Ecological Economics and Journal of Environmental Management.

7.3 Major international research projects

- PRATIQUE (enhancements of Pest Risk Analysis TechniQUEs), focusing on the development of methods to assess pest risk and analyze management options, https://secure.fera.defra.gov.uk/pratique/.
- POnTE (pest organisms threatening Europe) focusing on the biology, epidemiology, vector ecology and economic impacts of three main pathosystems that threaten strategic crops and natural landscapes in the EU, www.ponteproject.eu/.

- HOMED (holistic management of emerging forest pest and diseases), aiming at the development of a full panel of scientific knowledge and practical solutions for the management of emerging native and non-native pests threatening European forests, www.homed-project.eu/.

8 References

Aukema, J. E., Leung, B., Kovacs, K., Chivers, C., Britton, K. O., Englin, J., Frankel, S. J., Haight, R. G., Holmes, T. P., Liebhold, A. M., McCullough, D. G. and Von Holle, B. (2011), Economic impacts of non-native forest insects in the continental United States, *PLoS ONE*, 6(9), e24587.

Bragard, C., Dehnen-Schmutz, K., Di Serio, F., Gonthier, P., Jacques, M. A., Jaques Miret, J. A., Justesen, A. F., MacLeod, A., Magnusson, C. S., Milonas, P., Navas-Cortés, J. A., Potting, R., Reignault, P. L., Thulke, H. H., van der Werf, W., Vicent Civera, A., Yuen, J., Zappalà, L., Boscia, D., Chapman, D., Gilioli, G., Krugner, R., Mastin, A., Simonetto, A., Spotti Lopes, J. R., White, S., Abrahantes, J. C., Delbianco, A., Maiorano, A., Mosbach-Schulz, O., Stancanelli, G., Guzzo, M. and Parnell, S. (2019), Update of the scientific opinion on the risks to plant health posed by *Xylella fastidiosa* in the EU territory, *EFSA J.*, 17(5), e05665.

Boardman, A. E., Greenberg, D. H., Vining, A. R. and Weimer, D. L. (2006), *Cost Benefit Analysis: Concepts and Practice* (3rd edn.), New Jersey: Pearson Prentice Hall.

Bourguignon, F. and Pereira da Silva, L. (2003), *The Impact of Economic Policies on Poverty and Income Distribution, Evaluation Techniques and Tools*, Washington, DC: The World Bank and Oxford University Press.

Brockerhoff, E. G., Liebhold, A. M., Richardson, B. and Suckling, D. M. (2010), Eradication of invasive forest insects: concepts, methods, costs and benefits, *N. Z. J. For. Sci.*, 40, S117–S135.

Buhle, E. R., Margolis, M. and Ruesink, J. L. (2005), Bang for buck: cost-effective control of invasive species with different life histories, *Ecol. Econ.*, 52(3), 355–366.

Clark, C. W. (1990), *Mathematical Bioeconomics: The Optimal Management of Renewable Resources*, New York: John Wiley and Sons.

Dixon, P. B. and Parmenter, B. R. (1996), Computable general equilibrium modeling for policy analysis and forecasting, in Amman, H. M., Kendrick, D. A. and Rust, J. (Eds.), *Handbook of Computational Economics* (vol.-I), Amsterdam, North Holland: Elsevier science B.V.

European Food Safety Authority (EFSA) (2018), Update of the Xylella spp. host plant database, *EFSA J.* 16(9), e05408

Elliston, L., Hinde, R. and Yainshet, A. (2005), Plant disease incursion management, *Lect. Notes Comput. Sci.* 3415, 225–235.

Epanchin-Niell, R. S. (2017), Economics of invasive species policy and management, *Biol. Invasions* 19(11), 3333–3354.

Epanchin-Niell, R. S., Brockerhoff, E. G., Kean, J. M. and Turner, J. A. (2014), Designing cost-efficient surveillance for early detection and control of multiple biological invaders, *Ecol. Appl.*, 24(6), 1258–1274.

Evans, H. F., Mc Namara, D. G., Braasch, H., Chadoeuf, J. and Magnusson, C. (1996), Pest Risk Analysis (PRA) for the territories of the European Union (as PRA area) on

Bursaphelenchus Xylophilus and its vectors in the genus Monochamus, *EPPO Bull.*, 26(2), 199–249.

FAO (2004), Pest risk analysis for quarantine pests including analysis of environmental risks. *International Standards for Phytosanitary Measures*, Publication No. 11. Rev 1. FAO, R. (IT).

Fenichel, E. P., Castillo-Chavez, C., Ceddia, M. G., Chowell, G., Gonzalez Parra, P. A., Hickling, G. J., Holloway, G., Horan, R., Morin, B., Perrings, C., Springborn,M., Velazquez, L. and Villalobos, C. (2011), Adaptive human behavior in epidemiological models, *Proc. Natl. Acad. Sci. U. S. A.*, 108(15), 6306–6311.

Gatto, P., Zocca, A., Battisti, A., Barrento, M. J., Branco, M. and Paiva, M. R. (2009), Economic assessment of managing processionary moth in pine forests: a case-study in Portugal, *J. Environ. Manage.*, 90(2), 683–691.

Haight, R. G., Homans, F. R., Horie, T., Mehta, S. V., Smith, D. J. and Venette, R. C. (2011), Assessing the cost of an invasive forest pathogen: a case study with Oak Wilt, *Environ. Manage.*, 47(3), 506–517.

Hill, L., Jones, G., Atkinson, N., Hector, A., Hemery, G. and Brown, N. (2019), The £15 billion cost of ash dieback in Britain, *Curr. Biol.*, 29(9), R315–R316.

Hillier, D., Ross, S. A., Westerfield, R. W., Jaffe, J. and Jordan, B. D. (2020), Corporate finance, 4th European Edition, McGraw Hill. ISBN-9781526848086.

Holland, J. (2007), *Tools for Institutional, Political, and Social Analysis of Policy Reform. A source book for Development practitioners,* Washington, D.C.: The World Bank and Oxford University Press.

Holmes, T. P., Aukema, J. E., Von Holle, B., Liebhold, A. and Sills, E. (2009), Economic impacts of invasive species in forests: past, present, and future, *Ann. N. Y. Acad. Sci.*, 1162, 18–38.

Juliá, R., Holland, D. W. and Guenthner, J. (2007), Assessing the economic impact of invasive species: the case of yellow starthistle (Centaurea solsitialis L.) in the rangelands of Idaho, USA, *J. Environ. Manage.*, 85(4), 876–882.

Kehlenbeck, H., Cannon, R., Breukers, A., Battisti, A., Leach, A., Mumford, J. and MacLeod, A. (2012), A protocol for analysing the costs and benefits of phytosanitary measures, *EPPO Bull.* 42(1), 81–88.

Kovacs, K. F., Haight, R. G., McCullough, D. G., Mercader, R. J., Siegert, N. W. and Liebhold, A. M. (2010), Cost of potential emerald ash borer damage in US communities, 2009–2019, *Ecol. Econ.* 69(3), 569–578.

Kriticos, D. J., Leriche, A., Palmer, D. J., Cook, D. C., Brockerhoff, E. G., Stephens, A. E. A. and Watt, M. S. (2013), Linking climate suitability, spread rates and host-impact when estimating the potential costs of invasive pests, *PLoS ONE*, 8(2), e54861.

Leontief, W. (1986), *Input-Output Economics*, New York: Oxford University Press.

MacLeod, A., Head, J. and Gaunt, A. (2004), The assessment of the potential economic impact of Thrips palmi on horticulture in England and the significance of a successful eradication campaign, *Crop Prot.*, 23(7), 601–610.

Mas-Colell, A., Whinston, M. D. and Green, J. R. (1995), *Microeconomic Theory*, New York: Oxford University Press.

McDermott, S. M., Finnoff, D. C. and Shogren, J. F. (2013), The welfare impacts of an invasive species: endogenous versus exogenous price models, *Ecol. Econ.*, 85, 43–49.

Morelli, M., García-Madero, J. M., Jos, Á, Saldarelli, P., Dongiovanni, C., Kovacova, M., Saponari, M., Arjona, A. B., Hackl, E., Webb, S. and Compant, S. (2021), Xylella

fastidiosa in Olive: a review of control attempts and current management, *Microorganisms*, 9(8), 1771.
Miller, R. and Blair, P. (1985), *Input Output Analysis: Foundations and Extensions*, Englewood Cliffs, NJ: Prentice-Hall.
Oude Lansink, A. J. G. M. (Eds.) (2007), New approaches to the economics of plant health, Wageningen UR Frontis Series 20, Springer Netherlands.
Pacilly, F. C. A., Groot, J. C. J., Hofstede, G. J., Schaap, B. F. and van Bueren, E. T. L. (2016), Analysing potato late blight control as a social-ecological system using fuzzy cognitive mapping, *Agron. Sustain. Dev.*, 36(2), 35.
Petucco, C., Lobianco, A. and Caurla, S. (2019), Economic evaluation of an invasive forest pathogen at a large scale: the case of ash dieback in France, *Environ. Model. Assess.*, 25(1), 1–21.
Pimentel, D., Zuniga, R. and Morrison, D. (2005), Update on the environmental and economic costs associated with alien-invasive species in the United States, *Ecol. Econ.*, 52(3), 273–288.
Rich, K. M., Miller, G. Y. and Winter-Nelson, A. (2005), A review of economic tools for the assessment of animal disease outbreaks, *Rev. Sci. Tech.*, 24(3), 833–845.
Robinet, C., Van Opstal, N., Baker, R. and Roques, A. (2011), Applying a spread model to identify the entry points from which the pine wood nematode, the vector of pine wilt disease, would spread most rapidly across Europe, *Biol. Invas.*, 13(12), 2981–2995.
Robinet, C., Kehlenbeck, H., Kriticos, D. J., Baker, R. H. A., Battisti, A., Brunel, S., Dupin, M., Eyre, D., Faccoli, M., Ilieva, Z., Kenis, M., Knight, J., Reynaud, P., Yart, A. and Van der Werf, W. (2012), A suite of models to support the quantitative assessment of spread in pest risk analysis, *PLoS ONE*, 7(10), e43366.
Sadoulet, E. and de Janvry, A. (1995), *Quantitative Development Policy Analysis*, 341–372, Baltimore, MD: The Johns Hopkins University Press.
Schneider, K., Mourits, M., Van der Werf, W. and Oude Lansink, A. O. (2021), On consumer impact from *Xylella fastidiosa* subspecies *pauca*, *Ecol. Econ.*, 185, 107024.
Schneider, K., Van der Werf, W., Cendoya, M., Mourits, M., Navas-Cortés, J. A., Vicent, A. and Oude Lansink, A. (2020), Impact *Xylella fastidiosa* subspecies *pauca* in European olives, *Proc. Natl. Acad. Sci. U. S. A.*, 117(17), 9250–9259.
Sharov, A. A. and Liebhold, A. M. (1998), Model of slowing the spread of gypsy moth (*Lepidoptera: Lymantriidae*) with a barrier zone, *Ecol. Appl.*, 8(4), 1170–1179.
Soliman, T., Mourits, M. C. M., Oude Lansink, A. G. J. M. and Van der Werf, W. (2010), Economic impact assessment in pest risk analysis, *Crop Prot.*, 29(6), 517–524.
Soliman, T., Mourits, M. C. M., Oude Lansink, A. G. J. M. and Van der Werf, W. (2015), Quantitative economic impact assessment of invasive plant pests: what does it require and when is it worth the effort?, *Crop Prot.*, 69, 9–17.
Soliman, T., Mourits, M. C. M., Van der Werf, W., Hengeveld, G. M., Robinet, C. and Oude Lansink, A. G. J. M. (2012), Framework for modelling economic impacts of invasive species, applied to pine wood nematode in Europe, *PLoS ONE*, 7(9), e45505.
Strona, G., Carstens, C. J. and Beck, P. S. A. (2017), Network analysis reveals why Xylella fastidiosa will persist in Europe, *Sci. Rep.*, 7(1), 71.
Welsh, M. J., Turner, J. A., Epanchin-Niell, R. S., Monge, J. J., Soliman, T., Robinson, A. P., Kean, J. M., Phillips, C., Stringer, L. D., Vereijssen, J., Liebhold, A. M., Kompas, T., Ormsby, M. and Brockerhoff, E. G. (2021), Approaches for estimating benefits and costs of interventions in plant biosecurity across invasion phases, *Ecol. Appl.*, 31(5), e02319.

Chapter 10

Developing effective phytosanitary measures to prevent the introduction of invasive insect pests

Alan MacLeod and Dominic Eyre, DEFRA, UK

1 Introduction

Within the literature on invasion biology, the word 'invasive' is used in combination with a variety of other words to form terms such as 'invasive alien species', 'invasive species' and 'invasive pests' (e.g. Richardson and Pyšek, 2004; Mooney, 2005; Paini et al., 2016). As such, there can be confusion over terminology, and the word 'invasive' can be subjective (Colautti and MacIsaac, 2004; Guiaşu, 2016). In addition, 'alien', 'exotic', 'introduced', 'non-indigenous', 'non-native' and 'transboundary' also appear within the literature and have been used over recent decades (Williamson, 1996; National Research Council, 2002; Black and Bartlett, 2020), occasionally interchangeably. 'Harmful organism' and 'pest' can also appear synonymous. In this chapter, we consider an invasive insect pest as an insect species that is introduced to an area where it is not

http://dx.doi.org/10.19103/AS.2022.0113.14

native, and the area becomes overpopulated with the species, resulting in negative impacts (based on Davis and Thompson, 2000).

The number of invasive plant pest introductions is increasing (Chapman et al., 2017; Seebens et al., 2017). The introduction and spread of invasive insect pests into new areas is often attributed to increasing global connectivity (globalization) and to expanding international trade (Work et al., 2005; Liebhold et al., 2012; Boyd et al., 2013; Roy et al., 2014; Garnas et al., 2016). International trade has significant benefits, lifting millions of people out of poverty worldwide (Anderson et al., 2010). Trade expansion has been driven by advances in technology (Thiemann et al., 2012), including the development of faster and more efficient transport systems (Rodrigue et al., 2016), and by the World Trade Organization's (WTO) efforts to liberalize trade. The WTO aims to encourage international trade by building a non-discriminatory trading system. It recognizes that the development of increasing numbers of trade routes has opened new pathways for plant pests and invasive species which have increased the frequency of pest species being introduced to new geographic regions (Ruiz and Carlton, 2003; Caton et al., 2006; McCullough et al., 2006). Indeed, there is a strong positive relationship between new trade and the introduction of new species, and between the growth in trade volumes and the likelihood that species will enter, establish and spread (Dalmazzone, 2000; Tatem, 2009; Dehnen-Schmutz et al., 2010; Roques, 2010; Marini et al., 2011). The increased volume in trade of plant products and plants-for-planting (Hulme, 2009; Chapman et al., 2017) has created new opportunities for invasive species to spread globally and has facilitated the spread of plant pests. These invasive pests can threaten global production of foods and impact on food security (Flood, 2010), fuel and fibre resources, and the health of natural ecosystems (Perrings, 2016); harm environmental resources (Brasier, 2008; Liebhold et al., 2012); disrupt international trade (Alam and Rolfe, 2006); impact regional economies (Wittwer et al., 2005) and lead to significant social impacts (Tripathi et al., 2009). On a global scale, plant pest introductions into high-income countries are most likely to result from trade in plants and plant products. In contrast, pest introductions into least developed countries are correlated with general air travel rather than imports (Early et al., 2016). In addition to the increased movement of goods and people, the changing climate has allowed plant pests to expand the range in which they can establish (Yan et al., 2017; Bebber, 2015). This has altered the area suitable for production of some crops, facilitating the introduction of novel agricultural crops (Gardner et al., 2021) and changes in the selection of trees planted (Roloff et al., 2009).

2 International agreements and trade intervention

Bradshaw et al. (2016) estimated that, across the globe, invasive insects cost a minimum of US$70 billion per year, with associated human health costs exceeding US$6.9 billion per year. Whilst such impacts represent massive costs, they are recognized as gross underestimates. In economic terms, protection against invasive species is a 'public good' (Perrings et al., 2002), and market forces alone will underprovide such protection, warranting the need for government intervention.

Procedures to control the introductions of invasive species act against WTO ambitions for free trade. Nevertheless, recognizing that trade and the movement of goods can provide pathways for the spread of harmful pests, interventions to provide safer trade are permitted by the WTO Agreement on the Application of Sanitary and Phytosanitary Measures, the so-called SPS Agreement. This agreement sets out how governments can apply safety measures to inhibit the spread of threats to food safety, and animal and plant health, while facilitating international trade (Black and Bartlett, 2020). Specifically, Article 5 of the SPS Agreement allows WTO member countries to protect the life and health of their people, plants and animals from pest introductions by applying protective measures to trade pathways. Any measures applied to mitigate the risk of pest introduction should, amongst other things, be scientifically based and the least trade restrictive (Anderson et al., 2001). The SPS Agreement also notes that measures placed on trade should be based on the assessment of risk following international standards for risk assessment. The WTO recognizes the International Plant Protection Convention (IPPC) as the appropriate international standard setting body for the development of risk assessment techniques for plant health.

The IPPC is primarily a plant protection agreement and international treaty, which aims to secure common and effective action to prevent the spread of plant pests and to promote appropriate measures for pest control whilst having minimal impact on trade. It provides a framework for international cooperation, the origins of which can be traced back to the late nineteenth century (Box 1). The WTO SPS Agreement and the IPPC are reciprocal agreements, and although they have different emphases, both require risk-mitigating measures to be science based, align with appropriate risk assessment techniques and have minimal impact on trade. National Plant Protection Organizations (NPPOs) are the official government services responsible for implementing the functions specified within the IPPC such as the preventative and management activities relating to plant pests, including the assessment and management of pest risks, import inspections, the eradication of pest outbreaks and control of invasive pests (Magarey et al., 2009). In terms of preventing pest introductions, a hierarchical approach is often followed, which comprises prevention of the introduction of plant pests as the preferred aim followed by eradication at the earliest possible stage if an introduction has already taken place. As a last resort, if prevention or eradication is not feasible, suppression or containment of the introduced pest population can be employed (Schrader and Unger, 2003).

The Convention on Biological Diversity (CBD) is another international agreement that addresses invasive species. The aim of the CBD, to prevent the introduction of alien species, corresponds in large measure to the aim of the IPPC in relation to plant pests (Lopian, 2005). The IPPC is more focussed on plant pests, whilst the CBD has a broader remit covering invasive alien species in general. Article 8(h) of the CBD notes that Contracting Parties shall, as far as possible and as appropriate, prevent the introduction of, control or eradicate those alien species which threaten ecosystems, habitats or species. To avoid conflicting developments within the IPPC and the CBD regarding invasive alien species and plant pests, the secretariats of the two conventions established a Memorandum of Cooperation and collaborated with a joint work plan. Figure 1 illustrates how the remit of the three international agreements mentioned above overlap and support each other.

There are other interconnected international regulatory frameworks and organizations relevant to the inhibition of the movement of invasive species, including insect pests, such as the International Civil Aviation Organization, which aims to reduce the likelihood that civil air transportation will spread invasive alien species (Shine, 2007; Ormsby and Brenton-Rule, 2017).

Box 1: The origins of international cooperation against invasive insect pests

International cooperation and the development of the first agreement designed to inhibit the spread of an invasive insect plant pest dates back to the accidental introduction of grape phylloxera (*Daktulosphaira vitifoliae* (Fitch) Hemiptera: Sternorrhyncha) into Europe from North America via infested US grapevines in 1862. At the time, the pest was named *Phylloxera vastatrix*. Following its arrival in Europe it quickly spread. The pest forms galls on the mature storage roots of European grapevine varieties, allowing secondary pathogens to infest the plants, ultimately killing them. Over the next few years, the pest spread and caused huge losses in French vineyards (Ordish, 1972). Seven European countries signed the International Convention on Measures to be taken against *Phylloxera vastatrix* in 1878. The convention described early phytosanitary measures such as providing written assurance of the *Phylloxera* free status of host plant material traded internationally and powers to inspect plants and to take action on non-compliant material. The continued international spread of other plant pests and pathogens ultimately led to the development of the International Plant Protection Convention (IPPC), the principal multilateral treaty for international cooperation in plant protection. Its goals are to prevent the spread of pests of plants and plant products whilst promoting appropriate measures for their control (FAO, 2002; MacLeod et al., 2010). As of January 2023, there were 184 contracting parties to the IPPC.

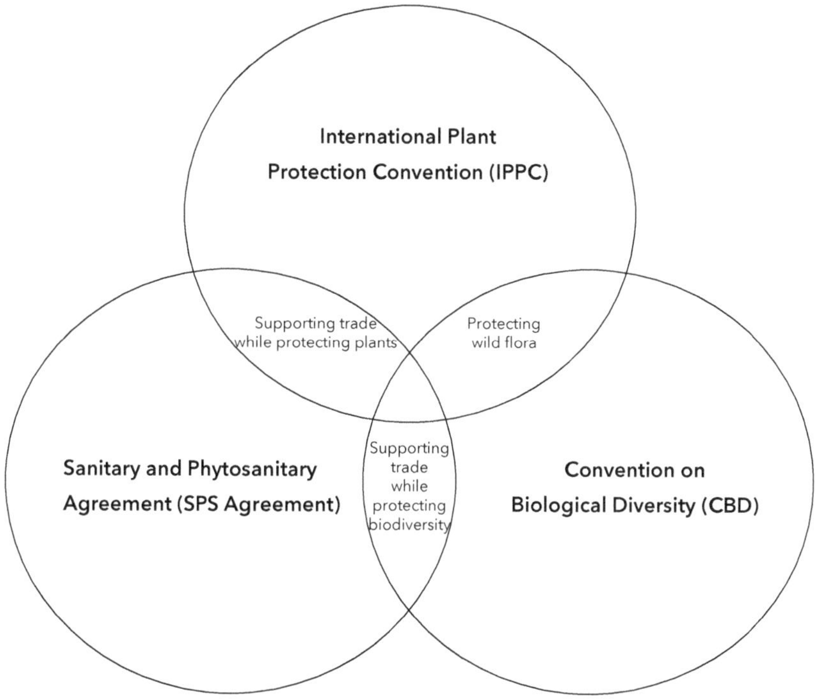

Figure 1 The overlapping aims of key international agreements seeking to act against invasive pests.

3 What are phytosanitary measures?

The term phytosanitary measure(s) relates to any legislation, regulation or official procedure with the purpose of preventing the introduction or spread of a quarantine pest, 'a pest of potential economic importance to the area endangered thereby and not yet present there, or present but not widely distributed and being officially controlled' (IPPC, 2021). The standard way of evaluating whether a potential quarantine pest is of economic importance is through a pest risk analysis (PRA) (Section 4). To qualify as a quarantine pest, a pest must be absent from the PRA area or if present, be under official control and not widely distributed. The implication of a pest being 'officially controlled' is that there will be some statutory (legal) measures in place to reduce the likelihood of introduction or spread.

The goals of phytosanitary measures vary according to circumstances. In the hierarchy of protection, phytosanitary measures are applied to a specified area with the aim of

- preventing the entry or establishment of a pest,
- eradicating a pest,

- suppressing a pest population, or
- containing a pest.

Phytosanitary import requirements are the specific phytosanitary measures established by the NPPO of an importing country concerning consignments moving into that country. In a new or unexpected phytosanitary situation presenting a pest risk, a phytosanitary measure can be established as a matter of urgency; the emergency measure may or may not be a provisional measure. Phytosanitary measures can take place at any stage of the production and shipment of plants or plant products, for example, in the export or import country, or any location in between. Phytosanitary measures can also take many forms, as discussed in Section 5.

4 Phytosanitary measures within the framework of pest risk analysis

To prescribe import requirements in the form of phytosanitary measures that minimize the likelihood of the introduction of plant pests via international trade, and to comply with international agreements, importing countries need to provide technical justification. Such justification is based on an assessment of risk and is very often in the form of a PRA (Schrader and Unger, 2003; IPPC, 2019). PRA incorporates both pest risk assessment and pest risk management and provides the rationale for phytosanitary decision-making, supporting decision-makers striving to protect plant resources from harmful organisms, such as invasive insect pests. The IPPC has established principles that form the basis of the standards and science-based guidelines for the assessment of risk and the development of import requirements (Allen et al., 2017). For example, International Standards for Phytosanitary Measures No. 11 (IPPC, 2019) describes factors to consider when conducting PRA. Typically, a PRA framework consists of three stages. These are:

1 Initiation: A PRA can be initiated due to (i) the identification of a pest that may require phytosanitary measures to protect plants in the PRA area, for example, following horizon scanning (Tepa-Yotto et al., 2021) or (ii) the identification of a pathway that provides the means of entry or spread of a pest, for example, the import of plants from a new trade partner creating a new pathway (MacLeod, 2015) or (iii) the review or revision of phytosanitary policies.
2 Risk assessment: This is the core of a PRA in which biological, economic and other information is gathered and systematically evaluated to inform judgements regarding the likelihood that the pest could enter the PRA area and establish, spread and cause unacceptable economic

impacts. Assessments can be qualitative (MacLeod et al., 2002; Venette and Gould, 2006), semi-quantitative (Favrin and Cree, 2016) or quantitative (Devorshak, 2012; Holt et al., 2017; EFSA Panel on Plant Health, 2018). Qualitative assessments are most common due to lack of information and the inability to quantify all sources of variability and uncertainty, although many assessments involve a mixture of qualitative and quantitative methods (Burgman et al., 2014).

3 Risk management: If the pest being assessed has the characteristics of a quarantine pest and could cause unacceptable risk, phytosanitary measures (Table 1) can be identified, considered and evaluated in terms of efficacy and feasibility to determine whether they could reduce the risk to an acceptable level.

5 Examples of phytosanitary measures

NPPOs are organizations responsible for plant health at a national level. To reduce the likelihood of quarantine pests being introduced, NPPOs can require specific phytosanitary measures. These measures can apply to one or more host plants and can apply to commodities from one or more exporting countries. Broadly, phytosanitary measures can be divided into two categories, although there can be some overlap between these categories. Firstly, plants can be grown in a location where the likelihood of them being infested or infected with a quarantine pest is low. Secondly, a phytosanitary procedure may be required to ensure that the likelihood of commodities being infested or infected is low. Such a procedure could involve a form of monitoring or inspection, or alternatively it could be a treatment of plants when they are growing or postharvest.

Before consignments of regulated plants or plant products are moved across international borders, exporting countries need to issue phytosanitary certificates (PCs). PCs are a physical or electronic declaration by the NPPO of the exporting country that a consignment of plants or plant products meets the requirements of an importing country (IPPC, 2011). Importing countries can require that the NPPOs of exporting countries include a statement of which option they have chosen to meet the phytosanitary requirements of an importing country, and this is called an 'Additional Declaration' (IPPC, 2011). Some phytosanitary measures are described in the remainder of Section 5, and there is a list of phytosanitary measures in Table 1.

5.1 Pest-free countries

One phytosanitary measure that importing countries can require is that the commodities come from a country where that pest is not present. NPPOs are responsible for determining the status (occurrence) of pests in their countries.

Table 1 Example phytosanitary measures

Phytosanitary measure	Stage of process	Rationale and limitations	Likely to be suitable as a stand-alone measure?
Pest-free area or pest-free country	Pre-production	• Only possible in certain export countries • May be expensive to manage	Yes
Pest-free place of production/ Pest-free production site	Pre-production or during production	• There needs to be a structural barrier (e.g. glass of a glasshouse) or a sufficient distance between the production site and pest populations • Likely to require regular inspections	Yes
Production in protected environments	During production	• Likely to be more expensive than outdoor production • Some pests may enter via open doorways, air vents or contamination on people, clothing or equipment	No
Resistant cultivars	During production	• Resistance is unlikely to provide 100% protection	No
Visual inspection of crop	During production	• Pests need to have visible symptoms at the time of inspection for this to be effective • Crop may become infested after inspection	No
Control measures applied to crop	During production	• Plant protection products or biological control agents can be applied to crops • Efficacy of control can be variable and is unlikely to approach 100%	No

Phytosanitary measure	Stage of process	Rationale and limitations	Likely to be suitable as a stand-alone measure?
Specified age/stage of crop	Harvest	• Some pests are likely on particular plant stages, e.g. mature trees as opposed to nursery plants • Risk of moving foliar pests can be removed if plants are moved without leaves or fruit	No
Visual inspection of plants or plant products	Post harvest/pre-export/in importing country	• Efficacy is dependent on the pests being detectable or symptoms being apparent • By inspecting a set number of plants or products, it is possible to ensure that the infestation level of apparent pests is below a set threshold to a set level of confidence	Possibly
Destructive sampling	Postharvest/pre-export/in importing country	• Destructive sampling can be an effective method for detecting pests that are hidden within plants or part plants, e.g. larvae of Tephritidae within fruit	Possibly
Removal of soil	Postharvest /pre-export	• Soil can harbour many invertebrate pests. Removing the soil can reduce the likelihood of these pests being moved with the plants or plant products	No

(*Continued*)

Table 1 (*Continued*)

Phytosanitary measure	Stage of process	Rationale and limitations	Likely to be suitable as a stand-alone measure?
Heat treatment	Pre-export	• Heat treatments such as the treatment described for ISPM 15 can significantly reduce wood boring pests in wood packaging material • Unlikely to be suitable for plants or fresh produce	Yes
Cold treatment	Pre-export/in transit	• Cold treatments can be used to kills eggs and larvae of Tephritidae within fruit • Treatments take a long time and require energy and labour	Yes
Irradiation	Pre-export	• Irradiation can be used to kill pests within fresh food • Establishing facilities for irradiation treatments may be expensive	Yes
Transport conditions	During transit	• Some commodities may become re-infested during transit. Risks can be reduced by enclosing the commodities, only transporting at safe times of year or avoiding travel through infested areas	No
Sampling and testing	On arrival in importing country	• If a reliable testing method is available, sampling and testing a sufficient number of plants can ensure the infestation level is below a certain level with a set level of confidence	Possibly

However, the NPPOs in importing countries may request evidence to support the determination of pest status by an exporting country (IPPC Secretariat, 2021). The amount of evidence required by an importing country to support the declaration of pest status may be influenced by the potential importance of the proposed trade and the threat posed by any pests that could be associated with that trade. The absence of records of a pest in a country may not be sufficient for concluding that a pest is absent (IPPC Secretariat, 2021). However, the absence of records of a pest may be sufficient to justify the pest status for pests that are not considered a major threat or for pests that are considered very unlikely to be present in an area. Some of the factors that might help to increase confidence in a declaration that a country is free of a pest are:

- If the country declaring pest freedom has a very different climate from countries where the pest is known to occur,
- If the country declaring pest freedom is a long way or geographically isolated from countries where the pest is known,
- If there have been recent pest-specific surveys for the pest at an intensity that is likely to detect the pest. Such surveys are likely to be more effective if there are effective tools (e.g. pheromone traps) for detecting the pest in question,
- If the country has a high level of general surveillance which would be likely to detect the pest. When the term 'general surveillance' is used in relation to plant health, it refers to the collation of evidence as to the status of a pest in a country from sources other than pest-specific surveillance. A range of sources can be used for general surveillance such as specimens in museum collections or pest records from members of the public, farmers or consultants (IPPC Secretariat, 2021) and
- If the pest or signs or pest symptoms are likely to be very noticeable when the pest is present. For example, large invertebrates such as adult Anoplophora beetles are very apparent to members of the public (Ciampitti and Cavagna, 2014).

5.2 Pest-free areas

If an area within a country is pest free, the NPPO may consider establishing a 'pest-free area' (PFA) - 'an area in which a specific pest does not occur as demonstrated by scientific evidence and in which, where appropriate, this condition is being officially maintained' (IPPC, 2017a). PFAs can be expensive to establish and maintain; hence, it is important to carry out a cost-benefit analysis before establishing a PFA to ensure that it is likely to offer sufficient benefits to justify the costs (FAO, 2019). The benefits are most likely to be economic but could also be environmental or social. A PFA is more likely to be implemented when there is a valuable export market that will be enabled by the PFA. Other factors to consider include the biology (including the mobility) of the pest, the technology available for surveillance, the capacity of the NPPO and the likely level of stakeholder compliance with measures. It is likely to be more feasible to establish a PFA in an area that has some form of geographical isolation, e.g. on an island, and in an area providing limited pathways for human-assisted spread, e.g. due to lack of roads and rail links with areas where the pest is present. PFAs are also more likely to be viable in areas which are sub-optimal for the establishment of a pest, for example if the climate is unsuitable or favoured hosts are absent. One example is the PFA for South American cucurbit fruit fly, *Anastrepha grandis* (Diptera: Tephritidae) in the states of Rio Grande and Ceará, Brazil. The establishment of this PFA has created an export trade for melons from these areas to the USA (FAO, 2019).

5.3 Pest-free place of production, pest-free production site and protected conditions

If it is not feasible for an NPPO to establish a PFA for a particular pest, an alternative could be the establishment of a smaller, more limited area, designating a particular plant-growing site as pest free. When this status applies to a complete premises or collection of fields run as a single production unit, it is called a pest-free place of production (PFPP). When the status applies to an even smaller area, e.g. a single field or glasshouse, this is called a pest-free production site (PFPS). An example of a PFPS in Ecuador is as part of a trade agreement - Ecuador exports dragon fruit, *Hylocereus undatus* (Haworth) to the USA from PFPSs which use specific monitoring systems for fruit flies (FAO, 2019).

One way of reducing the likelihood of insect pests moving from the wider environment on to cultivated plants is to grow the plants in protected conditions. For this to be effective, the protection provided by a structure must form an effective insect proof barrier for pests, or alternatively it could ensure that it creates an environment that is unsuitable for pests to establish. This is one of the phytosanitary measures introduced by the European Union (EU) in 2008 to reduce the likelihood of *Anoplophora chinensis* (Coleoptera: Cerambycidae), the citrus longhorn beetle, being introduced on plants-for-planting, as described in Section 7.

5.4 Phytosanitary treatments

Phytosanitary treatments are the most direct way of reducing the likelihood that plants or plant products are infested or infected with pests. Treatments can be applied during the production of plant products, as part of processing, pre-export, during transit or on arrival at the place of destination. Phytosanitary treatments can take a variety of forms and be physical or chemical. Treatments are applied to either kill or harm any pests that or present or alternatively reduce the susceptibility of a plant or plant product to future attack.

The intercontinental movement of wood packaging materials (WPM) has been a major pathway for the introduction of non-native forestry pests to new regions. The pathway has been shown to be the cause of numerous outbreaks of *Anoplophora glabripennis* (Coleoptera: Cerambycidae), the Asian longhorn beetle, in Europe and North America (Eyre and Haack, 2017). The risk associated with this trade has been greatly reduced by the introduction of an international phytosanitary standard, commonly referred to as ISPM 15 (IPPC, 2018), that sets out appropriate measures for treating WPM to reduce the likelihood of it harbouring pests (Haack et al., 2014). ISPM 15 provides four options for treating WPM: heat treatment, heat treatment by dielectric heating, fumigation with methyl bromide or fumigation with sulfuryl fluoride. The standard sets out the specifications and describes how wood packaging needs to be marked to show how and where it has been treated.

One relatively simple method for reducing the likelihood of certain pests occuring on deciduous trees is to require that they can only be imported when dormant. It may also be combined with a requirement that the plants are free from soil, i.e. bare rooted, and free from dead leaves and other plant debris. This can reduce the likelihood of introducing insects and other pests which may be sheltering and present in soil, on leaves or amongst plant debris.

Inspection can be used as a phytosanitary measure to check that consignments of plants or plant products for export meet the phytosanitary requirements of the importing country (IPPC, 2005). If the necessary criteria are met, a PC can be issued by the NPPO of the exporting country, indicating that the commodity satisfies the necessary requirements (IPPC, 2011, 2017b). The inspection process can involve a visual inspection of the goods. Inspections by the importing NPPO are essentially an audit of the system and can involve examining the commodity looking for pests and a check of associated documentation including PCs. For large consignments, the inspection is likely to be restricted to a sample of the whole consignment. If pests are detected on a consignment, it may be necessary for inspectors to submit a sample to an official laboratory to obtain an identification and determine whether the pest is regulated or not. Some pests may be present but difficult or impossible to detect by a visual examination.

Destructive sampling can be used as a means of detecting pests that are hard to detect by external examination and is recommended as a means of confirming the presence of fruit flies (Tephritidae) in fruit (EPPO, 2021). The detection of other pests, such as some microscopic invertebrates or plant pathogens, may only be possible by laboratory analysis. Great Britain's import regulations for some hosts of the bacterial plant pathogen *Xyella fastidiosa* include a requirement for sampling and laboratory testing to demonstrate that the proportion of infested plants is less than 5% with 99% confidence. ISPM 31 (IPPC, 2008) provides statistical guidance regarding sampling to achieve such requirements.

In addition to phytosanitary measures being used to reduce the likelihood of pests being introduced on consignments of plants or plant products in trade, they can also be applied to eradicate or contain a pest that has become locally established in a new area. Eradication of plant pests will normally require a surveillance programme and some pest management measures to eliminate the pest. Eradication programmes are often unsuccessful (Carnegie and Nahrung, 2019), and so it can be advisable to assess the likelihood of success with the tools and budget available at an early stage. Commonly, a range of phytosanitary measures need to be employed to achieve eradication, as seen in the eradication of *A. glabripennis* from the UK (Eyre and Barbrook, 2021).

Table 1 provides a summary of types of phytosanitary measures that may be applied in the production and trade chain, together with brief comments on the rationale for applying each measure and limitations of each. A judgement as to whether the phytosanitary measure is suitable as a single, stand-alone, measure that can provide appropriate quarantine protection is also provided. Combinations of measures in Table 1 that are not suitable as a stand-alone measure could be considered within a systems approach (see Section 8).

Having identified pest risks, selected phytosanitary measures can then be set in place, for example on high-risk pathways identified as likely routes for pest entry. Phytosanitary measures should accord with IPPC principles of necessity, managed risk, minimal impact on trade, transparency, harmonization, non-discrimination and technical justification (IPPC, 2006). Surveillance of pathways for a period before and after the implementation of phytosanitary measures can illustrate the effectiveness of the measures (see case studies explained in Sections 6 and 7).

The minimum phytosanitary measure applied to any pest is the declaration in NPPO phytosanitary regulations that the organism is a quarantine pest. This declaration prohibits both the entry of the pest in an isolated state and the import of consignments infested by the pest. It also enables an NPPO to act against the pest if it is detected during import inspections or when an outbreak is detected. Action can be taken against pests that are not listed as quarantine pests on an emergency and precautionary basis for a short time (Cannon et al., 1999) but can only be maintained against a pest once an assessment of risk has been conducted, the conclusion of which is that the pest satisfies the definition of being a quarantine pest.

6 Case study: strengthening phytosanitary measures against *Bemisia tabaci* and *Liriomyza huidobrensis* in the UK

Plants and plant products can be broadly grouped into three categories: (i) those that present a very low risk of introducing pest species and are imported without phytosanitary controls, (ii) those that can present some risk, but the risk is mitigated if phytosanitary measures are applied, such as requiring official inspection prior to export to ensure the commodity complies with phytosanitary regulations and (iii) those commodities that present such a high risk of introducing pests that the only satisfactory phytosanitary measure is to prohibit the import of the commodity entirely.

A proportion of material is inspected on arrival at airports and docks. Pest finds are recorded and data compiled and analysed to identify emerging risks. If a quarantine pest is found in a consignment, the plant material can be destroyed. Around the world, insects are the main taxonomic group encountered during import inspections (Turner et al., 2021). During the early years of the twenty-first century, analysis of insect pest interceptions on cut flowers by the UK NPPO revealed an emerging threat from the quarantine pests *B. tabaci* (Gennadius) (Hemiptera: Aleyrodidae), and *L. huidobrensis* (Blanchard) (Diptera: Agromyzidae) via imported *Solidago* and *Gypsophila* cut flowers. At that time, the cut flowers were not regulated by phytosanitary legislation and could be imported freely, given that they were perceived to be relatively short-lived, low-risk, commodities. The likelihood of pests transferring from the cut flowers to growing hosts was judged low. However, some producers of cut flowers supplement their own production by importing cut flowers and so increasing the likelihood of contaminating pests transferring to growing plants and establishing a population.

B. tabaci, the tobacco whitefly (Fig. 2), is a highly polyphagous pest with a wide range of protected ornamental and vegetable crops. It can introduce and vector over 120 plant viruses, almost none of which are present in the UK (Jones, 2003). *B. tabaci* is widely distributed in Africa, Asia, Oceania and Central America and also in several South American countries and in the southern parts of North America. It also occurs in Europe but is not established in the UK (EPPO global database). In the UK, *B. tabaci* could only survive on protected crops such as salad crops, including tomatoes, cucumbers and sweet peppers, and on a range of ornamentals grown in glasshouses. Whilst there have been outbreaks in UK glasshouses, they have always been eradicated.

L. huidobrensis, the serpentine or South American leaf miner (Fig. 3), originates from South and Central America but has spread to North America, Africa, Asia, Oceania and parts of Europe. The pest is a polyphagous species. In South America, it is a major pest of potatoes, although in other regions, it

Figure 2 *Bemsia tabaci* (adult, 1 mm long).

Figure 3 *Liriomyza huidobrensis* (adult, 2-3 mm long).

feeds on a range of vegetables and glasshouse grown ornamentals such as cucumbers and chrysanthemums; in the Mediterranean region, it feeds outdoors on lettuce and beet. Larvae mine and develop in the leaves and petioles of hosts, and the resulting damage can severely reduce the quality of ornamental plants. Yield can be lowered in vegetables. Whilst there have been outbreaks in UK glasshouses, they have always been eradicated.

The analysis of pest interceptions revealed that *B. tabaci* and *L. huidobrensis* could be found on imported *Solidago* and *Gypsophila* and led to the commodities being regulated, whereby they were required to be subject to official plant health inspection prior to export and should only be exported if found free from such quarantine pests. By comparing the percentage of *Solidago* and *Gypsophila* consignments contaminated in the months before and after phytosanitary measures were required, the effectiveness of the phytosanitary measure could be demonstrated. Following the requirements for official inspection prior to export, the mean level of consignments found to be contaminated with either of the quarantine pests during UK import inspections fell from ~7% to 3%, a statistically significant reduction (MacLeod et al., 2005).

7 Case study: EU measures against *Anoplophora chinensis*

A. chinensis (Fig. 4) is a species of longhorn beetle that is native to Asia. Most records of *A. chinensis* come from China, Japan and South Korea, but it has also been reported from Malaysia, North Korea, Vietnam, Taiwan, the Philippines, Indonesia and Myanmar (Lingafelter and Hoebeke, 2002). It is a pest of deciduous trees, especially fruit trees, notably citrus trees in Japan and China. Following the discovery of an outbreak of *A. chinensis* in Lombardy, Italy, in 2000, it has become one of the most important invasive plant pests in Europe. Interceptions have been discovered in 11 European countries, and in addition to Italy, outbreaks have been detected in Croatia, France and the Netherlands, with outbreaks persisting in Italy and Croatia (EPPO global database). In the UK, there were 12 confirmed finds of *A. chinensis* associated with *Acer palmatum* in 2008. The infested plants were part of a consignment of 46 000 *A. palmatum* imported from China, via the Netherlands and distributed around the UK by a mail-order company based in Guernsey. No outbreaks were detected after subsequent monitoring and publicity (Eyre et al., 2010).

The outbreak in Lombardy detected in 2000 was found at Parabiago, near Milan, by fieldwork students from an agricultural university monitoring invasive species in the surroundings of companies importing plants, especially bonsais from Asia. The demarcated areas covered approximately 357 km^2 in 2020

Figure 4 *Anoplophora chinensis* (Adults are 21 mm to 37 mm long, excluding antennae).

(Lombardy Region, 2020), and by 2017 over 50 000 trees had been felled to contain the infested area (Hérard and Maspero, 2019).

Figure 5 shows records of interceptions of *Anoplophora* spp. on consignments of plants recorded in EUROPHYT, a European Commission web-based network and database specifically concerned with plant health information. Between 2005 and 2019 there were 35 interceptions of *Anoplophora* spp. on plants-for-planting in the EU, of which 23 were diagnosed as *A. chinensis*. Given no other *Anoplophora* spp. was identified on plants-for-planting over this period, it is very likely that the remaining 12 interceptions were also *A. chinensis*. *Acer* spp. were recorded as the host for 32 out of 35 of the interceptions, and when the Acers were identified to species it was *A. palmatum* (14 interceptions), *Acer buergerianum* (2 interceptions) and *Acer rubrum* (1 interception). The exporting country was China for 33 interceptions, Japan (2 interceptions) and unknown (1 interception). These data showed that introducing effective phytosanitary measures for *Acer* spp. from China could have a significant impact on the number of interceptions. There are 18 interception records on the EUROPHYT database which appear to be an entry error, and these have been excluded from this analysis.

Following the numerous interceptions of *Anoplophora* spp. on plants-for-planting in Europe between 2000 and 2008, the EU agreed and published measures to reduce the risk of further incursions in November 2008 (European Commission, 2008). For countries where *A. chinensis* was known to be present, the measures included requirements that plants should come from a registered site, with two annual official inspections, and they should either be treated to prevent infestation or be grown within complete physical protection. There was also an obligation for the plants to be inspected before export. Despite the implementation of these requirements, there were further interceptions of *Anoplophora* spp. on plants from China in 2009 and 2010. An EU mission to

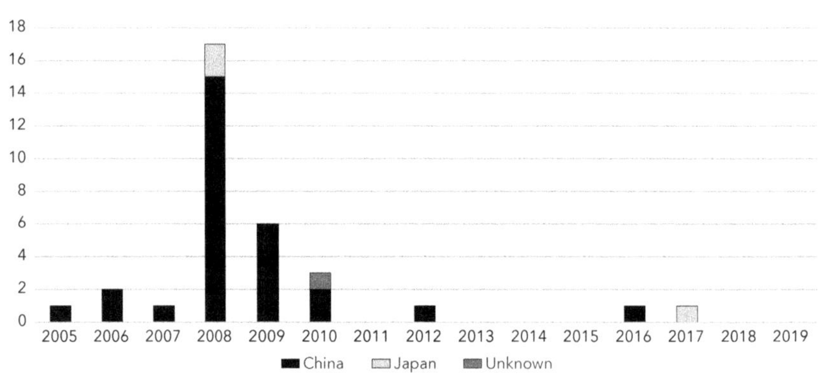

Figure 5 The annual number of interceptions of *Anoplophora* spp. on plants for planting in the EU from 2005 to 2019 and the origin of the infested plants.

China in 2009 concluded that the requirements of the EU's 2008 legislation had not been fully implemented. This demonstrated that phytosanitary measures created by importing countries may be ineffective if they are not followed in exporting countries. The conclusions of the EU mission and the further interceptions led to a 2-year prohibition of *Acer* spp. from China in 2010. Once this prohibition came to an end in 2012, imports of *Acer* spp. from China were subjected to new requirements for pre-export destructive sampling and destructive sampling on arrival in the EU (European Commission, 2012). Although it is not possible to statistically analyse the number of interceptions of *Anoplophora* spp. shown in Fig. 5 due to varying inspection rates and volume of trade between years, the results do indicate that the measures were relatively successful. In the 7 years leading up to 2012, there were 32 interceptions of *Anoplophora* on plants-for-planting in the EU, whereas in the 7 years after 2012, there were just two interceptions.

8 Combining measures: the systems approach

The integration of different phytosanitary measures, at least two of which act independently, and which cumulatively lower the pest threat to an appropriate level, is known as a 'systems approach' (IPPC, 2017c). The importance of independence is that if one measure fails, it will not affect the efficacy of other independent measures, although the overall performance of the system may be impaired (FAO/IAEA, 2011). Systems approaches are appropriate when a single-point risk management measure (i) does not satisfy phytosanitary import requirements alone or (ii) has some deleterious impact on the commodity, such as phytotoxicity or (iii) is not cost-effective or (iv) is overly trade restrictive and/or not feasible (Evans et al., 2012). Support for developing a systems approach is provided if (i) an effective system is already in place for a similar commodity with similar pests, (ii) the effectiveness of individual measures can be assessed, (iii) relevant growing, harvesting, packing, transport and distribution practices are known and standardized, (iv) the prevalence of target pests is known and can be monitored and (v) the system is cost-effective (Evans et al., 2012). Systems approaches arose from the need to support the increasing growth of horticultural exports in the mid-1980s. At the time, most quarantine treatments focused on postharvest treatment, with importers assuming the commodity would otherwise be infested by pests (Jang and Moffitt, 1994). In contrast, systems approaches enable pre- and postharvest measures to be used in combination to provide pest risk management. Historically, a combination of measures forming a systems approach would focus on a production and trade chain and could be agreed and used as the basis of a bilateral trade agreement. The methods for pest management were fixed in the agreement. It is important to consider systems approaches when managing pest risks because the integration of multiple measures might be less trade restrictive than other stand-alone phytosanitary measures, especially if the alternative is prohibition (IPPC, 2017c). There have been recent calls to include control points within systems approaches so that a system can be dynamic and respond to changes within the system (van Klinken et al., 2020), making the systems approach more flexible and effective. The ability to adjust within a system, either during real-time trade or after detection of an emerging pest risk, is a major advantage of the systems approach (Quinlan et al., 2021). Many NPPOs still focus on a single-point or stand-alone risk management measure. The uptake

Table 2 Countries exporting horticultural commodities to the USA with systems approaches in place to prevent the spread of pests

Exporting country	Commodity
Argentina	Citrus
Australia	Mango
Belize	Dragon fruit
Chile	Citrus, custard apple, pomegranate, Solanaceae
China	Pome fruit
Colombia	Avocado, gooseberry, papaya
Costa Rica	Dragon fruit, papaya, Solanaceae
Ecuador	Dragon fruit, papaya, Solanaceae
El Salvador	Dragon fruit, Papaya, Solanaceae
France	Solanaceae
Guatemala	Dragon fruit, papaya, Solanaceae
Honduras	Dragon fruit, papaya, Solanaceae
Japan	Citrus, persimmon
Korea	Citrus
Mexico	Avocado, citrus
Morocco	Solanaceae
Nicaragua	Dragon fruit, papaya, Solanaceae
New Zealand	Persimmon
Panama	Dragon fruit, papaya, Solanaceae
Peru	Avocado, citrus, Solanaceae
Philippines	Banana
South Korea	Pome fruit
Spain	Avocado, citrus, Solanaceae, stone fruit
Uruguay	Citrus
Vietnam	Mango
Western Sahara	Solanaceae

Source: Adapted from: van Klinken et al. (2020).

of systems approaches is hampered because there is not an agreed framework for understanding how independent measures individually contribute to lower risk, and there is no agreed method for comparing pest risk from trade using a systems approach to trade using a stand-alone measure although such methods have been proposed (e.g. Jamieson et al., 2014; Quinlan et al., 2016; Holt et al., 2018). The decision regarding the acceptability of systems approach lies with the importing country.

van Klinken et al. (2020) examined protocols for systems approaches for horticultural crops and found that the USA had protocols in place for a variety of commodities being imported from a range of countries (Table 2). van Klinken et al. (2020) also identified systems approaches facilitating horticultural imports into China, the EU, Japan, New Zealand and Thailand, but they were confined to a much smaller range of commodities from many fewer countries, compared to the USA. Fruit flies were often identified as pests of concern, and most protocols included phytosanitary measures within a systems approach to prevent their entry.

When developing effective phytosanitary measures to prevent the introduction of plant pests, it is desirable that options be provided to trade partners to allow some flexibility. Alternative measures should at least provide equivalent levels of protection (IPPC, 2017c). As per the WTO SPS Agreement and the IPPC principle of minimal interference, a selected (stand-alone) phytosanitary measure, or a combination of measures (systems approach), should be the least stringent and cause the minimum impact on trade whilst achieving the desired necessary level of phytosanitary protection.

9 Challenges

When developing effective phytosanitary measures to prevent the introduction of invasive insect pests, pest risk managers and phytosanitary policy makers face a variety of challenges, a few of which are outlined below:

- **Trade and biosecurity**: There are challenges and tensions between maintaining biosecurity protection whilst participating in global trade (Maye and Chan, 2021), and there have been calls to significantly improve inspection, detection and diagnosis efforts on traded plant material and limit or prohibit the import of some types of plants such as large trees for 'instant gardens' (Brasier, 2008). Nevertheless, opportunities exist for collaborative working between biosecurity natural scientists, economists and social scientists to improve biosecurity systems and design phytosanitary measures to prevent the introduction of invasive pests as plant protection authorities seek to develop integrated decision-making within an interdisciplinary framework (MacLeod et al., 2016).

- **Volume of materials in trade**: The volume of international trade in goods relevant to plant health is increasing, and this increase brings an associated increase in the likelihood of alien pests being transported. For example, using inflation adjusted figures from FAOSTAT, Eschen et al. (2015) reported that the value of plants imported into the EU had increased by 60% over the previous 15 years. Mwebaze et al. (2010) showed that imports of fresh fruit and vegetables increased by around 60% between 1988 and 2006, and there was a linear increase in the number of pests intercepted. NPPOs have not seen a commensurate increase in resources to conduct import inspections (Liebhold et al., 2012), so face the challenge of developing smarter ways of detecting invasive insects and other pests.
- **Loss of chemical treatments,** e.g. methyl bromide. Methyl bromide is a highly effective broad-spectrum biocidal fumigant used to treat soils, commodities, WPM and physical structures. It is effective against insect pests, nematodes, weeds, pathogens and rodents. It was previously widely used as a quarantine fumigant. Treatments typically took 2–3 h to kill insect pests. However, methyl bromide contributes to stratospheric ozone depletion. Following the implementation of the Montreal Protocol on Substances that Deplete the Ozone Layer (United Nations Ozone Secretariat, 2000), the use of methyl bromide in quarantine dramatically reduced, and researchers began to seek alternatives. Newer fumigants such as propylene oxide, sulfuryl fluoride and ethyl formate are being investigated for quarantine treatments and are beginning to be used, but to date, no alternative fumigant has been widely adopted which offers a comparable level of efficacy and practicality.
- **Difficulty in detecting pests during inspection**: The IPPC recognizes that there are difficulties in detecting pests during inspections. For example, inspection efficacy at entry is limited by the fact that latent infections cannot be detected without testing and pests within plants may require destructive sampling before they can be detected. A measure of inspection efficacy should be included when selecting sample size (IPPC, 2005, 2008). However, there has been limited research to determine the efficacy of inspection resulting in a paucity of data (MacLeod, 2015). One notable scientific paper by Liebhold et al. (2012) estimated that 72% of infested plant shipments passed through US ports undetected.
- **New pests**: Many invasive insect pests are well studied and may be recognized internationally as serious plant pests which cause significant economic or environmental damage. Agreed and established phytosanitary measures may be able to inhibit their spread, but new and emerging invasive insect pests can provide significant challenges for risk managers and policy makers who need to design and implement phytosanitary measures against them. For example, some invasive insects

are not recorded as pests in their native range and may not be known to science until they are introduced to a new country. For example, the European wood wasp Sirex noctilio Fabricius (Hymenoptera: Siricidae) is not a pest in its native European forests but became a serious pest of conifers in New Zealand, Australia, South Africa and South America following its arrival, establishment and spread in the Southern Hemisphere (Ayres et al., 2014). Similarly, the red turpentine beetle, Dendroctonus valens (LeConte) (Coleoptera: Scolytidae) is not a serious pest in North America but became a major pest in China where it is estimated to have killed 10 million pine trees since its accidental introduction on imported logs in the 1980s (Garnas et al., 2012; EPPO, 2014). Several other plant pests now subject to phytosanitary measures only became recognized as harmful once they escaped from their geographical centres of origin and caused impacts in other parts of the world (Brasier, 2005). Identifying which species will become a problematic pest and developing effective phytosanitary measures to prevent their introduction is challenging if the organism is little studied in its native range due to its lack of economic significance. This can lead to phytosanitary measures being reactive, coming into effect only after the introduction or outbreak of a pest and once a pest problem has been identified (Brasier, 2008) and escalated to a point where there is political willingness to spend on measures (Ward, 2016).

- **Measuring efficacy of measures inhibiting entry**: Phytosanitary measures are often designed to prevent the entry of pests into a country or a specific part of a country. However, the paucity of detailed international pest interception data impedes the ability of researchers to conduct thorough analyses of the effectiveness of changes to such phytosanitary policies designed to inhibit entry (Haack et al., 2014). This could be addressed if NPPOs recorded and reported inspection and sampling effort in relation to the number of consignments imported and inspected; such information would provide data that could greatly help the interpretation of pest interceptions. Moreover, it would better inform phytosanitary risk management and policy decision-making and would allow the efficacy of the changes in phytosanitary policies to be measured (MacLeod et al., 2005; MacLeod, 2015).
- **Lack of resources**: It can be challenging for NPPOs to implement effective phytosanitary measures due to their limited available resources, particularly in the global south. For example, the fast spread of Tuta absoluta (Meyrick) (Lepidoptera: Gelechiidae) across Africa could partly be due to the challenges faced by NPPOs to adequately implement phytosanitary measures (Tonnang et al., 2015). Such challenges could be partly addressed through regional and global collaboration and support

for countries that lack sufficient resources (Ormsby and Brenton-Rule, 2017). There may also be a role for citizen science to detect and report invasive species (Gallo and Waitt, 2011).

- **Applying systems approaches**: The interest in applying systems approaches is growing but is hampered by the perceived complexity of determining the combined consequences of multiple measures when the individual efficacy of each measure is not well known. For example, there is little variability in the efficacy of stand-alone measures such as fumigation when applied under controlled conditions. In contrast, some measures used within a systems approach are applied under variable field conditions, leading to greater variability of efficacy, creating challenges when designing a systems approach (van Klinken et al., 2021).
- **Climate change**: Global climate change is leading to changes in the scale and nature of the threat posed by non-native pests. A changing climate can lead to new crops being planted in areas which were previously climatically unsuitable for those crops. In the UK, the vineyard area increased by 148% between 2004 and 2013. This increase has been partly attributed to climate change increasing temperatures (Nesbitt et al., 2016). The changing suite of crops grown in a country will change the suite of pests that cause a threat it. This creates challenges for NPPOs who need to maintain an awareness of changing crops and growing practices and of the (new) pests that could threaten them when developing phytosanitary measures to prevent the introduction of invasive pests.

10 Conclusion

Globalization, partly due to increasing international trade, facilitates the spread of invasive insect pests. Recognizing the harm that such pests can cause to plants, countries follow international standards to assess pest risk and use a variety of phytosanitary measures to inhibit the introduction and spread of quarantine plant pests. In accordance with these standards, phytosanitary measures should only be applied when necessary and to a degree that sufficiently manages the risk whilst having the minimum impact on trade. A single, stand-alone phytosanitary measure may be sufficient, or a combination of measures may be necessary within a systems approach, to manage the pest risk. Further work is needed to reduce many of the challenges that risk managers and policy makers deal with when selecting phytosanitary measures. For example, technologies to improve pest detection and mechanisms to better evaluate the efficacy of systems approaches would greatly support the decision makers. The growth of globalization cannot be curbed but by assessing pest risks and applying appropriate phytosanitary measures NPPOs can minimize the likelihood that invasive insect pests are introduced and thereby protect their plant resources.

11 Where to look for further information

Although a rather specialist subject, there are a number of books, websites and research centres that readers can access to gain further insights and knowledge in relation to phytosanitary matters, pest risk analysis and phytosanitary measures, to prevent the introduction of invasive insect pests.

11.1 Books

- Devorshak, C. (Ed.) (2012). Plant Pest Risk Analysis Concepts and Application. CABI, Wallingford. 296pp.
- Ebbels, D. L. (2003). Principles of Plant Health and Quarantine, CAB International, Wallingford, UK. 203pp.
- Gordh, G. and McKirdy, S. (2014). The Handbook of Plant Biosecurity: Principles and Practices for the Identification, Containment and Control of Organisms that Threaten Agriculture and the Environment Globally. Springer, Netherlands. 723pp.
- Quinlan, M. M., Mengersen, K., Mumford, J., Leach, A., Holt, J. and Murphy, R. (Eds), (2016). Beyond compliance: a production chain framework for plant health risk management in trade. Chartridge Books Oxford.

11.2 Websites

- European and Mediterranean Plant Protection Organization (EPPO) http://www.eppo.int.
- International Plant Protection Convention (IPPC) https://www.ippc.int/en/.
- International Pest Risk Research Group (IPRRG) https://pestrisk.org/.
- USDA Plant Epidemiology and Risk Analysis Laboratory (USDA PERAL) https://www.aphis.usda.gov/aphis/ourfocus/planthealth/import-information/commodity-import-approval-process/risk-assessment/analysis.

11.3 Research centres

- Centre of Excellence for Biosecurity Risk Analysis (CEBRA) https://cebra.unimelb.edu.au/home.
- CSIRO https://www.csiro.au/en/research/animals/pests.
- Imperial College London, Centre for Environmental Policy https://www.imperial.ac.uk/environmental-policy/.

- USDA Agricultural Research Service https://www.ars.usda.gov/crop-production-and-protection/crop-protection-and-quarantine/.
- Wageningen University https://www.wur.nl/en/Research-Results/Research-Institutes/plant-research/Field-crops/Plant-Health.htm.

12 References

Alam, K. and Rolfe, J. (2006). Economics of plant disease outbreaks. *Agenda: A Journal of Policy Analysis and Reform* 13, 133–146.

Allen, E., Noseworthy, M. and Ormsby, M. (2017). Phytosanitary measures to reduce the movement of forest pests with the international trade of wood products. *Biological Invasions* 19(11), 3365–3376.

Anderson, K., Cockburn, J. and Martin, W. (2010). *Agricultural Price Distortions, Inequality and Poverty*. World Bank, Washington, DC.

Anderson, K., McRae, C. and Wilson, D. (2001). Introduction. In: Anderson, K., McRae, C. and Wilson, D. (Eds.), *The Economics of Quarantine and the SPS Agreement*. Centre for International Economic Studies, Adelaide, Australia.

Ayres, M. P., Pena, R., Lombardo, J. A. and Lombardero, M. J. (2014). Sirex noctilio. *PLoS ONE* 9(3), e90321. https://doi.org/10.1371/journal.pone.0090321.

Bebber, D. P. (2015). Range-expanding pests and pathogens in a warming world. *Annual Review of Phytopathology* 53, 335–356.

Black, R. and Bartlett, D. M. F. (2020). Biosecurity frameworks for cross-border movement of invasive alien species. *Environmental Science and Policy* 105, 113–119.

Boyd, I. L., Freer-Smith, P. H., Gilligan, C. A. and Godfray, H. C. (2013). The consequence of tree pests and diseases for ecosystem services. *Science* 342(6160), 1235773.

Bradshaw, C. J., Leroy, B., Bellard, C., Roiz, D., Albert, C., Fournier, A., Barbet-Massin, M., Salles, J. M., Simard, F. and Courchamp, F. (2016). Massive yet grossly underestimated global costs of invasive insects. *Nature Communications* 7(1), 12986.

Brasier, C. M. (2005). Preventing invasive pathogens: Deficiencies in the system. *The Plantsman* 4, 54–57.

Brasier, C. M. (2008). The biosecurity threat to the UK and global environment from international trade in plants. *Plant Pathology* 57(5), 792–808.

Burgman, M., Roberts, B., Sansford, C., Griffin, R. and Mengersen, K. (2014). The role of pest risk analysis in Plant Biosecurity. In: Gordh, G. and McKirdy, S. (Eds.), *The Handbook of Plant Biosecurity*. Springer, Dordrecht, pp. 235–267.

Cannon, R. J. C., Pemberton, A. W. and Bartlett, P. W. (1999). Appropriate measures for the eradication of unlisted pests. *EPPO Bulletin* 29(1–2), 29–36.

Carnegie, A. J. and Nahrung, H. F. (2019). Post-border forest biosecurity in Australia: Response to recent exotic detections, current surveillance and ongoing needs. *Forests* 10(4), 336.

Caton, B. P., Dobbs, T. T. and Brodel, C. F. (2006). Arrivals of hitchhiking insect pests on international cargo aircraft at Miami International Airport. *Biological Invasions* 8(4), 765–785.

Chapman, D., Purse, B. V., Roy, H. E. and Bullock, J. M. (2017). Global trade networks determine the distribution of invasive non-native species. *Global Ecology and Biogeography* 26(8), 907–917.

Ciampitti, M. and Cavagna, B. (2014). Public awareness: A useful tool for the early detection and a successful eradication of the longhorned beetles *Anoplophora chinensis* and *A. glabripennis*. *EPPO Bulletin* 44(2), 248–250.

Colautti, R. I. and MacIsaac, H. J. (2004). A neutral terminology to define 'invasive' species. *Diversity and Distributions* 10(2), 135–141.

Dalmazzone, S. (2000). Economic factors affecting vulnerability to biological invasions. In: Perrings, C., Williamson, M. and Dalmazzone, S. (Eds.), *The Economics of Biological Invasions*. Elgar Publishing, Cheltenham, pp. 17–30.

Davis, M. A. and Thompson, K. (2000). Eight ways to be a colonizer; two ways to be an invader: A proposed nomenclature scheme for invasion ecology. *Bulletin of the Ecological Society of America* 81(3), 226–230.

Dehnen-Schmutz, K., Holdenrieder, O., Jeger, M. J. and Pautasso, M. (2010). Structural change in the international horticultural industry: Some implications for plant health. *Scientia Horticulturae* 125(1), 1–15.

Early, R., Bradley, B. A., Dukes, J. S., Lawler, J. J., Olden, J. D., Blumenthal, D. M., Gonzalez, P., Grosholz, E. D., Ibañez, I., Miller, L. P., Sorte, C. J. and Tatem, A. J. (2016). Global threats from invasive alien species in the twenty-first century and national response capacities. *Nature Communications* 7(1), 12485.

EFSA Panel on Plant Health. (2018). Scientific opinion on the pest risk assessment of *Spodoptera frugiperda* for the European Union. *EFSA Journal* 16(8), 5351.

EPPO. (2014). *Dendroctonus valens: An Invasive Forest Pest in China*. EPPO Reporting Service, 085.

EPPO. (2021). PM 3/92 (1) Consignment inspection of fresh fruit and vegetables for fruit flies. *EPPO Bulletin* 51(3), 406–417.

Eschen, R., Gregoire, J. C., Hengeveld, G. M., de Hoop, B. M., Rigaux, L. and Potting, R. P. J. (2015). Trade patterns of the tree nursery industry in Europe and changes following findings of citrus longhorn beetle, *Anoplophora chinensis* Forster. *NeoBiota* 26, 1–20.

European Commission. (2008). Commission decision of 7 November 2008 on emergency measures to prevent the introduction into and the spread within the Community of *Anoplophora chinensis* (Forster). Official Journal of the European Union, L 300/36 11.11.2008.

European Commission. (2012). Commission Implementing Decision of 1 March 2012 as regards emergency measures to prevent the introduction into and the spread within the Union of *Anoplophora chinensis* (Forster). Official Journal of the European Union L64(38) 3.3.2012.

Evans, H., Brokerhoff, E., Baker, R. and Eschen, R. (2012). What role can the systems approach play in preventing the introduction of invasive alien species. In: *Meeting of IUFRO Working Group on Invasive Species and International Trade*, Tokyo, Japan, pp. 10–16.

Eyre, D. and Barbrook, J. (2021). The eradication of Asian longhorned beetle at Paddock Wood, UK. *CABI Agriculture and Bioscience* 2(1), 12.

Eyre, D., Cannon, R., McCann, D. and Whittaker, R. (2010). Citrus longhorn beetle, *Anoplophora chinensis*: An invasive pest in Europe. *Outlooks on Pest Management* 21(4), 195–198.

Eyre, D. and Haack, R. A. (2017). Invasive cerambycid pests and biosecurity measures. In: Wang, Q. (Ed.), *Cerambycidae of the World: Biology and Management*. CRC Press, Boca Raton, pp 563–607.

FAO. (2002). *Guide to the International Plant Protection Convention*. Food and Agriculture Organization, Rome, p. 20.

FAO. (2019). *Guide for Establishing and Maintaining Pest Free Areas*. Published by FAO on Behalf of the Secretariat of the International Plant Protection Convention (IPPC), Rome. Available at: https://www.fao.org/3/ca5844en/CA5844EN.pdf.

FAO/IAEA. (2011). Guidelines for implementing systems approaches for pest risk management of fruit flies working material (no: INIS-XA--12N0618). Joint FAO/ IAEA Division of Nuclear Techniques in Food and Agriculture.

Favrin, R. and Cree, L. (2016). Linking science and policy to better protect Canada from plant pests. *Food Security* 8(1), 71-75.

Flood, J. (2010). The importance of plant health to food security. *Food Security* 2(3), 215-231.

Gallo, T. and Waitt, D. (2011). Creating a successful citizen science model to detect and report invasive species. *BioScience* 61(6), 459-465.

Gardner, A. S., Gaston, K. J. and Maclean, I. M. D. (2021). Combining qualitative and quantitative methodology to assess prospects for novel crops in a warming climate. *Agricultural Systems* 190, 103083.

Garnas, J. R., Auger-Rozenberg, M. A., Roques, A., Bertelsmeier, C., Wingfield, M. J., Saccaggi, D. L., Roy, H. E. and Slippers, B. (2016). Complex patterns of global spread in invasive insects: Eco-evolutionary and management consequences. *Biological Invasions* 18(4), 935-952.

Garnas, J. R., Hurley, B. P., Slippers, B. and Wingfield, M. J. (2012). Biological control of forest plantation pests in an interconnected world requires greater international focus. *International Journal of Pest Management* 58(3), 211-223.

Guiașu, R. C. (2016). The troublesome matter of subjective definitions. In: *Non-Native Species and Their Role in the Environment*. Brill, Boston, pp. 8-37. https://doi.org/10.1163/9789047426134_003.

Haack, R. A., Britton, K. O., Brockerhoff, E. G., Cavey, J. F., Garrett, L. J., Kimberley, M., Lowenstein, F., Nuding, A., Olson, L. J., Turner, J. and Vasilaky, K. N. (2014). Effectiveness of the International Phytosanitary Standard ISPM No. 15 on reducing wood borer infestation rates in wood packaging material entering the United States. *PLoS ONE* 9(5), e96611.

Hérard, F. and Maspero, M. (2019). History of discoveries and management of the citrus longhorned beetle, *Anoplophora chinensis*, in Europe. *Journal of Pest Science* 92(1), 117-130.

Holt, J., Leach, A. W., Johnson, S., Tu, D. M., Nhu, D. T., Anh, N. T., Quinlan, M. M., Whittle, P. J. L., Mengersen, K. and Mumford, J. D. (2018). Bayesian networks to compare pest control interventions on commodities along agricultural production chains. *Risk Analysis: An Official Publication of the Society for Risk Analysis* 38(2), 297-310.

Holt, J., Leach, A. W., MacLeod, A., Tomlinson, D., Christodoulou, M. and Mumford, J. D. (2017). A quantitative model for trade pathway analysis of plant pest entry and transfer to a host in European Union territory. *EPPO Bulletin* 47(2), 220-226.

Hulme, P. E. (2009). Trade, transport and trouble: Managing invasive species pathways in an era of globalization. *Journal of Applied Ecology* 46(1), 10-18.

IPPC. (2005). *International Standards for Phytosanitary Measures, ISPM 23. Guidelines for Inspection. International Plant Protection Convention*. Food and Agriculture Organization, Rome.

IPPC. (2006). *International Standards for Phytosanitary Measures, ISPM 1. Phytosanitary Principles for the Protection of Plants and the Application of Phytosanitary Measures in International Trade. International Plant Protection Convention*. Food and Agriculture Organization, Rome.

IPPC. (2008). *International Standards for Phytosanitary Measures, ISPM 31. Methodologies for Sampling of Consignments. International Plant Protection Convention*. Food and Agriculture Organization, Rome.

IPPC. (2011). *International Standards for Phytosanitary Measures, ISPM 7: Phytosanitary Certification System. International Plant Protection Convention*. Food and Agriculture Organization, Rome.

IPPC. (2017a). *International Standards for Phytosanitary Measures, ISPM 4: Requirements for the Establishment of Pest Free Areas. International Plant Protection Convention*. Food and Agriculture Organization, Rome.

IPPC. (2017b). *International Standards for Phytosanitary Measures, ISPM 12: Phytosanitary Certificates. International Plant Protection Convention*. Food and Agriculture Organization, Rome.

IPPC. (2017c). *International Standards for Phytosanitary Measures, ISPM 14. The Use of Integrated Measures in a Systems Approach for Pest Risk Management. International Plant Protection Convention*. Food and Agriculture Organization, Rome.

IPPC. (2018). *International Standards for Phytosanitary Measures, ISPM 15: Regulation of Wood Packaging Material in International Trade. International Plant Protection Convention*. Food and Agriculture Organization, Rome.

IPPC. (2019). *International Standards for Phytosanitary Measures, ISPM 11. Pest Risk Analysis for Quarantine Pests. International Plant Protection Convention*. Food and Agriculture Organization, Rome.

IPPC. (2021). *International Standards for Phytosanitary Measures, ISPM 5. Glossary of Phytosanitary Terms. International Plant Protection Convention*. Food and Agriculture Organization, Rome.

IPPC Secretariat. (2021). *Pest Status Guide: Understanding the Principal Requirements for Pest Status Determination*. FAO on Behalf of the Secretariat of the International Plant Protection Convention, Rome.

Jamieson, L. E., DeSilva, H. N., Worner, S. P., Rogers, D. J., Hill, M. G. and Walker, J. T. S. (2014). A review of methods for assessing and managing market access and biosecurity risks using systems approaches. *New Zealand Plant Protection* 66, 1-9.

Jang, E. B. and Moffitt, H. R. (1994). Systems approaches to achieving quarantine security. In: Sharp, J. L. and Hallman, G. J. (Eds.), *Quarantine Treatments for Pests of Food Plants*. CRC Press, Boca Raton, pp. 225-237.

Jones, D. R. (2003). Plant viruses transmitted by whiteflies. *European Journal of Plant Pathology* 109(3), 195–219.

Liebhold, A. M., Brockerhoff, E. G., Garrett, L. J., Parke, J. L. and Britton, K. O. (2012). Live plant imports: The major pathway for forest insect and pathogen invasions of the US. *Frontiers in Ecology and the Environment* 10(3), 135–143.

Lingafelter, S. W. and Hoebeke, E. R. (2002). *Revision of the Genus Anoplophora (Cerambycidae)*. The Entomological Society of Washington, Washington, DC.

Lopian, R. (2005). "The international plant protection convention and invasive alien species". Identification of risks and management of invasive alien species using the IPPC framework. In: *Proceedings of the Workshop on Invasive Alien Species and the International Plant Protection Convention*, Braunschweig, Germany, 22–26 September 2003. Food and Agriculture Organization, Rome.

MacLeod, A. (2015). The relationship between biosecurity surveillance and risk analysis. In: Jarrad, F., Low Chow, S. and Mengersen, K. (Eds.), *Biosecurity Surveillance: Quantitative Approaches*. CABI Publishing, Wallingford, UK, pp. 109–122.

MacLeod, A., Baker, R. H. A. and Hoddinott, R. (2005). The analysis of detections in consignments to identify and target pests' entry pathways. In: *Proceedings of the British Crop Protection Council International Congress, Crop Science and Technology*, Glasgow, 31 October, pp. 1013–1018.

MacLeod, A., Evans, H. F. and Baker, R. H. A. (2002). An analysis of pest risk from an Asian longhorn beetle (Anoplophora glabripennis). *Crop Protection* 21(8), 635–645.

MacLeod, A., Jones, G. D., Anderson, H. M. and Mumford, R. A. (2016). Plant health and food security, linking science, economics, policy and industry. *Food Security* 8(1), 17–25.

MacLeod, A., Pautasso, M., Jeger, M. J. and Haines-Young, R. (2010). Evolution of the international regulation of plant pests and challenges for future plant health. *Food Security* 2(1), 49–70.

Magarey, R. D., Colunga-Garcia, M. and Fieselmann, D. A. (2009). Plant biosecurity in the United States: Roles, responsibilities, and information needs. *BioScience* 59(10), 875–884.

Marini, L., Haack, R. A., Rabaglia, R. J., Petrucco Toffolo, E., Battisti, A. and Faccoli, M. (2011). Exploring associations between international trade and environmental factors with establishment patterns of exotic Scolytinae. *Biological Invasions* 13(10), 2275–2288.

Maye, D. and Chan, K. W. (2021). National biosecurity regimes, plant and animal biopolitics in the UK and China. In: Barker, K. and Francis, R. A. (Eds.), *Routledge Handbook of Biosecurity and Invasive Species*. Routledge, London, pp. 243–260.

McCullough, D. G., Work, T. T., Cavey, J. F., Liebhold, A. M. and Marshall, D. (2006). Interceptions of nonindigenous plant pests at US ports of entry and border crossings over a 17-year period. *Biological Invasions* 8(4), 611–630.

Mooney, H. A. (2005). Invasive alien species: The nature of the problem. In: Mooney, H. A., Mack, R. N. and McNeely, J. A. (Eds.), *Invasive Alien Species: A New Synthesis*. Island Press, London, pp. 1–15.

Mwebaze, P., Monaghan, J., Spence, N., MacLeod, A., Hare, M. and Revell, B. (2010). Modelling the risks associated with the increased importation of fresh produce from emerging supply sources outside the EU to the UK. *Journal of Agricultural Economics* 61(1), 97–121.

National Research Council. (2002). Predicting invasions of non-indigenous plants and pests. Report from the Committee on the Scientific Basis for Predicting the Invasive Potential of Non-Indigenous Plants and Plant Pests in the United States. National Academy Press, Washington, DC.

Nesbitt, A., Kemp, B., Steele, C., Lovett, A. and Dorling, S. (2016). Impact of recent climate change and weather variability on the viability of UK viticulture – Combining weather and climate records with producers' perspectives. *Australian Journal of Grape and Wine Research* 22(2), 324–335.

Ordish, D. (1972). *The Great Wine Blight*. Scribner, New York.

Ormsby, M. and Brenton-Rule, E. (2017). A review of global instruments to combat invasive alien species in forestry. *Biological Invasions* 19(11), 3355–3364.

Paini, D. R., Sheppard, A. W., Cook, D. C., De Barro, P. J., Worner, S. P. and Thomas, M. B. (2016). Global threat to agriculture from invasive species. *Proceedings of the National Academy of Sciences of the United States of America* 113(27), 7575-7579.

Perrings, C. (2016). Options for managing the infectious animal and plant disease risks of international trade. *Food Security* 8(1), 27-35.

Perrings, C., Williamson, M., Barbier, E. B., Delfino, D., Dalmazzone, S., Shogren, J., Simmons, P. and Watkinson, A. (2002). Biological invasion risks and the public good: An economic perspective. *Conservation Ecology* 6(1). Available at: http://www.consecol.org/vol6/iss1/art1.

Quinlan, M. M., Leach, A. and Mumford, J. (2021). Classification of objectives in systems approaches to manage horticultural biosecurity risks for market access. *Crop Protection* 139, 105286.

Quinlan, M. M., Mengersen, K., Mumford, J., Leach, A., Holt, J. and Murphy, R. (Eds.). (2016). *Beyond Compliance: A Production Chain Framework for Plant Health Risk Management in Trade*. Chartridge Books, Oxford, UK.

Region, L. (2020). *Anoplophora chinensis*: Normative e cartografia 2020 [Online]. Available at: https://www.regione.lombardia.it/wps/portal/istituzionale/HP/DettaglioRedazionale/servizi-e-informazioni/Imprese/Imprese-agricole/servizio-fitosanitario-regionale/organismi-nocivi/anoplophora-chinensis/anoplophora-chinensis-glabripennis (Accessed 28 January 2022).

Richardson, D. M. and Pyšek, P. (2004). *What is an Invasive Species? CABI Crop Protection Compendium*. CABI Publishing, Wallingford, UK. Available at: https://www.cabi.org/isc/FullTextPDF/2009/20093238299.pdf.

Rodrigue, J. P., Comtois, C. and Slack, B. (2016). *The Geography of Transport Systems* (4th edn.). Routledge, Abingdon, UK.

Roloff, A., Korn, S. and Gillner, S. (2009). The climate-species-matrix to select tree species for urban habitats considering climate change. *Urban Forestry and Urban Greening* 8(4), 295-308.

Roques, A. (2010). Alien forest insects in a warmer world and a globalised economy: Impacts of changes in trade, tourism and climate on forest biosecurity. *New Zealand Journal of Forestry Science* 40 (Suppl.), S77-S94.

Roy, B. A., Alexander, H. M., Davidson, J., Campbell, F. T., Burdon, J. J., Sniezko, R. and Brasier, C. (2014). Increasing forest loss worldwide from invasive pests requires new trade regulations. *Frontiers in Ecology and the Environment* 12(8), 457-465.

Ruiz, G. M. and Carlton, J. T. (2003). Invasion vectors: A conceptual framework for management. In: Ruiz, G. M. and Carlton, J. T. (Eds.), *Invasive Species: Vectors and Management Strategies*. Island, Washington, DC, pp. 459-504.

Schrader, G. and Unger, J. G. (2003). Plant quarantine as a measure against invasive alien species: The framework of the International Plant Protection Convention and the plant health regulations in the European Union. *Biological Invasions* 5(4), 357-364.

Seebens, H., Blackburn, T. M., Dyer, E. E., Genovesi, P., Hulme, P. E., Jeschke, J. M., Pagad, S., Pyšek, P., Winter, M., Arianoutsou, M., Bacher, S., Blasius, B., Brundu, G., Capinha, C., Celesti-Grapow, L., Dawson, W., Dullinger, S., Fuentes, N., Jäger, H., Kartesz, J., Kenis, M., Kreft, H., Kühn, I., Lenzner, B., Liebhold, A., Mosena, A., Moser, D., Nishino, M., Pearman, D., Pergl, J., Rabitsch, W., Rojas-Sandoval, J., Roques, A., Rorke, S., Rossinelli, S., Roy, H. E., Scalera, R., Schindler, S., Štajerová, K., Tokarska-Guzik, B., van Kleunen, M., Walker, K., Weigelt, P., Yamanaka, T. and Essl, F. (2017). No saturation in the accumulation of alien species worldwide. *Nature Communications* 8(1), 14435.

Shine, C. (2007). Invasive species in an international context: IPPC, CBD, European strategy on invasive alien species and other legal instruments. *EPPO Bulletin* 37(1), 103–113.

Tatem, A. J. (2009). The worldwide airline network and the dispersal of exotic species: 2007–2010. *Ecography* 32(1), 94–102.

Tepa-Yotto, G. T., Gouwakinnou, G. N., Fagbohoun, J. R., Tamò, M. and Saethre, M. G. (2021). Horizon scanning to assess the bioclimatic potential for the alien species *Spodoptera eridania* and its parasitoids after the pests detection in West and Central Africa. *Pest Management Science* 77(10), 4437–4446.

Thiemann, F., Fleming, E. M. and Mueller, R. A. (2012). Impact of information and communication technology (ICT) on international trade in fruit and vegetables: A gravity model approach. In: *Proceedings of the International Association of Agricultural Economists (IAAE) Conference, August 18–24, 2012 (Vol. 2012)*, Foz do Iguaçu, Brazil. No. 1007-2016-79510.

Tonnang, H. E., Mohamed, S. A., Khamis, F. and Ekesi, S. (2015). Identification and risk assessment for worldwide invasion and spread of *Tuta absoluta* with a focus on Sub-Saharan Africa: Implications for phytosanitary measures and management. *PLoS ONE* 10(8), e0135283.

Tripathi, L., Mwangi, M., Abele, S., Aritua, V., Tushemereirwe, W. K. and Bandyopadhyay, R. (2009). Xanthomonas wilt: A threat to banana production in east and Central Africa. *Plant Disease* 93(5), 440–451.

Turner, R. M., Brockerhoff, E. G., Bertelsmeier, C., Blake, R. E., Caton, B., James, A., MacLeod, A., Nahrung, H. F., Pawson, S. M., Plank, M. J., Pureswaran, D. S., Seebens, H., Yamanaka, T. and Liebhold, A. M. (2021). Worldwide border interceptions provide a window into human-mediated global insect movement. *Ecological Applications: A Publication of the Ecological Society of America* 31(7), e02412.

United Nations Ozone Secretariat. (2000). *The Montreal Protocol on Substances That Deplete the Ozone Layer*. United Nations Environment Programme, Nairobi, Kenya.

van Klinken, R. D., Fiedler, K., Kingham, L. and Barbour, D. (2021). The importance of distinguishing between demonstrating the efficacy and implementation of phytosanitary systems approaches. *Crop Protection* 139, 105287.

van Klinken, R. D., Fiedler, K., Kingham, L., Collins, K. and Barbour, D. (2020). A risk framework for using systems approaches to manage horticultural biosecurity risks for market access. *Crop Protection* 129, 1–12.

Venette, R. C. and Gould, J. R. (2006). A pest risk assessment for *Copitarsia* spp., insects associated with importation of commodities into the United States. *Euphytica* 148(1–2), 165–183.

Ward, M. (2016). Action against pest spread–The case for retrospective analysis with a focus on timing. *Food Security* 8(1), 77–81.

Williamson, M. (1996). *Biological Invasions*. Chapman and Hall, London, UK.

Wittwer, G., McKirdy, S. and Wilson, R. (2005). Regional economic impacts of a plant disease incursion using a general equilibrium approach. *Australian Journal of Agricultural and Resource Economics* 49(1), 75-89.

Work, T. T., McCullough, D. G., Cavey, J. F. and Komsa, R. (2005). Arrival rate of nonindigenous insect species into the United States through foreign trade. *Biological Invasions* 7(2), 323-332.

Yan, Y., Wang, Y. C., Feng, C. C., Wan, P. M. and Chang, K. T. (2017). Potential distributional changes of invasive crop pest species associated with global climate change. *Applied Geography* 82, 83–92.

Chapter 11

Mitigating invasive insect species: eradication, long-term management, and the importance of sampling and monitoring

Amy Morey, University of Minnesota, USA; and Robert Venette, USDA Forest Service, USA

1 Introduction

The movement of species around the world has greatly increased with rising trade and commerce from globalization (Hulme, 2009; Pyšek et al., 2010) and rapidly changing environments due to climate change (Bebber et al., 2013; Walther et al., 2009). In new areas, species may experience different ecological processes from those that regulate their populations in their native geographic range and/or they may encounter an environment that is particularly sensitive to their presence (Colautti et al., 2004; Jeschke et al., 2012). Consequently, species can become disruptive to novel landscapes. For example, the ladybird beetle, *Harmonia axyridis* (Pallas), has been introduced from its native range in temperate Asia into many regions of the world both intentionally (for biological control of other insects) and accidentally. However, despite its effectiveness at pest suppression in some systems, it has become infamous as a pest in fruit production, a nuisance pest in human dwellings, and a predator/competitor of native insect species (Comont et al., 2014; Koch and Galvan, 2007; Roy et al., 2016). Other species can be problematic in all areas where they occur,

http://dx.doi.org/10.19103/AS.2022.0113.15
Published by Burleigh Dodds Science Publishing Limited, 2023.

including their native range. For example, the fall armyworm, *Spodoptera frugiperda* (J.E. Smith), is a serious pest of corn and other crops across its native migratory range in North and South America (Montezano et al., 2018). Following its spread into and within Africa and Asia, *S. frugiperda* has continued to cause significant damage to food crops and is an active target of regulation and concern in regions where it has not yet established (Baloch et al., 2020; Goergen et al., 2016; Jeger et al., 2017). Species that do not historically occur in an area (i.e. species that are exotic, non-indigenous, non-native, or alien) and are likely to negatively affect the environment, economy, or human are considered invasive.

Though a small proportion of the species that arrive and establish in new areas become problematic (Kolar and Lodge, 2001; Williamson, 1996), harms can be substantial. Insects are the most common animal to be documented as invasive (Pyšek et al., 2020). Globally, costs from invasive insects to goods and services (i.e. costs that do not include impacts on human health or ecosystem processes) are estimated to be at least US$70 billion per year (Bradshaw et al., 2016). The negative impacts of invasive species on native species and ecosystems can also be substantial, albeit they are more difficult to quantify (Bradshaw et al., 2016; Gutierrez-Arellano and Mulligan, 2018; Pejchar and Mooney, 2009; Pyšek et al., 2020). Invasive insects can displace native species, negatively impact genetic resources through hybridization, disrupt ecosystem processes, and transmit or facilitate disease (Kenis et al., 2009). The need to prevent and otherwise mitigate the negative impacts of invasive species remains a top priority across global jurisdictions (NISC, 2016; Noar et al., 2021; Peyton et al., 2020).

As the number of invasive insects continues to increase, managers have been forced to quickly adapt. Preventing biological invasions is widely recognized as the most economical long-term approach to mitigating negative impacts (Leung et al., 2002; Lodge et al., 2006); however, this strategy is not always effective. Eradication of newly arrived pests is the next best alternative. The purpose of this chapter is to highlight major elements of eradication programs for invasive insect species of agricultural importance and to review the role integrated pest management can play in successfully mitigating invasive insects when eradication efforts are untenable. We provide examples of programs applied to species in both circumstances (Table 1). Emphasis is placed on the role that surveillance and monitoring tools play as critical components of eradication and long-term mitigation efforts of invasive insects. Continued technological advancements, particularly involving molecular tools and large-scale data acquisition and analysis, are allowing for more thorough, precise, and effective monitoring and management tools. Nonetheless, the frequency and diversity of insect incursions into new areas continues to outpace the knowledge available for effective mitigation in many situations.

Table 1 Examples of eradication and integrated pest management (IPM) programs against invasive insects

Species	Program goal (targeted area)	Major tactic(s) used	Example(s) of sampling/ monitoring tools used	Outcome
Cactoblastis cactorum	Eradication (Mexico)	• Cultural (sanitation) • Genetic (SIT) • Regulatory (quarantines, binational coordination)	Monitoring using female sex pheromone-baited trap and sentinel host plants	Declared eradicated from Mexico in 2009
Halyomorpha halys	IPM (eastern USA)	• Biological (parasitoids) • Chemical (threshold-based spray, spatially targeted spray) • Physical (attract-and-kill)	Dual pheromone-baited traps for treatment decision-making; pheromones, black lights, and public reporting for surveillance	Ongoing; tactics being developed, with integrated management still in infancy
Lymantria dispar dispar	IPM (eastern USA)	• Biological (natural fungus, virus formulation) • Chemical (Btk aerial spray for eradication, mating disruption for suppression) • Regulatory (quarantines, interstate and agency coordination)	Pheromone-baited traps for detection and delimitation; visual sampling	Ongoing; considered model for IPM of invasive species in USA
Pectinophora gossypiella	Eradication (southern USA)	• Chemical • Genetic (SIT, Bt cotton) • Regulatory • Physical	Pheromone-baited traps to inform SIT adjustments; visual sampling and pheromone traps to confirm population absence	Declared eradicated from continental USA in 2018
Teia anartoides	Eradication (New Zealand)	• Chemical (Btk aerial spray) • Genetic (SIT) • Physical (host removal)	Delimitation surveys using live female moths as lures	Declared eradicated from NZ in 2006
Vespa mandarinia	Eradication (western USA and Canada)	• Physical (nest destruction, lethal trapping)	Surveillance using radio tags and public reports	Ongoing

SIT, sterile insect technique; *Btk*, *Bacillus thuringiensis* var. *kurstaki*.

Increased research investments in the biology and ecology of invasive insects and management tools are critical to limiting the harm caused by these species.

2 Overview of invasive species management

Biological invasions are characterized typically by three discrete stages: arrival, establishment, and spread (Blackburn et al., 2014; Venette and Carey, 1998; Williamson, 1996). Despite the demarcation, the invasion process is continuous (Davis, 2009). Strategies for managing invasive species generally reflect where a given species is in the invasion process: Prevention excludes species that have not yet arrived, rapid responses keep newly arrived species from establishing or spreading, suppression limits densities and damage from established pests, and restoration attempts to return sites to their pre-invasion ecological structure and function (e.g. Venette and Morey, 2019). Effective sampling and monitoring are fundamental for realizing the goals of any invasive species management strategy.

Prevention is highly reliant on biosecurity regulations, risk assessments, and disinfestation treatments to prevent unwanted organisms from moving within pathways of introduction (Lodge et al., 2006; Venette and Morey, 2019). Such efforts are typically the responsibility of national agencies and cooperating international partners. Prevention measures can occur prior to the movement of a species into a new area (i.e. pre-border). For example, importing certain commodities can be prohibited or transnational agreements can be established that articulate phytosanitary standards and expectations for commodity production and/or processing (e.g. pest-free areas and systems approaches) (Follett and Neven, 2006; Kalaris et al., 2014). At ports of entry, border biosecurity measures directly (e.g. rejecting or destroying cargo) or indirectly (e.g. culling host material) exclude unwanted organisms (Poland and Rassati, 2019).

Efforts to prevent the arrival of invasive insects through regulation and surveillance improve with the sensitivity of detection methods to locate invasive species that are *en route* to new areas. Molecular identification techniques, such as DNA barcoding, environmental DNA (eDNA) metabarcoding, and genomic sequencing, are greatly improving the accuracy and efficiency of detecting species in arrival pathways (Hamelin and Roe, 2020; Young et al., 2021). However, restricting pathways or maintaining effective quarantine require constant vigilance and cooperation from myriad stakeholder groups (Garnas et al., 2016).

Should prevention strategies fail, early detection tools are used to document an introduced species through targeted (i.e. species-specific surveys) or incidental (i.e. routine surveys intended for other purposes) means (Reaser et al., 2020). Detection ideally occurs soon after a species arrives to an area, though due to lack of effective sampling, small population size, and/or lack of awareness, species may be in the landscape many years before detection. For example, emerald ash borer (*Agrilus planipennis* Fairmaire) was likely present in southeastern Michigan as early as the mid-1990s before it was detected in 2002 (Siegert et al., 2014). Once a species is first detected, a rapid response typically includes eradication and/or containment efforts. While eradication seeks to eliminate a species from an area, containment seeks to limit the distribution and damages from a species. Early detection and rapid response (EDRR) can be combined as a single, iterative, management process, though clear operating definitions of EDRR and its components are often lacking despite the wide usage of the strategy in regulatory frameworks (Reaser et al., 2020). Early detection of invasive species has historically been site based in scope and heavily reliant on visual encounters (Reaser et al., 2020). However, detection tools have broadened with technological advancements and can include acoustic, chemical, and genetic detection and remote-sensing technologies (Martinez et al., 2020).

Once an invasive species has established and spread beyond an extent amenable to eradication or containment, the invading population may be managed with the goal of suppressing it below a desired level. Suppression strategies typically rely on sampling to enumerate established populations in much the same way as sampling is used for established populations of native pests. Economic thresholds, often represented by the number of individuals per sampling unit (e.g. per plant, per area, or per trap), are a fundamental component of informed decision-making in managed agricultural systems when mitigating native or invasive pest species (Pedigo et al., 2021). Similarly, the suppression of invasive insects can utilize many of the same tools employed against native pests, including chemical, physical, cultural, biological, and genetic tactics. Importantly, however, the relevant biological and ecological information necessary for using many of these tactics often is absent for new invasive species, at least initially. A strategy that can accompany the suppression of an invasive population is the restoration of the system or habitat negatively impacted. Restoring and rehabilitating a system that underwent undesired change due to the presence of an invasive species focuses on efforts that manage general ecosystem properties, not necessarily a single species (Venette and Morey, 2019). Incorporating resilience, or the general ability of a system to maintain certain functions and structure following disturbance (Chambers et al., 2019), and resistance to further disturbance in the area of interest are often sought (Koch et al., 2021). Building resilience in ecological systems is often aspirational, and it is difficult to know if or when it has been achieved.

Specific techniques to monitor and sample invasive insects are essential for most management strategies. For example, traps baited with semiochemical lures, such as those containing host volatiles or sex pheromones, are important tools at many points during invasion. They can be used to detect incursions at ports of entry and inform EDRR measures and are common sampling tools for threshold-based decision-making used to manage established species (Augustin et al., 2012; El-Sayed et al., 2009; Poland and Rassati, 2019). Monitoring is ultimately required to demonstrate the effectiveness of any invasive insect management strategy, whether to declare a population to be absent from an area or determine that it is below the desired level. For example, internationally recognized pest-free areas can be defined by regulatory authorities to indicate where a species is below an agreed-upon level sufficient for trade to continue (Follett and Neven, 2006; Kalaris et al., 2014). Establishing measurable definitions for effectiveness or success of management, however, remains a challenge for eradication and long-term efforts against invasive species (Venette and Morey, 2019).

3 Eradication of invasive insects

3.1 Concepts of eradication

Pest eradication implies the intentional elimination of an undesirable species from an area (FAO, 2017; Hendrichs et al., 2021). The Food and Agricultural Organization of the United Nations characterizes the process in three steps: (1) surveillance to determine the spatial distribution of an unwanted species, (2) containment procedures to prevent spread of that species, and (3) treatment to eliminate the targeted population (FAO, 2017). In the United States, eradication is typically described as part of federal policies in support of biosecurity (Reaser et al., 2020). EDRR programs typically identify eradication, sometimes in addition or alternative to containment, as the aspired response to the detection of a pest (Reaser et al., 2020). Eradication is implemented to avoid the higher economic impacts from long-term management across a region (Suckling et al., 2019); especially in agricultural systems invasive insects can reduce crop productivity and profitability. Certain species may also curtail international trade. For example, commodity exports may be banned or made too expensive due to quarantine regulations if a pest establishes in an area (e.g. Siebert and Cooper, 1995). As such, avoidance of trade restrictions is a major motivation for eradication programs against agriculturally important insects (Hendrichs et al., 2021).

Tools for eradication vary. Broad-spectrum insecticides are historically the most used tactic, often via large-scale application as part of a 'scorched earth' approach (Hendrichs et al., 2021). In an effort to reduce the many detrimental impacts that can result from scorched earth approaches, recent eradications have moved toward more targeted and/or localized insecticide use, such as using microbial insecticides (or biopesticides). For example, the soil bacterium *Bacillus thuringiensis* var. *kurstaki* (*Btk*) is a microbial insecticide used as a targeted treatment against Lepidopteran pests (Suckling et al., 2014b). A challenge of chemical control is that the efficacy of registered products may not be known for a new invasive species (e.g. Tripodi and Hardin, 2020). Additionally, products may not be labeled for the specific location or application rate needed at the time of pest detection; in the United States, off-label pesticide use is illegal. Other insect eradication tools include physical (e.g. host removal) and cultural (e.g. sanitation) methods. Genetic control, particularly the sterile insect technique (SIT), has been important to the success of insect eradications, and methodology has expanded significantly in recent years (Teem et al., 2020). Semiochemical treatments for mating disruption, mass trapping, or attract-and-kill methods are effective tools if the necessary pheromone(s) or kairomone(s) have been identified (Suckling et al., 2014b). Regulatory tactics, such as quarantine restrictions, can bring accountability and public awareness to limit further introduction and spread of a species during eradication.

Integrating multiple tactics can yield effective results (Blackwood et al., 2012; Suckling et al., 2014b). Lampert and Liebhold (2021) demonstrated that the integration of insecticides, SIT, and mating disruption was the most efficacious eradication scheme for univoltine insects based on the number of individuals removed per dollar invested. However, the complementarity of tactics must be understood. For example, if used simultaneously, SIT and mating disruption will be counterproductive (Lampert and Liebhold, 2021). Importantly, Allee effects (a density-dependent phenomenon of small populations where decreasing individual fitness is related to decreasing population size or density) can be exploited by control tactics (integrated or singly) to facilitate eradication without the need to kill all individuals in a landscape (Barron et al., 2020; Blackwood et al., 2012). In such circumstances, bringing the population below an Allee threshold density will drive the population to extinction, though relevant Allee thresholds are often not known.

Sampling and monitoring play crucial roles throughout an insect eradication program. Eradication is often infeasible when effective detection and sampling methods do not exist for a species. Sensitive tools are not only vital for detecting small, nascent populations that are more amenable to eradication but are critical for guiding eradication treatments as they occur and, ultimately, indicating whether eradication has been successful (Liebhold and Kean, 2019; Tobin et al., 2014). For insects, semiochemical tools, like pheromone-baited traps, and visual surveys are widely used for initial detection and surveillance. However, other techniques, such as sentinel hosts, sniffer dogs, visual inspections, and mechanical detectors, are also utilized (reviewed in Poland and Rassati, 2019). Once detected, delimitation surveys guide the selection of a treatment decision that is appropriate for the density and spatial extent of the initial population. Delimiting grids can indicate more precisely where, and at what intensity, to target for treatment (Tobin et al., 2013). While there is no universal agreement on how to verify eradication has occurred, the criteria used for a given program must be specified before eradication begins (FAO, 2017). Criteria often rely on the inability to detect a population after a defined period with ongoing detection/monitoring efforts. For example, Klassen (1989) states that an insect could be considered eradicated once no individuals are captured for the time equivalent of 10 generations, though fewer generations (e.g. three) are deemed sufficient by others (e.g. Kean et al., 2022; Venette and Morey, 2019).

3.2 Factors of eradication success

Retrospective reviews and analyses have identified multiple factors associated with successful eradication programs (Brockerhoff et al., 2010; Epanchin-Niell et al., 2012; Liebhold et al., 2016; Myers et al., 2000; Pluess et al., 2012b; Tobin et al., 2014). One influential and consistently identified contributor is the timely detection of a target species upon its arrival to a region (but see also Pluess et al., 2012a). Lodge et al. (2006) suggest that eradication efforts need to proceed within 2 years from detection to be successful, assuming the detected population is still small and localized. Eradication of well-established invasive species can occur (e.g. *Pectinophora gossypiella* example in Section 3.4.1), but efforts come with significant economic and environmental costs (Myers et al., 1998; Myers, 2002). Another factor common to successful eradication is the availability of targeted control tactics (Liebhold and Kean, 2019; Tobin et al., 2014), which ultimately rely on the availability of quality *a priori* information about the biology and ecology of the invasive species or similar taxon (Pluess et al., 2012b). Socioeconomic factors, such as stakeholder coordination and robust public engagement and community support, are crucial to the success of eradication programs, though how best to approach engagement is not well understood (Epanchin-Niell et al., 2010; Pluess et al., 2012b).

3.3 Challenges of eradication

Eradication is not always feasible, particularly if the species of concern is already established over a large spatial area, if detection tools are unavailable or inadequate, or if treatment options do not exist. The infeasibility of eradication may be decided the moment that the incursion is discovered, or the realization may come over time. Eradication against the codling moth, *Cydia pomonella* (L.), in British Columbia began in the 1990s but was eventually restructured as a broader, long-term management program in 2000 after costs and stakeholder coordination became intractable (Myers, 2002; Thistlewood and Judd, 2019).

Even for relatively well-studied species, eradication efforts are often expensive and ultimately carry no guarantees of success, especially in the long term. Acting early, when populations are recently detected and small, is less economically costly relative to anticipated impacts, but estimates may not adequately forecast long-term monitoring costs to ensure a species remains absent (Myers, 2002). Typical cost assessments from eradication treatments also ignore the loss of non-market values, such as ecosystem services (Pejchar and Mooney, 2009; Rosenberger et al., 2012). Paradoxically, the loss of ecosystem services could result if a species is allowed to establish and spread. Effectively estimating the costs and benefits of eradication compared to other strategies is challenging (Brockerhoff et al., 2010; Hanley and Roberts, 2019; Olson and Roy, 2002), and methods to account for uncertainty are ongoing. The difficulty in confirming the absence of a species and lack of a uniform standard for declaring eradication success further confound realistic estimates of the economic costs of a program.

Eradication programs can be controversial, including among scientists (Carey and Harder, 2013; Liebhold, 2014). Perceptions of high costs, undesirable side effects, and few demonstrated benefits can challenge eradication efforts (Crowley et al., 2017; Liebhold and Kean, 2019), especially in the presence of misinformation and/or conflicting values and risk acceptance (Keller et al., 2014). The failure to adequately engage relevant authorities (Thistlewood and Judd, 2019), communicate to public communities, and engender public support has undermined eradication campaigns (Lindeman, 2013). For example, an eradication campaign against *Epiphyas postvittana* (Walker) in California was halted following widespread communication failings that led to public outrage and lack of coordinated support (Berenbaum et al., 2009; Lindeman, 2013).

3.4 Examples of eradication campaigns for invasive insects in agriculture

Liebhold et al. (2016) suggest that the first formal insect eradication program was the attempt to eliminate *Lymantria dispar dispar* (Lepidoptera: Erebidae) from Boston, Massachusetts, in 1890. Though that eradication program was unsuccessful, success has been documented for hundreds of other programs. The Global Eradication and Response Database (GERDA), available online, documents 1321 terrestrial arthropod eradication programs across approximately 99 countries (Kean et al., 2022) (Fig. 1). Of the 1125 programs with known outcomes (with the remaining programs in progress or with unreported outcomes), 83% were considered successful or likely successful and 17% were considered failed. The greatest numbers of eradications were against Diptera (634; primarily Tephritidae), Lepidoptera (286; primarily *Lymantria* spp.), and Coleoptera (202; primarily wood-boring beetles). These insect orders also

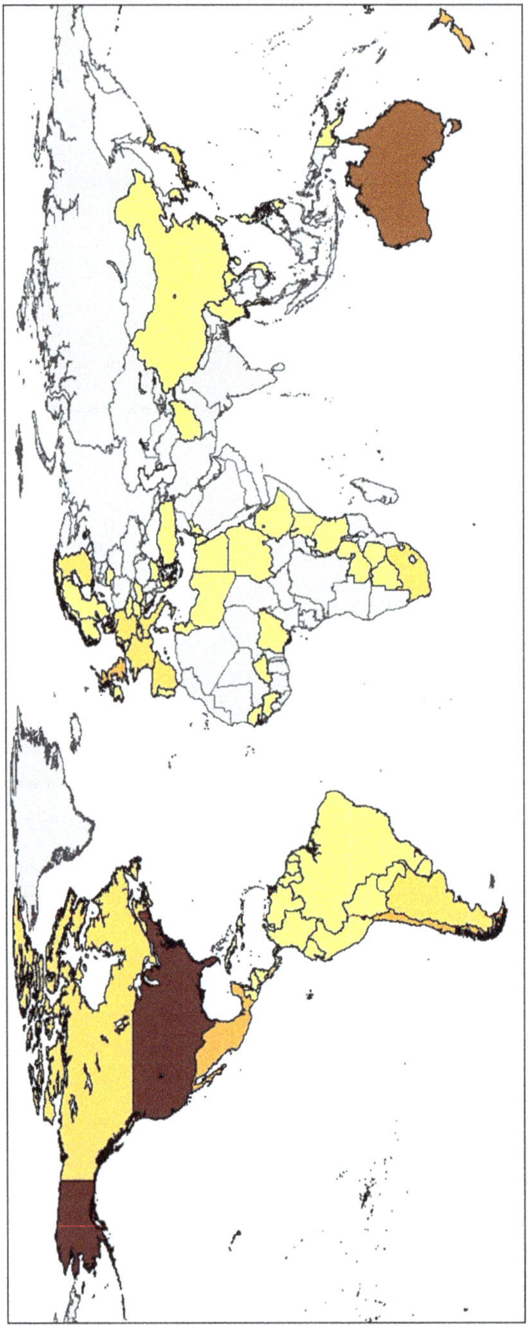

Figure 1 Global arthropod eradication programs (since 1890) reported in the global eradication and response database (GERDA) (Kean et al., 2022). Of the 1320 programs currently reported across ~99 countries, the regions with the five highest numbers of eradication programs (across all outcomes) are the USA (not including territories; 500), Australia (268), New Zealand (70), Chile (31), the United Kingdom (31), and Mexico (29). Colors indicate the number of reported arthropod eradications in GERDA, from zero (grey), <7 (yellow), 7-23 (light orange), 24-70 (orange), 71-268 (brown), and 269+ (dark brown). Data use permission and acknowledgement: John Kean and Better Border Biosecurity, New Zealand.

Published by Burleigh Dodds Science Publishing Limited, 2023.

have the highest numbers of successful eradication programs. For programs declared successful, most species (337) were first detected with targeted traps or lures (e.g. pheromones) or by public reports (126). Pesticides, host removal, bait sprays, and SIT were the most frequently reported tactics for successful arthropod eradications, though it is unclear how often programs used more than one tactic. While there are important caveats to these data and their interpretation (Liebhold et al., 2016; Tobin et al., 2014), GERDA can indicate meaningful trends of the eradication programs reported within it (Suckling et al., 2021).

Below, we provide brief examples of invasive insect eradication programs, including a description of the species' invasion history, summary of the eradication effort, and specific role of monitoring/sampling tools (see also Table 1). We have selected examples of agricultural importance that represent programs with declared success and those where eradication is currently ongoing. Many other examples exist, and though arthropod eradications occur globally, well-documented programs are heavily concentrated in certain areas of the world (Fig. 1). We draw heavily from examples in North America.

3.4.1 Pectinophora gossypiella (southwestern United States)

The pink bollworm, *Pectinophora gossypiella* (Lepidoptera: Gelechiidae), native to India or Australia/Indonesia, has invaded cotton production regions globally (Noble, 1969). Larval feeding within the developing cotton boll lowers seed and lint quantity and quality, causing significant economic losses. This insect was first reported in the United States in 1917 from Texas and subsequently spread to the southwestern United States, where it was particularly problematic. Invasion by this insect spurred intense federal quarantines and research. Concerns over spread to eastern cotton-producing states were long-standing and fueled by outbreaks in Louisiana, Arkansas, Florida, and Georgia, each eradicated from commercial cotton (Noble, 1969).

Strict management tactics have been applied to pink bollworm in the United States. Through the mid-1990s, most growers depended on physical and chemical methods to suppress pest densities locally, often relying on multiple insecticide applications (Henneberry, 2007). An SIT program was developed for pink bollworm in 1968, primarily to protect cotton in the San Joaquin Valley of southern California (Bartlett and Staten, 1996). Tens to hundreds of millions of sterilized female and male moths (dyed red for identification and tracking) were produced and released annually over this growing region (Staten and Walters, 2021). Pheromone baited traps captured wild and sterile moths, and wild moth counts were used to adjust rates and locations of sterile moth release. While the program was highly effective in lowering densities of wild pink bollworm, releasing enough sterile insects to extirpate wild populations proved difficult because wild moths still outnumbered sterile moths. In 1996, transgenic cotton expressing an insecticidal toxin from the bacterium *B. thuringiensis* was approved for release. Growers widely adopted the technology to manage pink bollworm and other lepidopteran pests, and pest densities declined dramatically. Computer simulations showed that sterile moths and *Bt* cotton could act synergistically to drive populations near zero; neither technology was capable of such an outcome alone (Tabashnik et al., 2021).

A formal eradication campaign against pink bollworm in the southwestern United States began in 2006 (Staten and Walters, 2021). From 2006 to 2014, most growers planted *Bt* cotton (e.g. 94% in Arizona), and through 2012, sterile moths were released throughout (approximately 1.6 billion moths/year in Arizona). Extensive pheromone trapping and boll sampling confirmed dramatic reductions in pink bollworm densities. In 2018, after multiple years without trapping a wild moth, the U.S. Department of Agriculture (USDA) declared pink bollworm to be eradicated from commercial cotton in the continental United States (Staten and Walters, 2021).

3.4.2 *Teia anartoides* (New Zealand)

Painted apple moth, *Teia anartoides* (Lepidoptera: Erebidae: Lymantriinae), native to southeastern Australia, was first detected in Auckland, New Zealand, in 1999. Larvae of this insect are polyphagous defoliators, capable of feeding on plants native to New Zealand in the families Solanaceae, Santalaceae, Myrtaceae, and Fabaceae, including Mimosoideae (Common, 1990) and non-native plants in another 15 families (Suckling et al., 2014a). Of these hosts, several are economically or ecologically significant, particularly apple, *Malus domestica*; gum trees, *Eucalyptus* spp.; Monterey pine, *Pinus radiata*; and wattles, *Acacia* spp. The Ministry of Agriculture and Forestry (MAF, 2002) estimated that NZ$58 million to NZ$356 million in losses could accrue over 20 years if the insect was not managed; nationally, more than 75% of these impacts were projected to occur in Monterey pine plantations.

The response to the incursion evolved rapidly from 1999 to 2004 (Suckling et al., 2007). The MAF Biosecurity Authority (i.e. 'New Zealand Biosecurity') coordinated a national management response, with an initial emphasis on surveying by visual inspections of potential hosts, followed by ground spraying and host removal, near where painted apple moths had been found. A breeding colony of the insect was established in 2001, and traps with caged female moths from the colony were used subsequently to monitor the distribution of insects until the end of the program. As the infested area in and around Auckland was relatively small (12 000 ha), 2000 traps were deployed, approximately every 500 m, though traps were less dense near the periphery and more dense near the core distribution or when wild moths were caught (Suckling et al., 2007). From

January 2002, aerial spraying of *Btk*-based insecticides occurred regularly and drove wild populations to about 1% of early trap catches but were controversial with residents. No breeding individuals were detected after 2003 and no wild males have been captured since January 2004 in west Auckland (Suckling et al., 2007).

A program using the SIT, the first of its kind in New Zealand, operated from January 2003 to April 2004 (Suckling et al., 2007). Male moths, irradiated as pupae, were dusted with fluorescent powder and released weekly; ultimately 350 000 moths were released at a rate of 100:1 (sterile:wild) males. Males were released from four points early in the program; later releases were within a few hundred meters of trap catches. The eradication program was declared successful in 2006.

3.4.3 *Cactoblastis cactorum* (Mexico)

The cactus moth, *Cactoblastis cactorum* (Berg) (Lepidoptera: Pyralidae), native to parts of South America (Argentina, Paraguay, Uruguay, and southern Brazil), was intentionally introduced into Australia in 1925 to control non-native prickly pear cacti (*Opuntia* spp.) infesting rangeland (Zimmermann et al., 2004). *Cactoblastis cactorum* larvae feed on a broad range of *Opuntia* species, including species outside the insect's native range, and primarily inflict damage by hollowing out young cactus pads (cladodes). Larval feeding also enables secondary microbial infections that can result in plant death (Zimmermann et al., 2004). Ninety percent of prickly pear was estimated to be destroyed in Queensland, Australia, following the release of *C. cactorum*.

The enormous success of *Opuntia* suppression by *C. cactorum* in Australia led to further introductions to South Africa, Hawaii, Mauritius, and the Caribbean Islands. In the latter program, *C. cactorum* was used to control native cacti deemed problematic in rangelands on multiple islands. Unfortunately, *C. cactorum* spread unintentionally throughout the Caribbean on weedy and non-weedy *Opuntia* (Zimmermann et al., 2004). It was found in southern Florida in 1989 and spread into coastal Louisiana and the Carolinas over the next 20 years (Bloem et al., 2007). Later molecular analyses confirmed that populations in the United States resulted from multiple introductions from populations outside of the insect's native range (Marsico et al., 2011). Arrival into the United States caused great concern over potential spread into southwestern states and especially into Mexico, where *C. cactorum* would have devastating impacts on the native desert ecosystems and commercial industries (Stiling, 2002; Zimmermann et al., 2004). Globally, Mexico is a biodiversity hot spot of *Opuntia* species, and the plants have significant cultural and agricultural value (Bello-Rivera et al., 2021).

Mexico initiated a surveillance campaign for *C. cactorum* in 2002, which detected two outbreaks in the state of Quintana Roo in 2006, and subsequently began an eradication program in collaboration with the United States (Bello-Rivera et al., 2021; Bloem et al., 2007). Detections were followed immediately by delimitation surveys. Intensive mechanical control through removal of aboveground host plant material and adult insect mass trapping then reduced the *C. cactorum* populations enough to allow for effective release of sterilized male and female moths; the targeted overflooding ratio was 10 sterile to 1 wild moth (Bello-Rivera et al., 2021).

Importantly, significant research into the biology of *C. cactorum* and development of potential management tactics had already occurred prior to the eradication efforts in Mexico (Bloem et al., 2007). For example, the characterization and synthesis of the female *C. cactorum* sex pheromone and validation of the trapping system allowed for effective area-wide monitoring of populations (Bloem et al., 2005). In addition, multiple national and international partners worked in support of the program, encouraging coordinated regulatory efforts, information exchange, and public awareness campaigns (Bello-Rivera et al., 2021). Transference of the implementation tools and insects required for effective SIT occurred through international collaboration (Bello-Rivera et al., 2021; Bloem et al., 2007).

In 2009, after the time required for three generations of insect development occurred without a detection, *C. cactorum* was declared eradicated from Mexico. Surveillance activities, including pheromone traps and sentinel host plants in areas where *Opuntia* is commercially grown, are ongoing to assure continued absence from the country (Bello-Rivera et al., 2021).

3.4.4 *Vespa mandarinia* (northwestern United States/southwestern Canada)

The northern giant hornet, *Vespa mandarinia* (Smith) (Hymenoptera: Vespidae), is the world's largest hornet (Makino, 2016) and widely distributed in its native range in East Asia (Makino, 2016; Tripodi and Hardin, 2020). The insect is most noted for its aggressive predation of honey bee colonies and potentially deadly stings to humans (Dooley, 2020; Makino, 2016; Matsuura and Koike, 2002; Matsuura and Sakagami, 1973; Tripodi and Hardin, 2020). European honey bee colonies (*Apis mellifera* L.) are particularly vulnerable, and without intervention, 100% of predated *A. mellifera* hives can be lost (Matsuura and Sakagami, 1973; Tripodi and Hardin, 2020).

The first North American detections of *V. mandarinia* occurred in British Columbia in the fall of 2019 from the discovery of a nest (Nanaimo, Vancouver Island) and single adult (White Rock). The first U.S. report came shortly thereafter of a dead specimen found near Blaine, WA, in December 2019, with additional reports of individual adult *V. mandarinia* observed earlier in the fall by local beekeepers (in Bellingham, WA). Whether *V. mandarinia* is established in the United States is uncertain. Preliminary genetic analysis suggests that the Canadian nest population was introduced separately from those in the United States (Wilson et al., 2020), and morphological comparisons show that the Washington specimens are most similar to forms found throughout the eastern portion of the insect's native range (Tripodi and Hardin, 2020). At this time, *V. mandarinia* has not been reported anywhere else outside of its native range. Unintentional (e.g. potted plants containing overwintering queens) and intentional (e.g. smuggling of live queens or brood comb as food and medicine) human-mediated spread pathways are highly feasible, though it is not known how *V. mandarinia* arrived into North America (Tripodi and Hardin, 2020).

Some regulatory measures have been enacted to prevent further arrival and spread of the hornet, especially into other states and provinces. In the United States, *V. mandarinia* is regulated as a quarantine pest (Tripodi and Hardin, 2020). U.S. federal allocations in 2021 were nearly US$1 million to support research and eradication efforts of *V. mandarinia* (USDA-APHIS, 2021). The Canadian Food Inspection Agency restricts the intentional importation of *V. mandarinia*, though it does not regulate *V. mandarinia* as a quarantine pest due to the lack of direct impacts on plants and uncertainty about arrival pathways (CFIA, 2020). To help guide further prevention and regional management decision-making, forecasts and assessments of risk are being produced (Norderud et al., 2021; Nuñez-Penichet et al., 2021; Tripodi and Hardin, 2020; Zhu et al., 2020).

Eradication efforts against *V. mandarinia* are focused on monitoring and trapping to delimit the extent of infestation and to locate nests for elimination. A major benefit to these is the preexisting literature on the species from its native range, including many aspects of the insect's biology and methods of control. For example, the life cycle and behavior of *V. mandarinia* colonies are well documented. Consequently, detection strategies have used both passive and active methods, depending on the estimated phenology of the colony and targeted life stage (Tripodi and Hardin, 2020). Sugar- and protein-based baits are typically used in traps due to their known effectiveness in Japan (Tripodi and Hardin, 2020). Further, information pertaining to related species that have also become invasive in other regions [e.g. *V. velutina* (Arca et al., 2015)] offer additional management insights. Radiotracking methods, which had shown success with *V. velutina* invasions in Europe (Kim et al., 2019), were used in 2020 to track an individual *V. mandarinia* back to its nest. This led to the first successful nest eradication in the United States (Rosenthal, 2021).

Contributions from the general public and citizen scientists have been crucial to the detection and continued monitoring of *V. mandarinia* in North America. The first detections in both Canada and the United States were by private citizens (Baker, 2020; Berube, 2020). Additional finds of *V. mandarinia* in Washington have been reported by private citizens, and the Washington State

Department of Agriculture has an active citizen scientist trapping network; over 770 traps were placed by citizen scientists in 2021 (WSDA, 2022).

The documented nest eradications in North America have used physical tactics, which included vacuuming out hornets and injecting the nest site with carbon dioxide (Berube, 2020; WSDA, 2020). Other control tactics may be used in the future (Tripodi and Hardin, 2020). Though individual nests have been eradicated, adult *V. mandarinia*, including mated queens, continue to be caught (WSDA, 2021). The eradication efforts in both the United States and Canada remain ongoing.

4 Long-term management of invasive insects

Established invasive species in agriculture can result in high levels of damage and high control costs. However, Simberloff et al. (2013) notes that, compared to past decades, technological advances enable 'long-term management of invaders [to be] neither futile nor necessarily damaging to non-targets'.

Once an invasive insect establishes within an area, devising management schemes to mitigate local harms may be sought, but often a parallel goal is to prevent or delay further spread (e.g. to other states or regions). In this way, eradication may be a component of the long-term management plan for an invasive insect (see *L. dispar dispar* example below). Alternatively, eradication may be the primary goal over the long term, as is the case in many SIT programs of agriculturally and medically important insect pests (Dyck et al., 2021). Epanchin-Niell et al. (2010) synthesized that economically optimal management for established invasive species relies on explicit consideration of both the control costs and economic measures of damage and the idiosyncratic invasion dynamics of a given species (e.g. population growth, spread, invasion extent, and likelihood for reinvasion). In recognition of the landscape-level influences on invasion dynamics, recent decades have seen a shift to area-wide management of invasive insect species, rather than the more typical field- or farm-level focus historically used for pest management (Hendrichs et al., 2007; Poland et al., 2021). Area-wide management targets a pest population within a delimited, large-scale (though not necessarily) area while considering the ecology of the species in that area (Hendrichs et al., 2007). In this way, area-wide management has similar goals to integrated pest management (the next section), though the two have historically been seen as opposing strategies (Hendrichs et al., 2007).

4.1 Integrated pest management for invasive species

Integrated pest management (IPM) has been a dominant paradigm in insect pest management for decades, particularly in agriculture (Kogan, 1998). The

Published by Burleigh Dodds Science Publishing Limited, 2023.

foundational aspirations of reducing the overreliance on chemical insecticides while encouraging natural population control from robust natural enemy presence persist as the stated goals of IPM programs today (Kogan, 1998). However, articulating the details of what defines IPM has proven ambiguous through time (Gray et al., 2009; Kogan, 1998), and few IPM programs have progressed into more sophisticated applications that manage multiple pests in diverse agroecosystems. Common themes of IPM are posed by the USDA's definition: '...a sustainable, science-based, decision-making process that combines biological, cultural, physical, and chemical tools to identify, manage and reduce risk from pests and pest management tools and strategies in a way that minimize overall economic health and environmental risks' (USDA-OPMP, 2018).

Using IPM for invasive insects is most feasible in later stages of invasion when a species is being managed with suppression and restoration strategies. Once a species has begun to integrate into a new environment and spread, it exhibits predictable population dynamics that make it equivalent to managing a resident pest with IPM (Venette and Koch, 2009). The major purpose of sampling and monitoring in IPM is to enumerate a population density or document the extent of its presence. This information is then used to determine whether management action is needed to suppress the population to a desired level. Typically, action is triggered when a pest population density reaches an economic threshold (ET). An ET is established to prevent a pest population from further reaching the economic injury level (EIL), a calculation that describes the lowest population density that will cause economic loss (Pedigo, 1986). Because of the extensive knowledge required about a given pest's biology and impact within a given cropping system, estimating an effective EIL and ET is often not possible for invasive insects early in the invasion process.

Managing invasive insects with IPM poses unique challenges. The research and knowledge base needed to devise an effective IPM plan may be incomplete for many years. For example, the soybean aphid (*Aphis glycines* Matsumura) was detected in the United States 6 years before the first ETs were calculated to reliably reduce economic loss and broad-spectrum insecticide use (Ragsdale et al., 2011). Though establishing thresholds was an important step, the viability of non-chemical tactics (e.g. biological control and host plant resistance) and their integration within soybean production are still active areas of research; soybean aphid management remains largely dependent on threshold-based applications of broad-spectrum insecticides (Koch et al., 2016). Implementing management for an invasive species in a system may also disrupt IPM programs already in place for other pests (e.g. Leskey et al., 2012). Lastly, defining IPM in a consistent and measurable way remains an important challenge for the field of pest management, broadly (Ehler, 2006; Peterson et al., 2018) and complicates applications of IPM to invasive species (Venette and Morey, 2019).

4.2 Examples of invasive agricultural insect pests being managed through integrated pest management

Below, we provide examples of invasive insect species being managed through IPM. Our intent is not to argue what qualifies as IPM *per se* but to show examples of species for which IPM has been an explicit strategy. We include a brief description of each species' invasion history, summary of the management effort, and role of specific monitoring/sampling tools (see also Table 1). As with the instances of eradication described previously, these descriptions are meant to convey common elements among programs deemed effective.

4.2.1 Lymantria dispar dispar (eastern United States)

Management of the spongy moth, *L. dispar dispar* (Lepidoptera: Erebidae: Lymantriinae), particularly through the Slow-the-Spread program, is 'the largest and most successful IPM program in the United States for managing the spread of an invasive forest pest... and is recognized as a model approach for managing invasive species' (Poland et al., 2021); it exemplifies how (local) eradication can become an integral component of area-wide IPM. This sub-species is native to Europe and has been spreading in the United States since 1868–9. Larvae can feed on more than 300 plant species, but aspen, basswood, birch, hawthorn, larch, oak, and willow are preferred. Defoliation becomes significant during outbreaks, and repeated defoliation events can kill trees. Outbreaks can also deter recreationists. Thus, the primary goal of Slow-the-Spread is to slow the expansion of generally infested areas and delay damages from *L. dispar dispar*.

A critical tool for *L. dispar dispar* IPM is the sex pheromone, (+) cis-7,8-epoxy-2-methyloctadecane, which is commercially synthesized and marketed as 'disparlure'. The compound is highly attractive to male *L. dispar dispar* moths (Carde et al., 1977), and its use with various trap designs, particularly the milk carton or delta trap, provides a sensitive way to detect low densities of the insect. A sprayable formulation also allows treated areas to be saturated with pheromone and impedes the ability of males to locate and mate with females (i.e. pheromone or mating disruption) (Leonhardt et al., 1996). Pheromone disruption is incompatible with pheromone-baited traps.

Slow-the-Spread (reviewed in Liebhold et al., 2021) focuses on the transition area between the leading edge, where the insect is detectable but at low densities, and areas with well-established populations (Fig. 2). Areas ahead of the leading edge are considered uninfested, and finds of the insect here are subject to eradication led by state agencies in partnership with the U.S. Department of Agriculture, Animal and Plant Health Inspection Service (USDA, APHIS). An extensive network of pheromone-baited traps is used to detect these populations, generally considered the result of inadvertent, long-distance

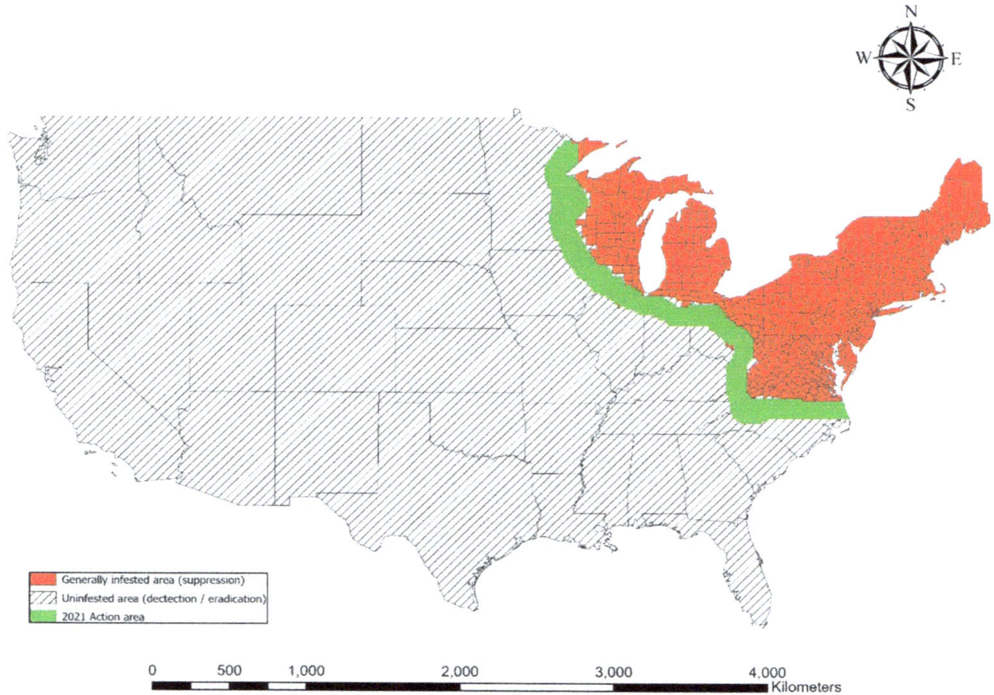

Figure 2 Distribution of *Lymantria dispar dispar* in the contiguous USA and status of the Slow-the-Spread program as of 2021. Green areas indicate the transition between the leading edge of the invasion and the areas with well-established populations (red). Low density populations in the transition zone are managed through Slow-the-Spread. Established populations are managed by suppression and uninfested areas (hatched) are the target of detection and eradication.Source: Map provided by Travis Perkins, Department of Entomology, Michigan State University, and the national *L. dispar* Slow-the-Spread Program.

movements of the insect (often as egg masses) by people. Trapping densities and frequencies vary by state and land use, generally in the range of 0.1–0.4 traps/km^2 every 2–4 years (Liebhold et al., 2021). Should one of these traps catch a moth, a more intensive grid of traps (~6 traps/km^2) is deployed the next year to confirm the population and delimit its spatial extent. Visual inspections of host material for egg masses, larvae, or pupae of *L. dispar dispar* are common to confirm local reproduction and define a 'core' population. Continued finds warrant treatment of the infested area in the following year, typically with sprays of *Btk* insecticides. In the year of treatment and one thereafter (i.e. 4 years after the initial detection), an intense trapping grid (again ~6 traps/km^2) is used to confirm treatment effectiveness or define new treatment areas (Liebhold et al., 2021). Two years with no moth captures is the minimum criteria to consider a population eradicated.

Slow-the-Spread has many similarities with the eradication program, but the goal is to slow the growth of the population, not eliminate it. A coordinated network of traps every 2–3 km in a semiregular grid in the transition area, ~100 km ahead of well-established populations, provides critical information for decision-making (Liebhold et al., 2021). Capturing a moth here triggers a series of actions similar to the eradication program, but treatments are likely to involve pheromone disruption, unless the catch is especially high (e.g. >100 moths/trap). Then, *Btk* treatments might be used. Traps are also placed every 5–8 km about 70 km behind the line of well-established populations to measure the progression of movement by *L. dispar dispar* (Liebhold et al., 2021). Slow-the-Spread is a coordinated effort between states in the transition area and USDA Forest Service.

USDA Forest Service also supports the suppression of *L. dispar dispar* populations in well-established areas during outbreaks. Areas may be treated with *Btk* or Gypcheck, a formulation of a nuclear polyhedrosis virus (NPV) that is highly host-specific and, therefore, particularly useful in environmentally sensitive areas. In addition, a fungal pathogen, *Entomophaga maimaiga*, also accidentally introduced, appears to be suppressing outbreaks of *L. dispar dispar* over large areas (Reilly et al., 2014).

4.2.2 *Halyomorpha halys* (United States)

The first confirmed United States record of the brown marmorated stink bug, *Halyomorpha halys* (Stål) (Hemiptera: Pentatomidae), was in 2001 in Pennsylvania, though the bug had been purportedly sighted in the area and intercepted at U.S. ports for many years prior. Native to eastern Asia, including Japan, Korea, and the temperate and subtropical regions of eastern China, *H. halys* likely arrived accidently in the United States associated with international travelers or commerce (Haye et al., 2015). This bug has a broad host range of over 170 plant species in North America, including many horticultural and row crops (Leskey and Nielsen, 2018) and is known to be an occasional economic pest in its native range (Lee et al., 2013). *Halyomorpha halys* can also become a severe nuisance pest due to its propensity to aggregate and overwinter in human structures (Hancock et al., 2019).

In 2010, outbreak populations in the mid-Atlantic United States caused severe to total losses in multiple crops (e.g. apples, peaches, and sweet corn) and resulted in tens of millions of dollars in damage. At the time, recommendations for management were nearly absent (Leskey and Nielsen, 2018). In its native range, mitigation of *H. halys* primarily relied on repeated insecticide applications, many of which were not registered for use in the affected crops in the United States (Kuhar and Kamminga, 2017). Laboratory and field research quickly identified insecticides for commercial U.S. applications, but without effective monitoring tools or alternative tactics, multi-fold increases in insecticide applications occurred in many crops (Leskey and Nielsen, 2018). In addition, most of the chemicals used were broad spectrum and had short residual activity, and the high mobility of *H. halys* made adequate application coverage difficult. Existing IPM programs were disrupted and secondary pest outbreaks occurred in many areas (Leskey et al., 2012, 2014).

Halyomorpha halys was known from its native range to be attracted to the pheromone of the stink bug, *Plautia stali* Scott, methyl (*E*,*E*,*Z*)-2,4,6 decatrienoate (MDT), but the semiochemical was not reliable for full-season monitoring (Nielsen et al., 2013). However, once the two-component male aggregation pheromone of *H. halys* [(3*S*,6*S*,7*R*,10*S*)-10,11-epoxy-1-bisabolen-3-ol and (3*R*,6*S*,7*R*,10*S*)-10,11-epoxy-1-bisabolen-3-ol] was isolated and found to be synergistic with MDT in baited traps, this combination provided season-long attraction to both adult and nymph bugs across multiple states (Leskey et al., 2015; Weber et al., 2014). Though less targeted, other sampling tools for *H. halys* have also been useful to monitor the spread and population buildup into new areas, including black light traps, beat and sweep net samples, and public reports (Hahn et al., 2016; Nielsen et al., 2013; Quarles, 2014).

Having effective, targeted semiochemical lures, which can be used in multiple trap designs (Acebes-Doria et al., 2018), allowed for the development of tools that have helped to significantly reduce insecticide inputs in some systems (e.g. apples). For example, using a provisional 10-bug threshold trap capture in apple orchards, insecticide applications were reduced by 40% (Short et al., 2017). Applications were further refined by exploiting the perimeter-driven aggregation behavior of *H. halys* in an attract-and-kill strategy, wherein movement and retention to crop perimeters are reinforced by pheromone lures and populations are removed through regular insecticide application (Morrison et al., 2019). Perimeter-targeted sprays against *H. halys* have also helped to begin restoring crop-based IPM programs in peaches (Blaauw et al.,

Published by Burleigh Dodds Science Publishing Limited, 2023.

2015) and apples (Akotsen-Mensah et al., 2020) and show some promise in soybeans (Aigner et al., 2017).

Other tools are actively being researched to support management of *H. halys* in agricultural settings, such as the use of insecticide-treated netting (Leskey et al., 2020), use of harmonic radar for monitoring the dispersal and movement behavior of individual bugs (Lee et al., 2014b), and use of canine detectors to characterize overwintering sites as possible targets of management (Lee et al., 2014a). A large, collaborative effort to characterize and evaluate the effectiveness of biological control in North America has been ongoing since the establishment of *H. halys* (Abram et al., 2017; Buffington et al., 2018). Generalist predators in the United States may be effective against *H. halys* in some areas (reviewed in Abram et al., 2017) but are often hampered by the insecticides used for *H. halys* (Leskey et al., 2020). Classical biological control evaluations were underway for parasitoid wasps found in the native range of *H. halys*, when adventive populations of one candidate species, *Trissolcus japonicus*, were discovered in multiple states (Ludwick et al., 2020). Within some of these states, efforts are underway to redistribute and enhance local *T. japonicus* populations (Ludwick et al., 2020). Work evaluating the interaction of insecticides commonly used against *H. halys* with *T. japonicus* indicates incompatibility of these tools (Lowenstein et al., 2019; Ribeiro et al., 2021), though some insecticides may be less detrimental in certain settings (Ludwick et al., 2020).

The most severe agricultural and nuisance problems have been reported from the mid-Atlantic and other portions of the eastern United States, particularly on specialty crops. Therefore, most management tactics have been developed within that context. However, *H. halys* is rapidly spreading throughout the United States and increasing in the extent of its impact (Ludwick et al., 2021). A critical part of the management efforts of *H. halys* in the USA has been the early coordination of research, extension, communication, and regulatory priorities through a national team of multidisciplinary members across 10 states (Ludwick et al., 2021). The immediate foci of the group were on mitigating impacts with chemical tactics, with the intention of integrating them with future tools (Leskey et al., 2012). The development of IPM programs for *H. halys* and reestablishment of programs disrupted by the insect's arrival are among the primary long-term goals of the group (Ludwick et al., 2021). Though many individual management tools have been, or are actively being, explored for *H. halys*, effective integration is an active area of research and remains to be demonstrated for most agroecosystems.

5 Conclusion

Invasive insects can be highly damaging in agriculture. For species that evade prevention attempts, eradication has been successful in stopping, or at least delaying, the widespread establishment of some populations. Success depends upon several factors, but common features include coordination among multiple partners and authorities, the ability to detect and enumerate populations, and the availability of multiple, complementary mitigation tactics. However, eradication is not feasible in many circumstances. Management of invasive insects during later stages of invasion using IPM is possible and has shown success. Consistently and clearly implementing IPM nonetheless continues to be challenging (Venette and Morey, 2019). Research continues to be needed to devise economically acceptable and optimal long-term management strategies for new invasive species where information is limited (Epanchin-Niell and Hastings, 2010).

6 Future trends

The number of attempted eradication programs for invasive insects continues to increase across the world (Suckling et al., 2021). Though the rate of new incursions dwarfs the available knowledge and tools needed to mitigate all harms, recent research and technology is making important progress in effective management and in highlighting areas where further works remains to be needed.

Tools to support eradication and long-term management of invasive insects have advanced in the past 20 years (Martinez et al., 2020). A notable example is the progress in genomic technology that can support decision-making and management throughout the invasion process. These genetic tools improve rapid, accurate diagnostics (e.g. eDNA (Allen et al., 2021; Valentin et al., 2018, 2020)) and the characterization of complex taxonomies (Garnas et al., 2016; Hamelin and Roe, 2020); identify population sources and spread pathways (Hamelin and Roe, 2020), which can help refine risk assessment components (e.g. Morey and Venette, 2020); and enhance targeted control through gene editing and engineering (Dyck et al., 2021; Harvey-Samuel et al., 2017; Maselko et al., 2020; Nikolouli et al., 2018; Teem et al., 2020). Automated monitoring and communications tools are improving spatial surveillance and area-wide management of invasive insects (Benavente-Sánchez et al., 2021; Schellhorn and Jones, 2021). The use of uncrewed aerial vehicles (i.e. drones) for monitoring or deploying population suppression technology is also an active subject of research (Iost Filho et al., 2020). For example, drones have been used in mark–release–recapture studies to track migrating *H. halys* (Stewart, 2021; Stumph et al., 2019).

Involvement of the public and citizen scientists to survey and monitor for invasive species is beneficial (Encarnação et al., 2021; Epanchin-Niell et al., 2021; Gallo and Waitt, 2011; Pyšek et al., 2020). Web-based applications that allow for mobile participation and information exchange have been particularly useful (Johnson et al., 2020). An estimated half of all new detections of invasive plant pests in New Zealand were reported from the public. In the United

States, all known infestations of *Anoplophora glabripennis* were first reported by private citizens (Tobin et al., 2014) and the recent incursion of the giant hornet, *V. mandarinia*, in the United States, was discovered by a private citizen (Baker, 2020). Effectively incorporating community surveillance in invasive species management, however, has received little research attention and is an important knowledge gap (Hester and Cacho, 2017). Additionally, alternatives to classical data analysis approaches continue to be needed to contend with the challenges of publicly collected data (Kamenova et al., 2017).

Lastly, research exploring societal issues and governance in invasion biology has increased in recent years, but challenges remain. Many emphasize the need for greater strides in engagement - that is, the incorporation of diverse knowledge, perspectives, and preferences - across management decisions phases, particularly for technologies that involve novel genetic and molecular tools (Kokotovich et al., 2020; Shackleton et al., 2019). Calls for REVALUATING and reframing approaches to invasive species management (Venette and Morey, 2019; Woodford et al., 2016) and policy (Keller et al., 2014) will undoubtedly require more effective integration of social and natural scientists (Marzano et al., 2017).

7 Acknowledgements

A portion of this project was supported by the Minnesota Invasive Terrestrial Plants and Pests Center through the Environment and Natural Resources Trust Fund, as recommended by the Legislative-Citizen Commission on Minnesota Resources. The findings and conclusions in this publication are those of the authors and should not be construed to represent any official USDA or U.S. Government determination of policy.

8 Where to look for further information

Educational texts and recent research summaries on the management of invasive insects and concepts of integrated pest management:

- Dyck, V. A., Hendrichs, J. and Robinson, A. S. (eds) (2021). Sterile Insect Technique - Principles and Practice in Area-Wide Integrated Pest Management. 2nd ed. Boca Raton, FL: CRC Press. DOI: 10.1201/9781003035572.
- Flint, M. L. (2012). IPM in Practice: Principles and Methods of Integrated Pest Management. 2nd ed. Oakland, CA: University of California Agriculture and Natural Resources.
- Hendrichs, J., Pereira, R. and Vreysen, M. J. B. (eds) (2021). Area-Wide Integrated Pest Management. 1st ed. Boca Raton, FL: CRC Press. DOI: 10.1201/9781003169239.

- Poland, T. M., Patel-Weynand, T., Finch, D. M., Miniat, C. F., Hayes, D. C. and Lopez, V. M. (eds) (2021). Invasive Species in Forests and Rangelands of the United States. Cham: Springer International Publishing. DOI: 10.1007/978-3-030-45367-1.
- Radcliffe, E. B., Hutchinson, W. D. and Cancelado, R. E. (eds) (2009). Integrated Pest Management: Concepts, Tactics, Strategies and Case Studies. Cambridge University Press.

Global organizations with expertise in, or regulatory authority of, invasive species and their management:

- CAB International, Invasive Species Compendium (https://www.cabi.org/isc/).
- Canadian Food Inspection Agency (https://inspection.canada.ca/).
- Centre of Excellence for Invasion Biology, Stellenbosch University (http://academic.sun.ac.za/cib/team_research.asp).
- CSIRO Australia (https://www.csiro.au/en/Research/BF).
- European Food Safety Authority, Panel on Plant Health (https://www.efsa.europa.eu/en/science/scientific-committee-and-panels/plh).
- European and Mediterranean Plant Protection Organization (https://www.eppo.int/).
- Instituto Chico Mendes de Conservação da Biodiversidade (https://www.gov.br/icmbio/pt-br).
- International Plant Protection Convention, Food and Agriculture Organization of the United Nations (https://www.ippc.int/en/).
- Ministry of the Environment, Government of Japan (https://www.env.go.jp/en/nature/).
- North American Invasive Species Management Association (https://naisma.org/).
- North American Plant Protection Organization (https://www.nappo.org/).
- United Kingdom Department for Environment, Food and Rural Affairs (https://secure.fera.defra.gov.uk/phiw/riskRegister/).
- United States Department of Agriculture, Animal and Plant Health Inspection Service, Plant Protection and Quarantine (https://www.aphis.usda.gov/aphis/ourfocus/planthealth).

The United States Department of Agriculture, Forest Service makes all research publications, which includes many on invasive insects, publicly available through TreeSearch (https://www.fs.usda.gov/treesearch/).

The GERDA (http://b3.net.nz/gerda/) is a free online resource that compiles information on global incursions and eradication programs of terrestrial arthropods and plant pathogens.

Geographic records of the occurrence of invasive species are publicly available through the Early Detection and Distribution Mapping System (https://www.eddmaps.org/) for the United States and Canada. The Global Biodiversity Information Facility (https://www.gbif.org/) is a global database of species occurrence records, including those that may be invasive in certain areas.

9 References

Abram, P. K., Hoelmer, K. A., Acebes-Doria, A., Andrews, H., Beers, E. H., Bergh, J. C., Bessin, R., Biddinger, D., Botch, P., Buffington, M. L., Cornelius, M. L., Costi, E., Delfosse, E. S., Dieckhoff, C., Dobson, R., Donais, Z., Grieshop, M., Hamilton, G., Haye, T., Hedstrom, C., Herlihy, M. V., Hoddle, M. S., Hooks, C. R. R., Jentsch, P., Joshi, N. K., Kuhar, T. P., Lara, J., Lee, J. C., Legrand, A., Leskey, T. C., Lowenstein, D., Maistrello, L., Mathews, C. R., Milnes, J. M., Morrison, W. R., Nielsen, A. L., Ogburn, E. C., Pickett, C. H., Poley, K., Pote, J., Radl, J., Shrewsbury, P. M., Talamas, E., Tavella, L., Walgenbach, J. F., Waterworth, R., Weber, D. C., Welty, C. and Wiman, N. G. (2017). Indigenous arthropod natural enemies of the invasive brown marmorated stink bug in North America and Europe. *Journal of Pest Science* 90(4): 1009–1020. DOI: 10.1007/s10340-017-0891-7.

Acebes-Doria, A. L., Morrison, W. R., Short, B. D., Rice, K. B., Bush, H. G., Kuhar, T. P., Duthie, C. and Leskey, T. C. (2018). Monitoring and biosurveillance tools for the brown marmorated stink bug, *Halyomorpha halys* (Stål) (Hemiptera: Pentatomidae). *Insects* 9(3): 82. DOI: 10.3390/insects9030082.

Aigner, B. L., Kuhar, T. P., Herbert, D. A., Brewster, C. C., Hogue, J. W. and Aigner, J. D. (2017). Brown marmorated stink bug (Hemiptera: Pentatomidae) infestations in tree borders and subsequent patterns of abundance in soybean fields. *Journal of Economic Entomology* 110(2): 487–490. DOI: 10.1093/jee/tox047.

Akotsen-Mensah, C., Blaauw, B., Short, B., Leskey, T. C., Bergh, J. C., Polk, D. and Nielsen, A. L. (2020). Using IPM-CPR as a management program for apple orchards. *Journal of Economic Entomology* 113(4): 1894–1902. DOI: 10.1093/jee/toaa087.

Allen, M. C., Nielsen, A. L., Peterson, D. L. and Lockwood, J. L. (2021). Terrestrial eDNA survey outperforms conventional approach for detecting an invasive pest insect within an agricultural ecosystem. *Environmental DNA* 3(6): 1102–1112. DOI: 10.1002/edn3.231.

Arca, M., Mougel, F., Guillemaud, T., Dupas, S., Rome, Q., Perrard, A., Muller, F., Fossoud, A., Capdevielle-Dulac, C., Torres-Leguizamon, M., Chen, X. X., Tan, J. L., Jung, C., Villemant, C., Arnold, G. and Silvain, J.-F. (2015). Reconstructing the invasion and the demographic history of the yellow-legged hornet, *Vespa velutina*, in Europe. *Biological Invasions* 17(8): 2357–2371. DOI: 10.1007/s10530-015-0880-9.

Augustin, S., De Kogel, W. J., Donner, P., Faccoli, M., Lees, D. C., Marini, L., Mori, N., Toffolo, E. P., Quilici, S., Roques, A., Yart, A. and Battisti, A. (2012). A list of methods to detect arthropod quarantine pests in Europe. *EPPO Bulletin* 42(1): 93–94. DOI: 10.1111/j.1365-2338.2011.02523.x.

Baker, M. (2020). Murder hornets' in the U.S.: The rush to stop the Asian giant hornet. *The New York Times*, 2 May. Available at: https://www.nytimes.com/2020/05/02/us/asian-giant-hornet-washington.html.

Baloch, M. N., Fan, J., Haseeb, M. and Zhang, R. (2020). Mapping potential distribution of *Spodoptera frugiperda* (Lepidoptera: Noctuidae) in Central Asia. *Insects* 11(3): 1-10. DOI: 10.3390/insects11030172.

Barron, M. C., Liebhold, A. M., Kean, J. M., Richardson, B. and Brockerhoff, E. G. (2020). Habitat fragmentation and eradication of invading insect herbivores. *Journal of Applied Ecology* 57(3): 590-598. DOI: 10.1111/1365-2664.13554.

Bartlett, A. C. and Staten, R. T. (1996). The sterile insect release method and other genetic control strategies. *Radcliffe's IPM World Textbook*. Available at: https://ipmworld.umn.edu/bartlett.

Bebber, D. P., Ramotowski, M. A. T. and Gurr, S. J. (2013). Crop pests and pathogens move polewards in a warming world. *Nature Climate Change* 3(11): 985-988. DOI: 10.1038/nclimate1990.

Bello-Rivera, A., Pereira, R., Enkerlin, W., Bloem, S., Bloem, K., Hight, S. D., Carpenter, J. E., Zimmermann, H. G., Sanchez-Anguiano, H. M., Zetina-Rodriguez, R. and Trujillo-Arriaga, F. J. (2021). Successful area-wide programme that eradicated outbreaks of the invasive cactus moth in Mexico. In: *Area-Wide Integrated Pest Management*. CRC Press, pp. 561-580. DOI: 10.1201/9781003169239-30.

Benavente-Sánchez, D., Moreno-Molina, J. and Argilés-Herrero, R. (2021). Prospects for remotely piloted aircraft systems in area-wide integrated pest management programmes. In: *Area-Wide Integrated Pest Management*. CRC Press, pp. 903-916. DOI: 10.1201/9781003169239-52.

Berenbaum, M., Bundy, T. E., Cash, S. B., Jones, V. P., Mills, N. J., Moffit, L. J., Powell, J. A., Simberloff, D. S. and Venette, R. C. (2009). Review of the U.S. Department of Agriculture's Animal and Plant Health Inspection Service Response to Petitions to Reclassify the Light Brown Apple Moth as a Non-actionable Pest. A Letter Report. Available at: http://www.nap.edu/catalog/12762.html. Washington, DC: National Academy of Science.

Berube, C. (2020). Giant alien insect invasion averted - Canadian beekeepers thwart apicultural disaster. *American Bee Journal (February)*: 209-214. Available at: https://www.beesforbabar.org/pdf/20200130abj_agh.pdf.

Blaauw, B. R., Polk, D. and Nielsen, A. L. (2015). IPM-CPR for peaches: Incorporating behaviorally-based methods to manage *Halyomorpha halys* and key pests in peach. *Pest Management Science* 71(11): 1513-1522. DOI: 10.1002/ps.3955.

Blackburn, T. M., Essl, F., Evans, T., Hulme, P. E., Jeschke, J. M., Kühn, I., Kumschick, S., Marková, Z., Mrugała, A., Nentwig, W., Pergl, J., Pyšek, P., Rabitsch, W., Ricciardi, A., Richardson, D. M., Sendek, A., Vilà, M., Wilson, J. R., Winter, M., Genovesi, P. and Bacher, S. (2014). A unified classification of alien species based on the magnitude of their environmental impacts. *PLoS Biology* 12(5): e1001850. DOI: 10.1371/journal.pbio.1001850.

Blackwood, J. C., Berec, L., Yamanaka, T., Epanchin-Niell, R. S., Hastings, A. and Liebhold, A. M. (2012). Bioeconomic synergy between tactics for insect eradication in the presence of Allee effects. *Proceedings. Biological Sciences* 279(1739): 2807-2815. DOI: 10.1098/rspb.2012.0255.

Bloem, K., Bloem, S., Carpenter, J., Hight, S., Floyd, J. and Zimmermann, H. (2007). Don't let cacto blast us: Development of a bi-national plan to stop the spread of the cactus moth *Cactoblastis cactorum* in North America. In: Vreysen, M. J. B., Robinson, A. S. and Hendrichs, J. (Eds) *Area-Wide Control of Insect Pests*. Springer Netherlands, pp. 337-344. DOI: 10.1007/978-1-4020-6059-5_31.

Bloem, S., Hight, S. D., Carpenter, J. E. and Bloem, K. A. (2005). Development of the most effective trap to monitor the presence of the cactus moth *Cactoblastis cactorum* (Lepidoptera: Pyralidae). *Florida Entomologist* 88(3): 300-306. DOI: 10.1653/0015-4040(2005)088[0300:DOTMET]2.0.CO;2.

Bradshaw, C. J. A., Leroy, B., Bellard, C., Roiz, D., Albert, C., Fournier, A., Barbet-Massin, M., Salles, J. M., Simard, F. and Courchamp, F. (2016). Massive yet grossly underestimated global costs of invasive insects. *Nature Communications* 7(1): 12986. DOI: 10.1038/ncomms12986.

Brockerhoff, E. G., Liebhold, A. M., Richardson, B. and Suckling, D. M. (2010). Eradication of invasive forest insects: Concepts, methods, costs and benefits. *New Zealand Journal of Forestry Science* 40: 117-135. Available at: www.scionresearch.com/nzjfs.

Buffington, M. L., Talamas, E. J. and Hoelmer, K. A. (2018). Team *Trissolcus*: Integrating taxonomy and biological control to combat the brown marmorated stink bug. *American Entomologist* 64(4): 224-232. DOI: 10.1093/ae/tmy057.

Carde, R. T., Doane, C. C., Granett, J., Hill, A. S., Kochansky, J. and Roelofs, W. L. (1977). Attractancy of racemic disparlure and certain analogues to male gypsy moths and the effect of trap placement. *Environmental Entomology* 6(6): 765-767. DOI: 10.1093/ee/6.6.765.

Carey, J. R. and Harder, D. (2013). Clear, present, significant, & imminent danger questions for the California light brown apple moth (*Epiphyas postvittana*) technical working group. *American Entomologist* 59(4): 240-247.

CFIA. (2020). Decision document: Vespa Mandarinia (Asian giant hornet). Available at: https://inspection.canada.ca/plant-health/invasive-species/insects/asian-giant-hornet/decision-document/eng/1593718645505/1593718645899. Canadian Food Inspection Agency.

Chambers, J. C., Allen, C. R. and Cushman, S. A. (2019). Operationalizing ecological resilience concepts for managing species and ecosystems at risk. *Frontiers in Ecology and Evolution* 7(July): 1-27. DOI: 10.3389/fevo.2019.00241.

Colautti, R. I., Ricciardi, A., Grigorovich, I. A. and MacIsaac, H. J. (2004). Is invasion success explained by the enemy release hypothesis? *Ecology Letters* 7(8): 721-733. DOI: 10.1111/j.1461-0248.2004.00616.x.

Common, I. F. B. (1990). *Moths of Australia*. Melbourne University Press.

Comont, R. F., Purse, B. V., Phillips, W., Kunin, W. E., Hanson, M., Lewis, O. T., Harrington, R., Shortall, C. R., Rondoni, G. and Roy, H. E. (2014). Escape from parasitism by the invasive alien ladybird, *Harmonia axyridis*. *Insect Conservation and Diversity* 7(4): 334-342. DOI: 10.1111/icad.12060.

Crowley, S. L., Hinchliffe, S. and McDonald, R. A. (2017). Conflict in invasive species management. *Frontiers in Ecology and the Environment* 15(3): 133-141. DOI: 10.1002/fee.1471.

Davis, M. A. (2009). *Invasion Biology*. Oxford University Press.

Dooley, B. (2020). In Japan, the 'murder hornet' is both a lethal threat and a tasty treat. *The New York Times*, 5 May. Available at: https://www.nytimes.com/2020/05/05/world/asia/murder-hornet-japan.html?action=click&auth=login-email&login=email&module=RelatedLinks&pgtype=Article.

Dyck, V. A., Hendrichs, J. and Robinson, A. S. (Eds). (2021). *Sterile Insect Technique - Principles and Practice in Area-Wide Integrated Pest Management* (2nd edn.). Boca Raton, FL: CRC Press. DOI: 10.1201/9781003035572.

Jeger, M., Bragard, C., Caffier, D., Candresse, T., Chatzivassiliou, E., Dehnen-Schmutz, K., Gilioli, G., Gregoire, J. C., Jaques Miret, J. A., Navarro, M. N., Niere, B., Parnell, S., Potting, R., Rafoss, T., Rossi, V., Urek, G., Van Bruggen, A., Van der Werf, W., West, J., Winter, S., Gardi, C., Aukhojee, M. and MacLeod, A. (2017). Pest categorisation of *Spodoptera frugiperda*. *EFSA Journal: European Food Safety Authority* 15(7): e04927. DOI: 10.2903/j.efsa.2017.4927.

Ehler, L. E. (2006). Integrated pest management (IPM): Definition, historical development and implementation, and the other IPM. *Pest Management Science* 62(9): 787–789. DOI: 10.1002/ps.1247.

El-Sayed, A. M., Suckling, D. M., Byers, J. A., Jang, E. B. and Wearing, C. H. (2009). Potential of "lure and kill" in long-term pest management and eradication of invasive species. *Journal of Economic Entomology* 102(3): 815–835.

Encarnação, J., Teodósio, M. A. and Morais, P. (2021). Citizen science and biological invasions: A review. *Frontiers in Environmental Science* 8: 1–13. DOI: 10.3389/fenvs.2020.602980.

Epanchin-Niell, R., Thompson, A. L. and Treakle, T. (2021). Public contributions to early detection of new invasive pests. *Conservation Science and Practice* 3(6): 1–15. DOI: 10.1111/csp2.422.

Epanchin-Niell, R. S., Haight, R. G., Berec, L., Kean, J. M. and Liebhold, A. M. (2012). Optimal surveillance and eradication of invasive species in heterogeneous landscapes. *Ecology Letters* 15(8): 803–812. DOI: 10.1111/j.1461-0248.2012.01800.x.

Epanchin-Niell, R. S. and Hastings, A. (2010). Controlling established invaders: Integrating economics and spread dynamics to determine optimal management. *Ecology Letters* 13(4): 528–541. DOI: 10.1111/j.1461-0248.2010.01440.x.

Epanchin-Niell, R. S., Hufford, M. B., Aslan, C. E., Sexton, J. P., Port, J. D. and Waring, T. M. (2010). Controlling invasive species in complex social landscapes. *Frontiers in Ecology and the Environment* 8(4): 210–216. DOI: 10.1890/090029.

FAO. (2017). Guidelines for pest eradication programmes (ISPM 9). Food and Agriculture Organization of the United Nations, International Plant Protection Convention Secretariat. Available at: https://www.fao.org/publications/card/en/c/30115bd4-b440-43e5-849d-ef425f343a41/.

Follett, P. A. and Neven, L. G. (2006). Current trends in quarantine entomology. *Annual Review of Entomology* 51: 359–385. DOI: 10.1146/annurev.ento.49.061802.123314.

Gallo, T. and Waitt, D. (2011). Creating a successful citizen science model to detect and report invasive species. *BioScience* 61(6): 459–465. DOI: 10.1525/bio.2011.61.6.8.

Garnas, J. R., Auger-Rozenberg, M.-A., Roques, A., Bertelsmeier, C., Wingfield, M. J., Saccaggi, D. L., Roy, H. E. and Slippers, B. (2016). Complex patterns of global spread in invasive insects: Eco-evolutionary and management consequences. *Biological Invasions* 18(4): 935–952. DOI: 10.1007/s10530-016-1082-9.

Goergen, G., Kumar, P. L., Sankung, S. B., Togola, A. and Tamò, M. (2016). First report of outbreaks of the fall armyworm *Spodoptera frugiperda* (J. E. Smith) (Lepidoptera, Noctuidae), a new alien invasive pest in West and Central Africa. *PLoS ONE* 11(10): e0165632. DOI: 10.1371/journal.pone.0165632.

Gray, M. E., Ratcliffe, S. T. and Rice, M. E. (2009). The IPM paradigm: Concepts, strategies and tactics. In: Radcliffe, E. B., Hutchison, W. D. and Cancelado, R. E. (Eds) *Integrated Pest Management: Concepts, Tactics, Strategies and Case Studies*. Cambridge University Press, pp. 1–13.

Gutierrez-Arellano, C. and Mulligan, M. (2018). A review of regulation ecosystem services and disservices from faunal populations and potential impacts of agriculturalisation on their provision, globally. *Nature Conservation* 30: 1–39. DOI: 10.3897/nature conservation.30.26989.

Hahn, N. G., Kaufman, A. J., Rodriguez-Saona, C., Nielsen, A. L., LaForest, J. and Hamilton, G. C. (2016). Exploring the spread of brown marmorated stink bug in New Jersey through the use of crowdsourced reports. *American Entomologist* 62(1): 36–45. DOI: 10.1093/ae/tmw007.

Hamelin, R. C. and Roe, A. D. (2020). Genomic biosurveillance of forest invasive alien enemies: A story written in code. *Evolutionary Applications* 13(1): 95–115. DOI: 10.1111/eva.12853.

Hancock, T. J., Lee, D., Bergh, J. C., Morrison, W. R. and Leskey, T. C. (2019). Presence of the invasive brown marmorated stink bug *Halyomorpha halys* (Stål) (Hemiptera: Pentatomidae) on home exteriors during the autumn dispersal period: Results generated by citizen scientists. *Agricultural and Forest Entomology* 21(1): 99–108. DOI: 10.1111/afe.12312.

Hanley, N. and Roberts, M. (2019). The economic benefits of invasive species management. *People and Nature* 1(2): 124–137. DOI: 10.1002/pan3.31.

Harvey-Samuel, T., Ant, T. and Alphey, L. (2017). Towards the genetic control of invasive species. *Biological Invasions* 19(6): 1683–1703. DOI: 10.1007/s10530-017-1384-6.

Haye, T., Gariepy, T., Hoelmer, K., Rossi, J., Streito, J., Tassus, X. and Desneux, N. (2015). Range expansion of the invasive brown marmorated stinkbug, *Halyomorpha halys*: An increasing threat to field, fruit and vegetable crops worldwide. *Journal of Pest Science* 88(4): 665–673. DOI: 10.1007/s10340-015-0670-2.

Hendrichs, J., Enkerlin, W. R. and Pereira, R. (2021). Invasive insect pests: Challenges and the role of the sterile insect technique in their prevention, containment, and eradication. In: Dyck, V. A., Hendrichs, J. and Robinson, A. S. (Eds) *Sterile Insect Technique: Principles and Practice in Area-Wide Integrated Pest Managment* (2nd edn.). CRC Press, pp. 885–922. DOI: 10.1201/9781003035572-27.

Hendrichs, J., Kenmore, P., Robinson, A. S. and Vreysen, M. J. B. (2007). Area-Wide Integrated Pest Management (AW-IPM): Principles, practice and prospects. In: Vreysen, M. J. B., Robinson, A. S. and Hendrichs, J. (Eds) *Area-Wide Control of Insect Pests*. Dordrecht: Springer Netherlands, pp. 3–33. DOI: 10.1007/978-1-4020-6059-5_1.

Henneberry, T. J. (2007). Integrated systems for control of the pink bollworm *Pectinophora gossypiella* in cotton. In: Vreysen, M. J. B., Robinson, A. S. and Hendrichs, J. (Eds) *Area-Wide Control of Insect Pests*. Dordrecht: Springer Netherlands, pp. 567–579. DOI: 10.1007/978-1-4020-6059-5_53.

Hester, S. M. and Cacho, O. J. (2017). The contribution of passive surveillance to invasive species management. *Biological Invasions* 19(3): 737–748. DOI: 10.1007/s10530-016-1362-4.

Hulme, P. E. (2009). Trade, transport and trouble: Managing invasive species pathways in an era of globalization. *Journal of Applied Ecology* 46(1): 10–18. DOI: 10.1111/j.1365-2664.2008.01600.x.

Iost Filho, F. H., Heldens, W. B., Kong, Z. and de Lange, E. S. (2020). Drones: Innovative technology for use in precision pest management. *Journal of Economic Entomology* 113(1): 1–25. DOI: 10.1093/jee/toz268.

Jeschke, J., Gómez Aparicio, L., Haider, S., Heger, T., Lortie, C., Pyšek, P. and Strayer, D. (2012). Support for major hypotheses in invasion biology is uneven and declining. *NeoBiota* 14: 1–20. DOI: 10.3897/neobiota.14.3435.

Johnson, B. A., Mader, A. D., Dasgupta, R. and Kumar, P. (2020). Citizen science and invasive alien species: An analysis of citizen science initiatives using information and communications technology (ICT) to collect invasive alien species observations. *Global Ecology and Conservation* 21: e00812. DOI: 10.1016/j.gecco.2019.e00812.

Kalaris, T., Fieselmann, D., Magarey, R., Colunga-Garcia, M., Roda, A., Hardie, D., Cogger, N., Hammond, N., Martin, P. A. T. and Whittle, P. (2014). The role of surveillance methods and technologies in Plant Biosecurity. In: Gordh, G. and McKirdy, S. (Eds) *The Handbook of Plant Biosecurity*. Dordrecht: Springer Science+Business Media. DOI: 10.1007/978-94-007-7365-3_11.

Kamenova, S., Bartley, T. J., Bohan, D. A., Boutain, J. R., Colautti, R. I., Domaizon, I., Fontaine, C., Lemainque, A., Le Viol, I., Mollot, G., Perga, M.-E., Ravigné, V. and Massol, F. (2017). Invasions toolkit: Current methods for tracking the spread and impact of invasive species. In: *Advances in Ecological Research* (1st edn.). Elsevier Ltd., pp. 85–182. DOI: 10.1016/bs.aecr.2016.10.009.

Kean, J. M., Suckling, D. M., Sullivan, N. J., Tobin, P. C., Stringer, L. D., Smith, G. R., Kimber, B., Lee, D. C., Flores Vargas, R., Fletcher, J., Macbeth, F., McCullough, D. G., Herms, D. A. et al. (2022). Global eradication and response database (GERDA). Available at: http://b3.net.nz/gerda.

Keller, R. P., Cadotte, M. W. and Sandiford, G. (Eds). (2014). *Invasive Species in a Globalized World: Ecological, Social, and Legal Perspectives on Policy*. University of Chicago Press. DOI: 10.7208/chicago/9780226166216.001.0001.

Kenis, M., Auger-Rozenberg, M. A., Roques, A., Timms, L., Péré, C., Cock, M. J. W., Settele, J., Augustin, S. and Lopez-Vaamonde, C. (2009). Ecological effects of invasive alien insects. *Biological Invasions* 11(1): 21–45. DOI: 10.1007/s10530-008-9318-y.

Kim, S., Ju, C., Kim, J. and Son, H. I. (2019). A tracking method for the invasive Asian hornet: A brief review and experiments. *IEEE Access* 7: 176998–177008. DOI: 10.1109/ACCESS.2019.2958153.

Klassen, W. (1989). *Eradication of Introduced Arthropod Pests: Theory and Historical Practice. Miscellaneous Publication No. 73*. Entomological Society of America.

Koch, J., Pearson, D. E., Huebner, C. D., Young, M. K. and Sniezko, R. A. (2021). Restoration of landscapes and habitats affected by established invasive species. In: Poland, T. M., Patel-Weynand, T., Finch, D. M., et al. (Eds) *Invasive Species in Forests and Rangelands of the United States*. Cham: Springer, pp. 185–202. DOI: 10.1007/978-3-030-45367-1_8.

Koch, R. L. and Galvan, T. L. (2007). Bad side of a good beetle: The North American experience with *Harmonia axyridis*. In: Roy, H. E. and Wajnberg, E. (Eds) *From Biological Control to Invasion: The Ladybird Harmonia axyridis as a Model Species*. Springer Netherlands, pp. 23–35. DOI: 10.1007/978-1-4020-6939-0_3.

Koch, R. L., Potter, B. D., Glogoza, P. A., Hodgson, E. W., Krupke, C. H., Tooker, J. F., DiFonzo, C. D., Michel, A. P., Tilmon, K. J., Prochaska, T. J., Knodel, J. J., Wright, R. J., Hunt, T. E., Jensen, B., Varenhorst, A. J., McCornack, B. P., Estes, K. A. and Spencer, J. L. (2016). Biology and economics of recommendations for insecticide-based management of soybean aphid. *Plant Health Progress* 17(4): 265–269. DOI: 10.1094/PHP-RV-16-0061.

Kogan, M. (1998). Integrated pest management: Historical perspectives and contemporary developments. *Annual Review of Entomology* 43: 243–270.

Kokotovich, A. E., Delborne, J. A., Elsensohn, J. and Burrack, H. (2020). Emerging technologies for invasive insects: The role of engagement. *Annals of the Entomological Society of America* 113(4): 266–279. DOI: 10.1093/AESA/SAZ064.

Kolar, C. S. and Lodge, D. M. (2001). Progress in invasion biology: Predicting invaders. *Trends in Ecology and Evolution* 16(4): 199–204. Available at: http://www.ncbi.nlm.nih.gov/pubmed/11245943.

Kuhar, T. P. and Kamminga, K. (2017). Review of the chemical control research on *Halyomorpha halys* in the USA. *Journal of Pest Science* 90(4): 1021-1031. DOI: 10.1007/s10340-017-0859-7.

Lampert, A. and Liebhold, A. M. (2021). Combining multiple tactics over time for cost-effective eradication of invading insect populations. *Ecology Letters* 24(2): 279–287. DOI: 10.1111/ele.13640.

Lee, D. H., Cullum, J. P., Anderson, J. L., Daugherty, J. L., Beckett, L. M. and Leskey, T. C. (2014a). Characterization of overwintering sites of the invasive brown marmorated stink bug in natural landscapes using human surveyors and detector canines. *PLoS ONE* 9(4): e91575. DOI: 10.1371/journal.pone.0091575.

Lee, D.-H., Park, C.-G., Seo, B. Y., Boiteau, G., Vincent, C. and Leskey, T. C. (2014b). Detectability of *Halyomorpha halys* (Hemiptera: Pentatomidae) by portable harmonic radar in agricultural landscapes. *Florida Entomologist* 97(3): 1131–1138.

Lee, D. H., Short, B. D., Joseph, S. V., Bergh, J. C. and Leskey, T. C. (2013). Review of the biology, ecology, and management of *Halyomorpha halys* (Hemiptera: Pentatomidae) in China, Japan, and the Republic of Korea. *Environmental Entomology* 42(4): 627–641. DOI: 10.1603/EN13006.

Leonhardt, B. A., Mastro, V. C., Leonard, D. S., McLane, W., Reardon, R. C. and Thorpe, K. W. (1996). Control of low-density gypsy moth (Lepidoptera: Lymantriidae) populations by mating disruption with pheromone. *Journal of Chemical Ecology* 22(7): 1255–1272. DOI: 10.1007/BF02266964.

Leskey, T. C., Agnello, A., Bergh, J. C., Dively, G. P., Hamilton, G. C., Jentsch, P., Khrimian, A., Krawczyk, G., Kuhar, T. P., Lee, D. H., Morrison, W. R., Polk, D. F., Rodriguez-Saona, C., Shearer, P. W., Short, B. D., Shrewsbury, P. M., Walgenbach, J. F., Weber, D. C., Welty, C., Whalen, J., Wiman, N. and Zaman, F. (2015). Attraction of the invasive *Halyomorpha halys* (Hemiptera: Pentatomidae) to traps baited with semiochemical stimuli across the United States. *Environmental Entomology* 44(3): 746–756. DOI: 10.1093/ee/nvv049.

Leskey, T. C., Hamilton, G. C., Nielsen, A. L., Polk, D. F., Rodriguez-Saona, C., Bergh, J. C., Herbert, D. A., Kuhar, T. P., Pfeiffer, D., Dively, G. P., Hooks, C. R. R., Raupp, M. J., Shrewsbury, P. M., Krawczyk, G., Shearer, P. W., Whalen, J., Koplinka-Loehr, C., Myers, E., Inkley, D., Hoelmer, K. A., Lee, D. and Wright, S. E. (2012). Pest status of the brown marmorated stink bug, *Halyomorpha halys* in the USA. *Outlooks on Pest Management* 23(5): 218–226. DOI: 10.1564/23oct07.

Leskey, T. C. and Nielsen, A. L. (2018). Impact of the invasive brown marmorated stink bug in North America and Europe: History, biology, ecology, and management. *Annual Review of Entomology* 63(1): 599–618. DOI: 10.1146/annurev-ento-020117-043226.

Leskey, T. C., Short, B. D. and Lee, D. H. (2014). Efficacy of insecticide residues on adult *Halyomorpha halys* (Stål) (Hemiptera: Pentatomidae) mortality and injury in apple and peach orchards. *Pest Management Science* 70(7): 1097–1104. DOI: 10.1002/ps.3653.

Leskey, T. C., Short, B. D. and Ludwick, D. (2020). Comparison and refinement of integrated pest management tactics for *Halyomorpha halys* (Hemiptera: Pentatomidae) management in apple orchards. *Journal of Economic Entomology* 113(4): 1725-1734. DOI: 10.1093/jee/toaa067.

Leung, B., Lodge, D. M., Finnoff, D., Shogren, J. F., Lewis, M. A. and Lamberti, G. (2002). An ounce of prevention or a pound of cure: Bioeconomic risk analysis of invasive species. *Proceedings. Biological Sciences* 269(1508): 2407-2413. DOI: 10.1098/rspb.2002.2179.

Liebhold, A. M. (2014). Responses to 'Clear, present, significant, and imminent danger: Questions for the California Light Brown Apple Moth (*Epiphyas postvittana*) Technical Working Group'. *American Entomologist* 60(4): 244-248.

Liebhold, A. M., Berec, L., Brockerhoff, E. G., Epanchin-Niell, R. S., Hastings, A., Herms, D. A., Kean, J. M., McCullough, D. G., Suckling, D. M., Tobin, P. C. and Yamanaka, T. (2016). Eradication of invading insect populations: From concepts to applications. *Annual Review of Entomology* 61(1): 335-352. DOI: 10.1146/annurev-ento-010715-023809.

Liebhold, A. M. and Kean, J. M. (2019). Eradication and containment of non-native forest insects: Successes and failures. *Journal of Pest Science* 92(1): 83-91. DOI: 10.1007/s10340-018-1056-z.

Liebhold, A. M., Leonard, D., Marra, J. L., et al. (2021). Area-wide management of invading gypsy moth (*Lymantria dispar*) populations in the USA. In: Hendrichs, J., Pereira, R. and Vreysen, M. J. B. (Eds) *Area-Wide Integrated Pest Management*. CRC Press, pp. 551-560. DOI: 10.1201/9781003169239-29.

Lindeman, N. (2013). Subjectivized knowledge and grassroots advocacy: An analysis of an environmental controversy in Northern California. *Journal of Business and Technical Communication* 27(1): 62-90. DOI: 10.1177/1050651912448871.

Lodge, D. M., Williams, S., MacIsaac, H. J., Hayes, K. R., Leung, B., Reichard, S., Mack, R. N., Moyle, P. B., Smith, M., Andow, D. A., Carlton, J. T. and McMichael, A. (2006). Biological invasions: Recommendations for U.S. policy and management. *Ecological Applications : A Publication of the Ecological Society of America* 16(6): 2035-2054. DOI: 10.1890/1051-0761(2006)016[2035:BIRFUP]2.0.CO;2.

Lowenstein, D. M., Andrews, H., Mugica, A. and Wiman, N. G. (2019). Sensitivity of the egg parasitoid *Trissolcus japonicus* (Hymenoptera: Scelionidae) to field and laboratory-applied insecticide residue. *Journal of Economic Entomology* 112(5): 2077-2084. DOI: 10.1093/jee/toz127.

Ludwick, D., Morrison, W. R., Acebes-Doria, A. L., Agnello, A. M., Bergh, J. C., Buffington, M. L., Hamilton, G. C., Harper, J. K., Hoelmer, K. A., Krawczyk, G., Kuhar, T. P., Pfeiffer, D. G., Nielsen, A. L., Rice, K. B., Rodriguez-Saona, C., Shearer, P. W., Shrewsbury, P. M., Talamas, E. J., Walgenbach, J. F., Wiman, N. G. and Leskey, T. C. (2021). Invasion of the brown marmorated stink bug (Hemiptera: Pentatomidae) into the United States: Developing a national response to an invasive species crisis through collaborative research and outreach efforts. *Journal of Integrated Pest Management* 11(1): 4. DOI: 10.1093/jipm/pmaa001.

Ludwick, D. C., Patterson, J., Leake, L. B., Carper, L. and Leskey, T. C. (2020). Integrating *Trissolcus japonicus* (Ashmead, 1904) (Hymenoptera: Scelionidae) into management programs for *Halyomorpha halys* (Stål, 1855) (Hemiptera: Pentatomidae) in apple orchards: Impact of insecticide applications and spray patterns. *Insects* 11(12): 833. DOI: 10.3390/insects11120833.

MAF (2002). Painted apple moth: Reassessment of potential economic impacts. *Report for the Ministry of Agriculture and Forestry New Zealand*. Available at: https://citeseerx.ist.psu.edu/viewdoc/download?doi=10.1.1.730.5398&rep=rep1&type=pdf.

Makino, S. (2016). Post-hibernation ovary development in queens of the Japanese giant hornet *Vespa mandarinia* (Hymenoptera: Vespidae). *Entomological Science* 19(4): 440–443. DOI: 10.1111/ens.12205.

Marsico, T. D., Wallace, L. E., Ervin, G. N., Brooks, C. P., McClure, J. E. and Welch, M. E. (2011). Geographic patterns of genetic diversity from the native range of *Cactoblastis cactorum* (Berg) support the documented history of invasion and multiple introductions for invasive populations. *Biological Invasions* 13(4): 857–868. DOI: 10.1007/s10530-010-9874-9.

Martinez, B., Reaser, J. K., Dehgan, A., Zamft, B., Baisch, D., McCormick, C., Giordano, A. J., Aicher, R. and Selbe, S. (2020). Technology innovation: Advancing capacities for the early detection of and rapid response to invasive species. *Biological Invasions* 22(1): 75–100. DOI: 10.1007/s10530-019-02146-y.

Marzano, M., Allen, W., Haight, R. G., Holmes, T. P., Keskitalo, E. C. H., Langer, E. R. L., Shadbolt, M., Urquhart, J. and Dandy, N. (2017). The role of the social sciences and economics in understanding and informing tree biosecurity policy and planning: A global summary and synthesis. *Biological Invasions* 19(11): 3317–3332. DOI: 10.1007/s10530-017-1503-4.

Maselko, M., Feltman, N., Upadhyay, A., Hayward, A., Das, S., Myslicki, N., Peterson, A. J., O'Connor, M. B. and Smanski, M. J. (2020). Engineering multiple species-like genetic incompatibilities in insects. *Nature Communications* 11(1): 4468. DOI: 10.1038/s41467-020-18348-1.

Matsuura, M. and Koike, K. (2002). Studies on the ecology of social wasps and bees in urban environments, 1. Records on aerial nests of the giant hornet, *Vespa mandarinia* japonica (Hymenoptera: Vespidae) within human buildings. *Medical Entomology and Zoology* 53(3): 183–186.

Matsuura, M. and Sakagami, S. (1973). A bionomic sketch of the giant hornet, *Vespa mandarinia*, a serious pest for Japanese apiculture. *Zoology: Journal of the Faculty of Science Hokkaido University Series VI* 19(1): 125–162. Available at: http://hdl.handle.net/2115/27557.

Montezano, D. G., Specht, A., Sosa-Gómez, D. R., Roque-Specht, V. F., Sousa-Silva, J. C., Paula-Moraes, S. V., Peterson, J. A. and Hunt, T. E. (2018). Host plants of *Spodoptera frugiperda* (Lepidoptera: Noctuidae) in the Americas. *African Entomology* 26(2): 286–300. DOI: 10.4001/003.026.0286.

Morey, A. C. and Venette, R. C. (2020). Minimizing risk and maximizing spatial transferability: challenges in constructing a useful model of potential suitability for an invasive insect. Annals of the Entomological Society of America 113(2): 100–113. https://doi.org/10.1093/aesa/saz049.

Morrison, W. R., Blaauw, B. R., Short, B. D., Nielsen, A. L., Bergh, J. C., Krawczyk, G., Park, Y. L., Butler, B., Khrimian, A. and Leskey, T. C. (2019). Successful management of *Halyomorpha halys* (Hemiptera: Pentatomidae) in commercial apple orchards with an attract-and-kill strategy. *Pest Management Science* 75(1): 104–114. DOI: 10.1002/ps.5156.

Myers, J. H. (2002). Eradication: Is it ecologically, financially, environmentally, and realistically possible? *Managing for Healthy Ecosystems*, pp. 533–540.

Myers, J. H., Savoie, A. and Randen, E. van (1998). Eradication and pest management. *Annual Review of Entomology* 43(1): 471–491. DOI: 10.1146/annurev.ento.43.1.471.

Myers, J. H., Simberloff, D., Kuris, A. M. and Carey, J. R. (2000). Eradication revisited: Dealing with exotic species. *Trends in Ecology and Evolution* 15(8): 316–320. DOI: 10.1016/S0169-5347(00)01914-5.

Nielsen, A. L., Holmstrom, K., Hamilton, G. C., Cambridge, J. and Ingerson-Mahar, J. (2013). Use of black light traps to monitor the abundance, spread, and flight behavior of *Halyomorpha halys* (Hemiptera: Pentatomidae). *Journal of Economic Entomology* 106(3): 1495–1502. DOI: 10.1603/EC12472.

Nikolouli, K., Colinet, H., Renault, D., Enriquez, T., Mouton, L., Gibert, P., Sassu, F., Cáceres, C., Stauffer, C., Pereira, R. and Bourtzis, K. (2018). Sterile insect technique and *Wolbachia* symbiosis as potential tools for the control of the invasive species *Drosophila suzukii*. *Journal of Pest Science* 91(2): 489–503. DOI: 10.1007/s10340-017-0944-y.

NISC. (2016). NISC management plan 2016–2018. Available at: https://www.doi.gov/sites/doi.gov/files/uploads/2016-2018-nisc-management-plan.pdf%0Apapers3://publication/uuid/BD8DF32C-8243-4664-A88F-BF1ABC969960. United States Department of Interior, National Invasive Species Council.

Noar, R. D., Jahant-Miller, C. J., Emerine, S. and Hallberg, R. (2021). Early warning systems as a component of integrated pest management to prevent the introduction of exotic pests. *Journal of Integrated Pest Management* 12(1): 16. DOI: 10.1093/jipm/pmab011.

Noble, L. (1969). Fifty years of research on the pink bollworm in the United States. *Agriculture Handbook*. United States Department of Agriculture 357: 1–58. Available at: https://naldc.nal.usda.gov/download/CAT10309959/PDF.

Norderud, E. D., Powell, S. L. and Peterson, R. K. D. (2021). Risk assessment for the establishment of *Vespa mandarinia* (Hymenoptera: Vespidae) in the Pacific Northwest, United States. *Journal of Insect Science* 21(4): 1–34. DOI: 10.1093/jisesa/ieab052.

Nuñez-Penichet, C., Osorio-Olvera, L., Gonzalez, V. H., Cobos, M. E., Jiménez, L., DeRaad, D. A., Alkishe, A., Contreras-Díaz, R. G., Nava-Bolaños, A., Utsumi, K., Ashraf, U., Adeboje, A., Peterson, A. T. and Soberon, J. (2021). Geographic potential of the world's largest hornet, *Vespa mandarinia* Smith (Hymenoptera: Vespidae), worldwide and particularly in North America. *PeerJ* 9: e10690. DOI: 10.7717/peerj.10690.

Olson, L. J. and Roy, S. (2002). The economics of controlling a stochastic biological invasion. *American Journal of Agricultural Economics* 84(5): 1311–1316.

Pedigo, L. (1986). Economic injury levels in theory and practice. *Annual Review of Entomology* 31(1): 341–368. DOI: 10.1146/annurev.ento.31.1.341.

Pedigo, L. P., Rice, M. E. and Krell, R. K. (2021). *Entomology and Pest Management* (7th edn.). Waveland Press, Inc.

Pejchar, L. and Mooney, H. A. (2009). Invasive species, ecosystem services and human well-being. *Trends in Ecology and Evolution* 24(9): 497–504. DOI: 10.1016/j.tree.2009.03.016.

Peterson, R. K. D., Higley, L. G. and Pedigo, L. P. (2018). Whatever happened to IPM? *American Entomologist* 64(3): 146–150. DOI: 10.1093/ae/tmy049.

Peyton, J. M., Martinou, A. F., Adriaens, T., Chartosia, N., Karachle, P. K., Rabitsch, W., Tricarico, E., Arianoutsou, M., Bacher, S., Bazos, I., Brundu, G., Bruno-McClung, E., Charalambidou, I., Demetriou, M., Galanidi, M., Galil, B., Guillem, R., Hadjiafxentis, K., Hadjioannou, L., Hadjistylli, M., Hall-Spencer, J. M., Jimenez, C., Johnstone, G.,

Kleitou, P., Kletou, D., Koukkoularidou, D., Leontiou, S., Maczey, N., Michailidis, N., Mountford, J. O., Papatheodoulou, A., Pescott, O. L., Phanis, C., Preda, C., Rorke, S., Shaw, R., Solarz, W., Taylor, C. D., Trajanovski, S., Tziortzis, I., Tzirkalli, E., Uludag, A., Vimercati, G., Zdraveski, K., Zenetos, A. and Roy, H. E. (2020). Horizon scanning to predict and prioritize invasive alien species with the potential to threaten human health and economies on Cyprus. *Frontiers in Ecology and Evolution* 8: 566281. DOI: 10.3389/fevo.2020.566281.

Pluess, T., Cannon, R., Jarošík, V., Pergl, J., Pyšek, P. and Bacher, S. (2012a). When are eradication campaigns successful? A test of common assumptions. *Biological Invasions* 14(7): 1365–1378. DOI: 10.1007/s10530-011-0160-2.

Pluess, T., Jarošík, V., Pyšek, P., Cannon, R., Pergl, J., Breukers, A. and Bacher, S. (2012b). Which factors affect the success or failure of eradication campaigns against alien species? *PLoS ONE* 7(10): e48157. DOI: 10.1371/journal.pone.0048157.

Poland, T. M. and Rassati, D. E. (2019). Improved biosecurity surveillance of non-native forest insects: A review of current methods. *Journal of Pest Science* 92(1): 37–49. DOI: 10.1007/s10340-018-1004-y.

Poland, T. M., Juzwik, J., Rowley, A., et al. (2021). Management of landscapes for established invasive species. In: Poland, T. M., Patel-Weynand, T., Finch, D. M., et al. (Eds) *Invasive Species in Forests and Rangelands of the United States*. Cham: Springer International Publishing, pp. 133–184. DOI: 10.1007/978-3-030-45367-1_7.

Pyšek, P., Hulme, P. E., Simberloff, D., Bacher, S., Blackburn, T. M., Carlton, J. T., Dawson, W., Essl, F., Foxcroft, L. C., Genovesi, P., Jeschke, J. M., Kühn, I., Liebhold, A. M., Mandrak, N. E., Meyerson, L. A., Pauchard, A., Pergl, J., Roy, H. E., Seebens, H., van Kleunen, M., Vilà, M., Wingfield, M. J. and Richardson, D. M. (2020). Scientists' warning on invasive alien species. *Biological Reviews of the Cambridge Philosophical Society* 95(6): 1511–1534. DOI: 10.1111/brv.12627.

Pyšek, P., Jarosik, V., Hulme, P. E., Kühn, I., Wild, J., Arianoutsou, M., Bacher, S., Chiron, F., Didziulis, V., Essl, F., Genovesi, P., Gherardi, F., Hejda, M., Kark, S., Lambdon, P. W., Desprez-Loustau, M. L., Nentwig, W., Pergl, J., Poboljsaj, K., Rabitsch, W., Roques, A., Roy, D. B., Shirley, S., Solarz, W., Vilà, M. and Winter, M. (2010). Disentangling the role of environmental and human pressures on biological invasions across Europe. *Proceedings of the National Academy of Sciences of the United States of America* 107(27): 12157–12162. DOI: 10.1073/pnas.1002314107.

Quarles, W. (2014). IPM for the brown marmorated stink bug. *The IPM Practitioner* 4(3): 1–8.

Ragsdale, D. W., Landis, D. A., Brodeur, J., Heimpel, G. E. and Desneux, N. (2011). Ecology and management of the soybean aphid in North America. *Annual Review of Entomology* 56(1): 375–399. DOI: 10.1146/annurev-ento-120709-144755.

Reaser, J. K., Burgiel, S. W., Kirkey, J., Brantley, K. A., Veatch, S. D. and Burgos-Rodríguez, J. (2020). The early detection of and rapid response (EDRR) to invasive species: A conceptual framework and federal capacities assessment. *Biological Invasions* 22(1): 1–19. DOI: 10.1007/s10530-019-02156-w.

Reilly, J. R., Hajek, A. E., Liebhold, A. M. and Plymale, R. (2014). Impact of *Entomophaga maimaiga* (Entomophthorales: Entomophthoraceae) on outbreak gypsy moth populations (Lepidoptera: Erebidae): The role of weather. *Environmental Entomology* 43(3): 632–641. DOI: 10.1603/EN13194.

Ribeiro, A. V., Holle, S. G., Hutchison, W. D. and Koch, R. L. (2021). Lethal and sublethal effects of conventional and organic insecticides on the parasitoid *Trissolcus japonicus*, a biological control agent for *Halyomorpha halys*. *Frontiers in Insect Science* 1(June): 1–12. DOI: 10.3389/finsc.2021.685755.

Rosenberger, R. S., Bell, L. A., Champ, P. A. and Smith, E. L. (2012). Nonmarket economic values of forest insect pests: An updated literature review. General Technical Report RMRS-GTR-275WWW. Available at: https://www.fs.usda.gov/treesearch/pubs/40491.

Rosenthal, G. (2021). USDA's cutting-edge methods help deliver a victory against Asian giant hornet. Available at: https://www.usda.gov/media/blog/2020/10/29/usdas-cutting-edge-methods-help-deliver-victory-against-asian-giant-hornet.

Roy, H. E., Brown, P. M. J., Adriaens, T., Berkvens, N., Borges, I., Clusella-Trullas, S., Comont, R. F., De Clercq, P., Eschen, R., Estoup, A., Evans, E. W., Facon, B., Gardiner, M. M., Gil, A., Grez, A. A., Guillemaud, T., Haelewaters, D., Herz, A., Honek, A., Howe, A. G., Hui, C., Hutchison, W. D., Kenis, M., Koch, R. L., Kulfan, J., Lawson Handley, L., Lombaert, E., Loomans, A., Losey, J., Lukashuk, A. O., Maes, D., Magro, A., Murray, K. M., Martin, G. S., Martinkova, Z., Minnaar, I. A., Nedved, O., Orlova-Bienkowskaja, M. J., Osawa, N., Rabitsch, W., Ravn, H. P., Rondoni, G., Rorke, S. L., Ryndevich, S. K., Saethre, M., Sloggett, J. J., Soares, A. O., Stals, R., Tinsley, M. C., Vandereycken, A., van Wielink, P., Viglášová, S., Zach, P., Zakharov, I. A., Zaviezo, T. and Zhao, Z. (2016). The harlequin ladybird, *Harmonia axyridis*: Global perspectives on invasion history and ecology. *Biological Invasions* 18(4): 997–1044. DOI: 10.1007/s10530-016-1077-6.

Schellhorn, N. A. and Jones, L. K. (2021). Real-time insect detection and monitoring: Breaking barriers to area-wide integrated management of insect pests. In: Hendrichs, J., Pereira, R. and Vreysen, M. J. B. (Eds) *Area-Wide Integrated Pest Management*. CRC Press, pp. 889–902. DOI: 10.1201/9781003169239-51.

Shackleton, R. T., Adriaens, T., Brundu, G., Dehnen-Schmutz, K., Estévez, R. A., Fried, J., Larson, B. M. H., Liu, S., Marchante, E., Marchante, H., Moshobane, M. C., Novoa, A., Reed, M. and Richardson, D. M. (2019). Stakeholder engagement in the study and management of invasive alien species. *Journal of Environmental Management* 229: 88–101. DOI: 10.1016/j.jenvman.2018.04.044.

Short, B. D., Khrimian, A. and Leskey, T. C. (2017). Pheromone-based decision support tools for management of *Halyomorpha halys* in apple orchards: Development of a trap-based treatment threshold. *Journal of Pest Science* 90(4): 1191–1204. DOI: 10.1007/s10340-016-0812-1.

Siebert, J. B. and Cooper, T. (1995). Embargo on California produce would cause revenue, job loss. *California Agriculture* (July–August): 7–12.

Siegert, N. W., McCullough, D. G., Liebhold, A. M. and Telewski, F. W. (2014). Dendrochronological reconstruction of the epicentre and early spread of emerald ash borer in North America. *Diversity and Distributions* 20(7): 847–858. DOI: 10.1111/ddi.12212.

Simberloff, D., Martin, J. L., Genovesi, P., Maris, V., Wardle, D. A., Aronson, J., Courchamp, F., Galil, B., García-Berthou, E., Pascal, M., Pyšek, P., Sousa, R., Tabacchi, E. and Vilà, M. (2013). Impacts of biological invasions: What's what and the way forward. *Trends in Ecology and Evolution* 28(1): 58–66. DOI: 10.1016/j.tree.2012.07.013.

Staten, R. T. and Walters, M. L. (2021). Technology used by field managers for pink bollworm eradication with its successful outcome in the United States and Mexico. In: Hendrichs, J., Pereira, R. and Vreysen, M. J. B. (Eds) *Area-Wide Integrated Pest Management*. Boca Raton, FL: CRC Press, pp. 51–92. DOI: 10.1201/9781003169239-5.

Stewart, S. D. (2021). Detecting invasive insects using uncrewed aerial vehicles and variational autoencoders. *Marquette University, Masters of Science*.

Stiling, P. (2002). Potential non-target effects of a biological control agent, prickly pear moth, *Cactoblastis cactorum* (Berg) (Lepidoptera: Pyralidae), in North America, and possible management actions. *Biological Invasions* 4(3): 273–281. DOI: 10.1023/A:1020988922746.

Stumph, B., Virto, M. H., Medeiros, H., et al. (2019). Detecting invasive insects with unmanned aerial vehicles. In: *International Conference on Robotics and Automation (ICRA)*, May 2019 (vol. 2019). IEEE Publications, pp. 648–654. DOI: 10.1109/ICRA.2019.8794116.

Suckling, D. M., Barrington, A. M., Chhagan, A., Stephens, A. E. A., Burnip, G. M., Charles, J. G. and Wee, S. L. (2007). Eradication of the Australian painted apple moth *Teia anartoides* in New Zealand: Trapping, inherited sterility, and male competitiveness. In: *Area-Wide Control. of Insect Pests*. Springer Netherlands, pp. 603–615. DOI: 10.1007/978-1-4020-6059-5_56.

Suckling, D. M., Stringer, L. D., Baird, D. B. and Kean, J. M. (2019). Will growing invasive arthropod biodiversity outpace our ability for eradication? *Ecological Applications: A Publication of the Ecological Society of America* 29(8): e01992. DOI: 10.1002/eap.1992.

Suckling, D. M., Stringer, L. D. and Kean, J. M. (2021). Trends in arthropod eradication programmes from the global eradication database, Gerda. In: Hendrichs, J., Pereira, R. and Vreysen, M. J. B. (Eds) *Area-Wide Integrated Pest Management*. CRC Press, pp. 505–518. DOI: 10.1201/9781003169239-26.

Suckling, D. M., Charles, J. G., Kay, M. K., Kean, J. M., Burnip, G. M., Chhagan, A., Noble, A. and Barrington, A. M. (2014a). Host range testing for risk assessment of a sexually dimorphic polyphagous invader, painted apple moth. *Agricultural and Forest Entomology* 16(1): 1–13. DOI: 10.1111/afe.12028.

Suckling, D. M., Stringer, L. D., Stephens, A. E., Woods, B., Williams, D. G., Baker, G. and El-Sayed, A. M. (2014b). From integrated pest management to integrated pest eradication: Technologies and future needs. *Pest Management Science* 70(2): 179–189. DOI: 10.1002/ps.3670.

Tabashnik, B. E., Liesner, L. R., Ellsworth, P. C., Unnithan, G. C., Fabrick, J. A., Naranjo, S. E., Li, X., Dennehy, T. J., Antilla, L., Staten, R. T. and Carrière, Y. (2021). Transgenic cotton and sterile insect releases synergize eradication of pink bollworm a century after it invaded the United States. *Proceedings of the National Academy of Sciences of the United States of America* 118(1): e2019115118. DOI: 10.1073/pnas.2019115118.

Teem, J. L., Alphey, L., Descamps, S., Edgington, M. P., Edwards, O., Gemmell, N., Harvey-Samuel, T., Melnick, R. L., Oh, K. P., Piaggio, A. J., Saah, J. R., Schill, D., Thomas, P., Smith, T. and Roberts, A. (2020). Genetic biocontrol for invasive species. *Frontiers in Bioengineering and Biotechnology* 8(May): 452. DOI: 10.3389/fbioe.2020.00452.

Thistlewood, H. M. A. and Judd, G. J. R. (2019). Twenty-five years of research experience with the sterile insect technique and area-wide management of codling moth, *Cydia pomonella* (L.), in Canada. *Insects* 10(9): 292. DOI: 10.3390/insects10090292.

Tobin, P. C., Blackburn, L. M., Gray, R. H., Lettau, C. T., Liebhold, A. M. and Raffa, K. F. (2013). Using delimiting surveys to characterize the spatiotemporal dynamics facilitates the management of an invasive non-native insect. *Population Ecology* 55(4): 545–555. DOI: 10.1007/s10144-013-0382-5.

Tobin, P. C., Kean, J. M., Suckling, D. M., McCullough, D. G., Herms, D. A. and Stringer, L. D. (2014). Determinants of successful arthropod eradication programs. *Biological Invasions* 16(2): 401–414. DOI: 10.1007/s10530-013-0529-5.

Tripodi, A. and Hardin, T. (2020). *Vespa mandarinia. Asian Giant Hornet. USDA-APHIS New Pest Response Guidelines*. United States Department of Agriculture, Animal and Plant Health Inspection Service.

USDA. (2018). OPMP. *A National Road Map for Integrated Pest Management*. Available at: https://www.ars.usda.gov/ARSUserFiles/OPMP/IPM Road Map FINAL.pdf. United States Department of Agriculture, Office of Pest Management Policy.

USDA. (2021). Aphis. *USDA Provides More Than $70 Million in Fiscal Year 2021 to Protect Agriculture and Natural Resources From Plant Pests and Diseases. News and Announcements*. Available at: https://www.aphis.usda.gov/aphis/newsroom/news/sa_by_date/sa-2021/national-ppa7721. United States Department of Agriculture, Animal and Plant Health Inspection Service, 5 January.

Valentin, R. E., Fonseca, D. M., Gable, S., Kyle, K. E., Hamilton, G. C., Nielsen, A. L. and Lockwood, J. L. (2020). Moving eDNA surveys onto land: Strategies for active eDNA aggregation to detect invasive forest insects. *Molecular Ecology Resources* 20(3): 746–755. DOI: 10.1111/1755-0998.13151.

Valentin, R. E., Fonseca, D. M., Nielsen, A. L., Leskey, T. C. and Lockwood, J. L. (2018). Early detection of invasive exotic insect infestations using eDNA from crop surfaces. *Frontiers in Ecology and the Environment* 16(5): 265–270. DOI: 10.1002/fee.1811.

Venette, R. C. and Carey, J. R. (1998). Invasion biology: Rethinking our response to alien species. *California Agriculture* 52(2): 13–17. DOI: 10.3733/ca.v052n02p13.

Venette, R. C. and Koch, R. L. (2009). IPM for invasive species. In: Radcliffe, E. B., Hutchison, W. D. and Cancelado, R. E. (Eds) *Integrated Pest Management: Concepts, Tactics, Strategies, and Case Studies*. Cambridge University Press, pp. 424–436.

Venette, R. C. and Morey, A. (2019). Advances in understanding the ecology of invasive crop insect pests and their impact on IPM. In: Kogan, M. and Heinrichs, E. (Eds) *Integrated Management of Insect Pests: Current and Future Developments*. Cambridge, UK: Burleigh Dodds Science Publishing, pp. 161–190. DOI: 10.19103/AS.2019.0047.06.

Walther, G. R., Roques, A., Hulme, P. E., Sykes, M. T., Pysek, P., Kühn, I., Zobel, M., Bacher, S., Botta-Dukát, Z., Bugmann, H., Czúcz, B., Dauber, J., Hickler, T., Jarosík, V., Kenis, M., Klotz, S., Minchin, D., Moora, M., Nentwig, W., Ott, J., Panov, V. E., Reineking, B., Robinet, C., Semenchenko, V., Solarz, W., Thuiller, W., Vilà, M., Vohland, K. and Settele, J. (2009). Alien species in a warmer world: Risks and opportunities. *Trends in Ecology and Evolution* 24(12): 686–693. DOI: 10.1016/j.tree.2009.06.008.

Weber, D. C., Leskey, T. C., Walsh, G. C. and Khrimian, A. (2014). Synergy of aggregation pheromone with methyl (E,E,Z)-2,4,6-decatrienoate in attraction of *Halyomorpha halys* (Hemiptera: Pentatomidae). *Journal of Economic Entomology* 107(3): 1061–1068. DOI: 10.1603/EC13502.

Williamson, M. (1996). *Biological Invasions*. Chapman and Hall.

Wilson, T. M., Takahashi, J., Spichiger, S.-E., Kim, I. and van Westendorp, P. (2020). First reports of *Vespa mandarinia* (Hymenoptera: Vespidae) in North America represent two separate maternal lineages in Washington State, United States, and British Columbia, Canada. *Annals of the Entomological Society of America* 113(6): 468–472. DOI: 10.1093/aesa/saaa024.

Woodford, D. J., Richardson, D. M., MacIsaac, H. J., Mandrak, N. E., van Wilgen, B. W., Wilson, J. R. U. and Weyl, O. L. F. (2016). Confronting the wicked problem of managing biological invasions. *NeoBiota* 31: 63–86. DOI: 10.3897/neobiota.31.10038.

WSDA. (2020). *WSDA Entomologists Eradicate First Asian Giant Hornet Nest*. News Release, 26 October. Washington State Department of Agriculture. Available at: https://agr.wa.gov/about-wsda/news-and-media-relations/news-releases?article=31892.

WSDA. (2021). *State and Federal Entomologists Confirm New Asian Giant Hornet Detection in Snohomish County*. News Release, 16 June. Washington State Department of Agriculture. Available at: https://agr.wa.gov/about-wsda/news-and-media-relations/news-releases?article=32455.

WSDA (2022). Asian giant hornet. Available at: https://agr.wa.gov/departments/insects-pests-and-weeds/insects/hornets. Washington State Department of Agriculture.

Young, R. G., Milián-García, Y., Yu, J., Bullas-Appleton, E. and Hanner, R. H. (2021). Biosurveillance for invasive insect pest species using an environmental DNA metabarcoding approach and a high salt trap collection fluid. *Ecology and Evolution* 11(4): 1558–1569. DOI: 10.1002/ece3.7113.

Zhu, G., Gutierrez Illan, J., Looney, C. and Crowder, D. W. (2020). Assessing the ecological niche and invasion potential of the Asian giant hornet. *Proceedings of the National Academy of Sciences of the United States of America* 117(40): 24646–24648. DOI: 10.1073/pnas.2011441117.

Zimmermann, H., Bloem, S. and Klein, H. (2004). Biology, history, threat, surveillance and control of the cactus moth, *Cactoblastis cactorum*. *Joint FAO/IAEA Programme of Nuclear Techniques in Food and Agriculture*. Available at: https://www-pub.iaea.org/MTCD/Publications/PDF/faobsc_web.pdf.

Index

www.ingramcontent.com/pod-product-compliance
Ingram Content Group UK Ltd.
Pitfield, Milton Keynes, MK11 3LW, UK
UKHW021825100926
8556IPUK00003B/168

9 781801 461078